Edited by R. Přikryl

DIMENSION
STONE

DIMENSION STONE 2004

PROCEEDINGS OF THE INTERNATIONAL CONFERENCE ON DIMENSION STONE 2004,
14–17 JUNE 2004, PRAGUE, CZECH REPUBLIC

Dimension Stone 2004

New Perspectives for a Traditional Building Material

Edited by

R. Přikryl
*Institute of Geochemistry, Mineralogy and Mineral Resources,
Charles University – Faculty of Science, Prague, Czech Republic*

CRC Press
Taylor & Francis Group
Boca Raton London New York

CRC Press is an imprint of the
Taylor & Francis Group, an **informa** business
A BALKEMA BOOK

Published by:
CRC Press/Balkema
P.O. Box 447, 2300 AK Leiden, The Netherlands
e-mail: Pub.NL@taylorandfrancis.com
www.crcpress.com – www.taylorandfrancis.com

© 2004 by Taylor & Francis Group, LLC
CRC Press/Balkema is an imprint of the Taylor & Francis Group, an informa business

No claim to original U.S. Government works

ISBN 13: 978-90-5809-675-3 (hbk)

Visit the Taylor & Francis Web site at
http://www.taylorandfrancis.com

and the CRC Press Web site at
http://www.crcpress.com

Cover picture: The use of this stone type is documented only from the Czech area. It has been widely used namely for sacral purposes (altars in baroque churches) during 18th century. Its use was very scarce due to the occurrence in very limited area. The main body of the rock is black due to the high content of black organic matter between fine-grained calcite matter. The white parts are fossilized remnants of organic bodies later filled with whitish calcite.

Although all care is taken to ensure the integrity and quality of this publication and the information herein, no responsibility is assumed by the publishers nor the author for any damage to property or persons as a result of operation or use of this publication and/or the information contained herein.

Dimension Stone 2004, Přikryl (ed)
© *2004 Taylor & Francis Group, London, ISBN 90 5809 675 0*

Table of Contents

Rock fabric and rock mechanics studies

Computers, databases and GIS applied to dimension stone studies

Dimension stone decay: from rock weathering studies to diagnostic and conservation efforts

Bowing of natural stone cladding (TEAM session)

Environmental aspects and technological research of dimension stone

Dimension Stone 2004, Přikryl (ed)
© 2004 Taylor & Francis Group, London, ISBN 90 5809 675 0

Preface

This book contains papers presented at the International Conference Dimension Stone 2004, held in Prague, the Czech Republic, 14–17 June 2004. The theme of the conference, "New perspectives for a traditional building material", represents the essence to provide an international forum for sharing ideas of stone traditional aspects and for finding new ways of stone use in modern world.

The invited lectures of the conference reflect the most recent trends related to dimension stone utilisation. In this respect, invited speaker Prof. E. Reynard highlighted controversies regarding stone extraction and landscape protection and invited speaker Prof. B.J. Smith focused on the actual problems related to climatic impact on dimension stone weathering.

The themes discussed cover various aspects of geology of traditional stone types but also of new deposits. The use of modern analytical techniques is advantageous, not only in academic research, but also when finding a solution for practical problems like natural stone bowing. This trend is supported by numerous papers in this volume, also by the specialised session on TEAM research (Testing and Assessment of Marble and Limestone). The importance of the application of digital techniques was reflected in the specialised session on geographical information systems, databases and computers in dimension stone research, exploration, extraction, and marketing. Dimension stone, as each building material, changes its properties during service; thus the deterioration and conservation of natural stones play a role in both research and in cultural, economic and political considerations. Waste produced during stone extraction presents another public-sensitive aspect. The reduction of waste and its practical utilisation form a challenge for future research as is demonstrated by several presentations in this volume.

The conference organisers and editor of this volume would like to thank all contributors for the timely submission of their manuscripts. The international conference DIMENSION STONE 2004 was organised by the Institute of Geochemistry, Mineralogy and Mineral Resources (Faculty of Science, Charles University in Prague), together with the Academy of Fine Arts in Prague, the Czech national group of the International Society for Rock Mechanics (ISRM), the Czech national group of the International Association for Engineering Geology and the Environment (IAEG), the Society for Geology Applied to Mineral Deposits (SGA), and the Czech Geological Survey. The conference planning and organisation would not have realised without the highly acknowledged financial support from projects MSM 520000001 "Research and development project in restoration of sculptural artistic monuments based on historical and current knowledge" and MSM 113100005 "Material and energy flows in the upper parts of the Earth".

The editor would like to express his sincere thanks to his family for their understanding and support during the preparation of this book. The help and patience from the Publisher's staff is highly acknowledged as well. The staff and students from the Institute of Geochemistry, Mineralogy and Mineral Resources (Faculty of Science, Charles University in Prague) and from the School of Sculptural Works of Art Restoration (Academy of Fine Arts in Prague) are kindly thanked for their help during the conference preparation.

Prague, June 2004   R. Přikryl
                                                                                                        (organizer)

Invited lectures

Dimension Stone 2004, Přikryl (ed)
© 2004 Taylor & Francis Group, London, ISBN 90 5809 675 0

Protecting stones: conservation of erratic blocks in Switzerland

E. Reynard

Institute of Geography, University of Lausanne, Switzerland

ABSTRACT: Geosites are geological and geomorphological objects that present a particular importance for the comprehension of the Earth's history. Because of the multitude of uses that can be made of geosites (exploitation, tourist uses, protection, etc.), they may be particularly vulnerable and the most important of them should be protected by legal measures. Because of their high value in reconstructing former glacial extensions in the plains surrounding the Alps, erratic blocks are considered as important geosites in Switzerland. In the 19th century, most of them were endangered by the granite extraction industry. Several actions were carried out by groups of scientists, Natural sciences associations, and some political authorities. This paper recalls the principal steps that permit the conservation of the major erratic blocks in Switzerland, and some recent actions for conserving and promoting them are discussed.

1 GEOSITE CONSERVATION

1.1 *Definitions*

Geosites (or geotopes) are portions of the geosphere that present a particular importance for the comprehension and reconstruction of the history of the Earth, climate and life (Reynard 2003). They can be single objects or large systems. Active geosites allow the visualization of geological and geomorphological processes in action, whereas passive geosites testify past processes.

The value of a geosite may be assessed on the basis of four groups of criteria: scientific, cultural/historical, aesthetic and/or social/economic (Panizza & Piacente 1993, 2003). Some types of geosites also have a particular value for the reconstruction of the history of geosciences and/or for educative purposes.

1.2 *Geosite vulnerability and conservation*

Geosites may be modified, damaged, and even destroyed, by natural processes and anthropogenic actions (Cavallin et al. 1994). In order to avoid damage and destruction, geosites need to be protected. Protection strategies generally follow three steps: the assessment of the quality of geosites (Panizza & Piacente 1993, Rivas et al. 1997, Grandgirard 1999, Coratza & Giusti, in press), the selection of sites worthy of being protected (Grandgirard 1999), and the implementation of legal protection rules (Reynard, in press a,b). Legal protection may be obtained by using

property rights instruments, as well as public policies (Reynard, in press a, b).

1.3 *Erratic blocks as geosites*

Erratic blocks are boulders that were removed from their bedrock and transported by glaciers to areas of other lithology (Benn & Evans 1998). These *exotic blocks* may have travelled very far (up to 1200 km in northern Europe), and they can be used for reconstructing former transportation pathways. *Indicator erratics* are terrains for which a definite source area is known (Benn & Evans 1998). The current position of an erratic boulder may be due to several glacial cycles.

Erratic blocks can be considered as geosites. They are geological objects that allow the reconstruction of former glacier positions, and more generally they participate in the reconstruction of the history of the Earth's climate. Most of them have therefore a high scientific value. Several blocks, especially in the areas surrounding the Alps and in northern Europe, were also concerned with scientific controversies during the 19th century; they have therefore a high value for the history of geosciences. Large boulders and stones with particular shapes created an impression on the local populations, several blocks have a name, and others were used as religious sites; this is what creates the cultural, historical and religious value of erratic blocks. They may also have an aesthetic value. In some regions, erratic blocks had a high economic value as sites for building material extraction. In Switzerland,

3

erratic blocks have also a particular value for the history of nature conservation: they were some of the first natural objects to be protected at a national level (Vischer 1946, Bachmann 1999).

In the following paragraphs, the principal steps that permitted the conservation of the major erratic blocks in Switzerland are recalled, and some recent actions for conserving and promoting them are discussed.

2   ERRATIC BLOCK CONSERVATION IN THE 19TH CENTURY

2.1   *Discussions on the origin of erratic blocks*

Since the first works of Horace Benedict de Saussure in the 18th century, the question of the origin of erratic blocks has occupied Earth scientists for several decades. Schaer (2000) accurately studied the respective role of different scientists in this debate. Several hypotheses were proposed to explain the displacement of large blocks of granite from the Alps to the Central Plateau of Switzerland and the flanks of the Jura Range: flow streams (diluvian hypothesis, de Saussure, von Buch), ice rafts (Darwin), ice plans tilting from the Alps to the Jura (Agassiz), gas eruptions (de Luc, von Buch), and glaciers (glacial hypothesis) (Schaer 2000).

The glacial hypothesis was proposed rather simultaneously by Ignaz Venetz in 1827 (Berchtold & Bumann 1990), Goethe in 1829 (von Engelhardt 1999), and other scientists like Hutton, Playfair, Esmark (de Charpentier 1841, Schaer 2000), sometimes on the basis of discussions with peasants. Jean de Charpentier in 1834 and Louis Agassiz in 1837 diffused the theory in the scientific community (Schaer 2000).

2.2   *Scientific value of erratic blocks*

De Charpentier was first sceptical about the hypothesis proposed by Venetz in 1827, but some years later, he agreed completely, and in his *Essay on glaciers and erratic terrains in the Rhone basin* (1841), he based his argumentation for the glacial origin of erratic blocks on three explanatory factors: there is no reduction of the block size with the distance from the Alps; the blocks are sharp, which is not the case of boulders transported by rivers; erratic terrains are not stratified.

Around 1850, erratic blocks were therefore recognized, with moraines and glacial striae, as the principal indices used for the reconstruction of former alpine glaciers. They were particularly important for deducing the width and elevation of the alpine glacial streams and for mapping their extension (Favre 1876).

The scientific value of erratic blocks was therefore recognized; but it was not sufficient for insuring their protection.

2.3   *Call to the Swiss population for the conservation of erratic blocks*

Because of the industrialization and progressive urbanisation in some areas of the Central Plateau of Switzerland, several large blocks were exploited for stone extraction. Limestones were used for lime production, whereas crystalline rocks were extracted for building material. This economic use of erratic blocks was particularly interesting in areas where the bedrock was formed of loose Tertiary material.

The rapid destruction of large and numerous blocks, and its negative impacts for the scientific development of Earth sciences particularly dismayed some scientists. In 1866, Prof. Alphonse Favre of Geneva proposed to take measures in order to protect the most important blocks to the Helvetic Society for Natural Sciences (HSNS) (currently: Swiss Academy of Sciences). The HSNS charged its geological commission to make propositions. One year later, the commission proposed to write a *Call to Swiss people for asking them to conserve erratic blocks* (Favre & Studer 1867).

The conclusions of the commission were that it was a too large enterprise for a small scientific society to protect blocks disseminated throughout the country. The idea was to involve the Swiss population in the process. But, the scientific value of blocks was not sufficient to diffuse a spirit of conservation in the population. The authors focused therefore their argumentation on the patriotic character of a conservation programme of erratic blocks in Switzerland. The country would have been reinforced by the action; it would obtain international recognition, as it had been the case with the discovery and protection of lacustrine archaeological remains in Switzerland (Favre & Studer 1867).

The two scientists gave different examples of existing protected blocks in several parts of Switzerland: the *Pierre-à-Bot* in Neuchâtel was the first block to be protected in 1838 (Vischer 1946); the Canton of Valais offered two blocks of the moraine of Monthey in 1853 to Jean de Charpentier (Renevier 1877). The *Call* was sent to all the cantons, to regional societies of natural sciences, to several other organizations like the Swiss Alpine Club, to various municipalities and different individuals. It was also published in national newspapers (Favre & Soret 1868). The call was diffused in France and Germany as well.

2.4   *First cases of boulder conservation*

The *Call* obtained a great popular success (Favre & Soret 1868, 1869, Favre 1871, 1872, 1876 Vischer 1946, Bachmann 1999). Scientific commissions, working groups and directives for the forestry administration were created by several cantons to organize an inventory of erratic blocks on the cantonal territory. Some cantons and local municipalities issued bans of exploitation and protection decrees. This was the case

4

of the city of Solothurn that protected more than 200 blocks on its properties.

Blocks situated on private properties were more difficult to conserve because of the rights of the owners to dispose freely of their land (Reynard in press a,b). In that case, the blocks and the terrain where they were lying were bought by Natural science societies. Because of the excessive prices that were asked by the owners, several popular subscriptions were organised. This was the case, for example, of the *Luegiboden Block* in the Interlaken valley, bought by the Natural History Museum of Berne with the benefices of a subscription (Favre & Soret 1868), or the *Steinhof Block*, bought by subscription by the HSNS (Favre & Soret 1869). Some private owners offered their blocks to the HSNS: Mr Briganti offered one of the blocks of the famous moraine of Monthey (Favre & Soret 1869); in 1875, two other blocks of the group of Monthey, the *Pierre des Muguets* and *Pierre à Dzo*, were given by de Charpentier's heirs to the Society of Natural Sciences of Canton Vaud (Renevier 1877).

In 1871, Favre published a list of 69 already protected erratic blocks. But a large number of blocks were still in danger, and in 1872, the Geological commission wrote to all the cantonal governments asking for two types of measures: to decide the protection of all the blocks situated in cantonal forests; and to put pressure on local municipalities to induce them to preserve the more important blocks. Criteria for selecting blocks worthy of being protected were their beauty, the presence of a local name, and the report of legends related to the blocks (Favre 1872). Du Pasquier (1891) proposed five general criteria for selecting the blocks to be protected: the blocks situated at high relative elevations (important for deducing the ice streams width); the indicator blocks; large blocks and perched boulders; blocks with archaeological value; blocks with ecological value (presence of plants transported by the erratic stone).

2.5 Map of Swiss erratic blocks (1884)

The *Call to the Swiss population* (1867) was accompanied by a project to compile all the findings and to map them on the recent Dufour Map at 1:100,000 scale (Favre & Studer 1867). Tens of scientists and individuals) carried out the inventory of erratic blocks in their area. The engineers that were preparing the Siegfried Map also contributed to the inventory, especially in the Jura Range. Finally, in 1884, Favre published a synthesis map covering the whole country at 1:250,000 scale.

2.6 Endangered stones

Large circles of the scientific community and large parts of the Swiss population were interested in the erratic blocks and were sensitive to their scientific, aesthetic and historical value. Natural science associations bought some blocks; others were put under protection by the State. However, economic speculation was important and tens of blocks continued to be destroyed. They were also no legal means to force private owners and local municipalities to put blocks under protection. Two cases are emblematic of the difficulties of protection.

In the canton of Neuchâtel, the commune of Bôle decided in 1891 to use its *Bloc du Mont de Boudry*, one of the most famous in the area, to extract granite to build steps for the municipal school, even though the block had been protected since 1864 by a municipal decree (Favre 1871). To save the block, the committee of the Natural Sciences Society of Neuchâtel had to find the sum of 700 francs in less than one week. With the help of the State and the Historical society of the Canton, it was possible and the block was saved.

In 1905, it was the turn of the largest block in Switzerland to be sold to an extraction company: the *Pierre des Marmettes* in Monthey (Schardt 1908). The sale provoked a large official and popular movement against the destruction of the *King of the erratic blocks* (Bachmann 1999). The block was situated on a private property. Because of the absence of legal rules for protecting erratic blocks, the municipality of Monthey called the Swiss government for help. After three years of procedures, the block was able to be bought from the proprietor for the tremendous sum of 31,500 Swiss francs (Swiss Confederation: 12,000 francs; Canton of Valais: 5000 francs; municipality of Monthey: 5500 francs; subscription to the Swiss population organised by the Helvetic society of natural sciences: 9000 francs). The success of the operation is explained by the size of the block (the largest in Switzerland), the sentiment that the country was to loose an important part of its natural and national heritage, and the scientific fame of the block (it was one of the blocks that had been used by de Charpentier in his argumentation for the glacial theory).

3 ERRATIC BLOCK CONSERVATION IN THE 21ST CENTURY

3.1 Necessity of conserving and promoting geosites in Switzerland

After these pioneering times of nature conservation focusing on geological heritage protection, conservation activities progressively tilted towards biological protection. In 1966 the Nature Conservation Act was adopted, and in the 1990s several articles were added for the protection of moors and moor landscapes.

The protection of erratic blocks now seems to be part of history. Because of the use of concrete in the

5

construction sector, the blocks are no longer endangered. But, since the end of the 1980s, the geoscientific community has become aware of the poor knowledge of the population concerning the Earth's history, geological phenomena and geomorphological processes. A working group for the protection of geosites was created in 1994 within the Swiss Academy of Sciences. The working group's objectives are to improve the public awareness of the concept of geosite and geological heritage. In 1995, a strategic report was published (Strasser et al. 1995), and four years later, the first unofficial Swiss inventory of geosites was presented (SAS 1999).

3.2 The Swiss inventory of geosites

The Swiss inventory of geosites was carried out by a group of experts under the auspice of the Swiss Academy of Science (SAS 1999). It lists 401 sites, but has no legal incentive for the political authorities. It is just an indicative list of objects with high geoscientific value. Geosites are therefore not explicitly protected in the Swiss environmental legislation (Jordan 1999, Reynard, in press a, b). This situation may reduce the possibilities of protection of relevant geological objects, which have a high scientific value, but a low aesthetic, cultural and/or ecological value. Within the 401 selected geosites, 15 are erratic blocks or groups of erratics.

Examples are: the glacial landscape of Steinhof, the *Pierre Agassiz* on Mount Vully, the *Pierre-à-Bot* (Neuchâtel; first protected block in 1838); the *Glacier Garden* in Lucerne (several blocks), the group of blocks in Monthey (Rhone valley), and several Verrucano-blocks in the canton of Zurich.

3.3 Assessing erratic blocks as geosites

The Swiss geosite inventory is not a systematic inventory based on transparent assessment methods. The selected erratic blocks are not representative of the 19th century controversies. They do not correspond to all the blocks that were protected by the end of the 19th century. For these reasons, a more systematic inventory based on transparent criteria should be carried out. Grandgirard & Schneuwly (1997) made an inventory of erratic blocks in the canton of Fribourg. Around 400 blocks were counted, from which 64 blocks were selected as erratic blocks of cantonal significance. Selection was based on two criteria: the volume and the lithology. Each criterion was divided into three classes.

This method only assesses the displacement and the glaciological significance of the blocks. It does not permit the evaluation of the geocultural (erratic blocks as archaeological heritage), geo-ecological (erratic blocks as exotic ecosystems), and/or geohistorical values (erratic blocks that were involved in the

19th century glacial theory controversies). A new method combining geoscientific and geocultural criteria is in preparation (Reynard & Lugon, in prep.).

4 CONCLUSIONS

Erratic blocks are some of the geological objects that played a predominant role in the advances in Earth sciences in the 19th century. The scientists first paid attention to their exotic presence in areas of different lithologies. Several hypotheses were proposed by known scientists (de Saussure, Darwin) to explain their transport. After it was clear that former glacier advances were at the origin of their exotic location, some of them became famous as indices of glacier positions in remote times. Various inventories carried out in the second part of the 19th century, and especially the presence of indicator blocks, permit the reconstruction of the former principal glacial ice streams (Rhone glacier, Aare glacier, Reuss glacier, Rhine glacier). Several erratic blocks have therefore a high *scientific* value for climate and glacial history reconstruction in Switzerland. They are also of great *geohistorical* significance because of the role they played in the glaciological and quaternary studies in the 19th century. Their *geocultural* value is poorly known and assessed.

Erratic block conservation began in 1838 (*Pierre-à-Bot*, Neuchâtel). Several additional blocks were protected in the second part of the 19th century, especially in relation with the *Call to the Swiss population* published by Favre & Studer (1867). The examination of two cases in the area of Neuchâtel (*Mont de Boudry*, 1891) and Monthey (*Pierre des Marmettes*, 1905) shows that because of economic speculation combined with the absence of specific legislation insuring their protection, several blocks were endangered by granite extraction activities. Property status was important for determining the degree of vulnerability (Reynard, in press a, b): blocks located on private terrains (case of the *Pierre des Marmettes*) were much more vulnerable than erratics situated on State territories. Because of the federalist organization of the Swiss Confederation, even the public property of terrains was not a guarantee for the long-term conservation of the blocks (case of the *Mont de Boudry* block). The sale of granite blocks was considered as an interesting income for relatively poor small municipalities. These are examples of the conflicts generated by the multifunctional characteristics of geosites (Reynard, in press a).

The informal geosite inventory carried out by the Swiss Academy of Sciences demonstrates the value of erratic blocks as geosites: 15 geosites out of 400 are erratics. Nevertheless, the inventory was not based on rational and transparent criteria, and several blocks protected at the end of the 19th century are not included.

6

Moreover, the archaeological value of erratic blocks is poorly known. For these various reasons, systematic inventories like that carried out in the Canton of Fribourg (Grandgirard & Schneuwly 1997) should be encouraged.

REFERENCES

Bachmann, S. 1999. *Zwischen Patriotismus und Wissenschaft. Die schweizerischen Naturschutzpioniere (1900–1938).* Zürich: Chronos Verlag.

Benn, D.I. & Evans, D.J.A. 1998. *Glaciers and glaciation.* London: Arnold.

Berchtold, S. & Bumann, P. 1990. *Ignaz Venetz, 1788–1859. Ingenieur und Naturforscher.* Visp: Mengis Verlag.

Cavallin, A., Marchetti, M., Panizza, M. & Soldati, M. 1994. The role of geomorphology in environmental impact assessment. *Geomorphology* 9: 143–153.

Coratza, P. & Giusti, C. in press. Methodological proposal for the assessment of the scientific quality of the geomorphosites. *Il Quaternario.*

De Charpentier, J. 1841. *Essai sur les glaciers et sur le terrain erratique du bassin du Rhône.* Lausanne: Ducloux.

Favre, A. & Soret, L. 1868. Rapport sur l'étude et la conservation des blocs erratiques en Suisse. In *Actes de la Société helvétique des sciences naturelles.* Einsiedeln: 143–151.

Favre, A. & Soret, L. 1869. Troisième rapport sur l'étude et la conservation des blocs erratiques en Suisse. In *Actes de la Société helvétique des sciences naturelles.* Soleure: 169–181.

Favre, A. & Studer, B. 1867. *Appel aux Suisses pour les engager à conserver les blocs erratiques.* Rheinfelden: Actes de la Société helvétique des sciences naturelles.

Favre, A. 1871. Quatrième rapport sur l'étude et la conservation des blocs erratiques en Suisse. In *Actes de la Société helvétique des sciences naturelles.* Frauenfeld: 193–217.

Favre, A. 1872. Cinquième rapport sur l'étude et la conservation des blocs erratiques en Suisse. In *Actes de la Société helvétique des sciences naturelles.* Fribourg: 162–171.

Favre, A. 1876. Notice sur la conservation des blocs erratiques et sur les anciens glaciers du revers septentrional des Alpes suisses. *Arch. des Sciences* 57: 181–205.

Favre, A. 1884. *Carte du phénomène erratique et des anciens glaciers du versant nord des Alpes suisses et de la chaîne du Mont-Blanc.* Berne: Commission géologique.

Grandgirard, V. & Schneuwly, D. 1997. Auswahl und Schutz bedeutender Findlinge im Kanton Freiburg (Schweiz). *Geowissenschaften* 12: 402–407.

Grandgirard, V. 1999. L'évaluation des géotopes. *Geol. Insubr.* 4(1): 59–66.

Jordan, P. 1999. Die Geotopschutz: die rechtliche Situation in der Schweiz. *Geol. Insubr.* 4(1): 55–58.

Panizza, M. & Piacente, S. 1993. Geomorphological assets evaluation. *Zeitschr. für Geomorphologie N.F.*, Suppl. Bd. 87: 13–18.

Panizza, M. & Piacente, S. 2003. *Geomorfologia culturale.* Bologna: Pitagora Editrice.

Renevier, E. 1877. Notice sur les blocs erratiques de Monthey. *Bull. Soc. Vaud. Sc. Nat.* 15(78): 105–116.

Reynard, E. & Lugon, R. in prep. The assessment of geocultural geosites within the Geosite Inventory of Canton Valais (Switzerland). In *Proceedings IUGS International Geological Congress 2004*: Florence.

Reynard, E. 2003. Geosites. In Andrew S. Goudie (ed.), *Encyclopedia of Geomorphology*: 440. London: Routledge.

Reynard, E. (in press a). Geomorphological sites, public policies and property rights. Conceptualization and examples from Switzerland. *Il Quaternario.*

Reynard, E. (in press b). Öffentliche Politik, Eigentumsverhältnisse und Schutz von Geomorphologischen Geotope. In *Proceedings Tagung Geotope Bad-Ragaz 2003*: Schriftenreihe der Deutsche Geologische Gesellschaft.

Rivas, V., Rix, K., Francés, E., Cendrero, A. & Brunsden, D. 1997. Geomorphological indicators for environmental impact assessment: consumable and non-consumable geomorphological resources. *Geomorphology* 18: 169–182.

SAS. 1999. Inventory of Geotopes of national significance. *Geol. Insubr.* 4(1): 31–48.

Schaer, J.P. 2000. Agassiz et les glaciers: sa conduite de la recherche et ses mérites. *Eclogae Geol. Helv.* 93: 231–256.

Schardt, H. 1908. La Pierre des Marmettes et la grande moraine de Monthey. *Eclogae Geol. Helv.* 10: 555–566.

Strasser, A., Heitzmann, P., Jordan, P., Stapfer, A., Stürm, B., Voger, A. & Weidmann, M. 1995. *Géotopes et la protection des objets géologiques en Suisse: un rapport stratégique.* Fribourg: Groupe suisse pour la protection des géotopes.

Vischer, W. 1946. *Naturschutz in der Schweiz.* Basel: SBN Verlag.

Von Engelhardt, W. 1999. Did Goethe discover the ice age? *Eclogae Geol. Helv.* 92: 123–128.

Dimension Stone 2004, Přikryl (ed)
© 2004 Taylor & Francis Group, London, ISBN 90 5809 675 0

Implications of climate change and increased 'time-of-wetness' for the soiling and decay of sandstone structures in Belfast, Northern Ireland

B.J. Smith & P.A. Warke
School of Geography, Queen's University Belfast, UK

J.M. Curran
Stone Conservation Services, Belfast, UK

ABSTRACT: Climate change scenarios for Northern Ireland propose a trend towards wetter, warmer and longer winter conditions that could translate into an increase in the time-of-wetness for stone buildings across the Province. Studies from Belfast suggest that the effects of this could already be manifesting themselves through the 'greening' of sandstone buildings, as surfaces are colonised by algal and fungal growths, and the deep penetration of mobile salts into saturated stonework – possibly by ion diffusion. Such changes have major implications for building aesthetics, decay patterns and, ultimately, conservation strategies.

1   INTRODUCTION AND BACKGROUND

The performance in use of building stone results from complex interactions between stone properties and exposure environment that determine susceptibility to decay and ultimately any necessary conservation strategies. In this context the understanding of stone decay and conservation within the UK has been strongly conditioned by the dominance of southeast of England as a source of conservation case studies. In turn, this has meant that research has tended to centre on limestones, rather than the quartz sandstones that are more important to the north and west. Thinking on decay processes has likewise been strongly influenced by the decrease in mean annual rainfall towards the south and east and the consequent emphasis on a weathering environment characterised by relatively infrequent surface wetting followed by thorough drying of building facades. This facilitates the surface and near-surface crystallization of salts that soil buildings with black gypsum crusts and can physically disrupt porous stone of any mineralogy through flaking, scaling and granular disaggregation. Eventually, this thinking resulted in the development of stone durability tests, such as the sodium sulphate test, which reflect these assumptions. Performance in this test is now effectively accepted within the UK as an industry standard for the specification of new and replacement stone.

All of the above cornerstones of our understanding of stone decay are ultimately dependent upon the assumptions that the same weathering conditions pertain across the UK and that climate is stable. There is, however, increasingly compelling evidence that not only do weathering conditions vary, but that climatic regimes are changing.

2   CLIMATE CHANGE

The climate change scenarios for the northwest of the UK typically project both increased short-term uncertainty in day-to-day weather conditions and an underlying trend towards wetter, warmer and longer winter conditions (Betts 2001, Hulme & Jenkins 1998). In relation to the decay and conservation of stone structures, there are a number of implications for how buildings might respond to a possible increase in the time-of-wetness. As Viles (2002) has indicated, one possible consequence is a likely enhancement of chemical weathering in the northwest of the UK compared to an increase in crystallization damage in the southeast. However, prolonged, and presumably more deeply penetrating, wetness must also be considered in the context of how it will affect other agents of decay, including the internal distribution and impact of salts and the biological colonization of stone – especially if winter wetness is to be associated with warmth. Any such changes must in turn dictate how those with a duty of care for these structures approach their future cleaning and conservation. This brief paper will examine

9

the implications of these changes in environmental conditions through research studies carried out in the City of Belfast.

3 THE HISTORY OF STONE DECAY IN BELFAST

Belfast has a long history of stone decay problems related to high levels of atmospheric pollution (especially sulphur and particulates) that are accentuated in winter by anticyclonic conditions that trap pollutants over the city beneath a temperature inversion. This, when combined with a cool temperate maritime location, creates a particularly wet (c. 200 rain days per annum according to Betts (1997)), salt rich urban environment in which widely used quartz sandstones are prone to rapid soiling by gypsum crusts in areas sheltered from rainfall, and eventually to salt-induced decay (Smith et al. 1991). The rapidity with which new stone soils was demonstrated in exposure trials carried out in the centre of Belfast in the early 1990s by Moses (1996), who recorded salt accumulation and the beginning of gypsum crusts formation on sheltered stone tablets within six months of exposure. In contrast, tablets exposed to rainwash remained largely gypsum free and were characterised by the opening out of the pore structure. There is also much evidence of chemical modification in the extensive case hardening of sandstones by iron-rich crusts formed by the mobilisation and outward migration of iron oxides that occur as cements and/or nodules within the original stone (McAlister et al. 2003). In terms salt induced decay, the commonly used Triassic sandstones such as the local Scrabo and Dumfries from southwest Scotland appear to be particularly susceptibility to damage (Smith et al. 1994). For this reason recent renovations have sought replacement stone from further afield, such as Carboniferous Dunhouse sandstone from northeast England, that are perceived to be more resistant to salt weathering.

4 PREDICTED EFFECTS OF INCREASED TIME-OF-WETNESS

The fact that building stones, especially sandstones, already react so adversely and so rapidly to the moist, salt rich environment of Belfast must inevitably raise fears of what could happen following any projected increase in time-of-wetness. Indeed, there is already anecdotal and some photographic evidence that changes in damage patterns have already begun. The most obvious of these being the widespread 'greening' of sandstone buildings through algal growth on rainwashed surfaces (Figs 1 & 2). However, if stonework were to remain wetter for longer and not dry out either

Figure 1. A section of renovated sandstone wall from St Mark's Church East Belfast photographed in May 1999 immediately after it was cleaned and replacement stone inserted.

Figure 2. The same section of wall from St Mark's Church as in Figure 1. Photographed in October 2001 to show the extent of post-conservation biological colonisation.

as frequently or as thoroughly as before, there are numerous possible implications.

4.1 *Possible changes in the nature of salt-related stone decay*

1. Prolonged saturation of complete stone blocks.
2. Irregular and possibly incomplete surface drying for much of year.
3. Deeper penetration of salts (especially those of high mobility).
4. Movement of salt through stone by ion diffusion rather than in solution.

10

5. Presence of a store of 'deep salt' that may facilitate continued retreat of individual blocks following initial delamination or removal of surface layers during conservation.

4.2 Possible effects of increased biological colonisation of rainwashed stonework

1. Visual disamenity.
2. Active biological decay, both chemical and physical.
3. Decrease in surface porosity/permeability.
4. Continued wetting of stone by frequent rainfall and condensation.
5. Prolonged saturation following reduced evaporation.
6. Facilitation of dust deposition and the formation of complex bio-mineralogic crusts.

To examine the validity of some these projections, research has recently been conducted in Belfast to examine in detail salt concentration and associated decay through the case study of a badly weathered Scrabo sandstone church (St. Matthew's) close to the city centre. In addition, surface alteration to sandstones was examined through a programme of 'exposure trials' in which blocks of commonly used building stone were placed at sites across the City to monitor changes to blocks sheltered from and exposed to rainwash. Selected results from this research are presented in the following section.

5 RESEARCH FINDINGS

5.1 Exposure trials

The specific objectives were to assess adjustment on exposure of cleaned and replacement stone, to evaluate soiling rates and to assess relationships between permeability and soiling/crust formation on three sandstone types: Dunhouse, Scrabo and Dumfries. Exposure blocks were $150 \times 150 \times 100$ mm and a grid of 16 surface permeability measurements were carried out prior to exposure using probe permeametry (see Carey and Curran (2000) for methodology). A fixed grid was use to ensure accurate re-positioning of the probe tip for subsequent measurements. Each exposure rack held 6 inclined samples, one of each stone type exposed to rainwash and another set sheltered beneath an open shelf. Trials ran from April 1999 to March 2001 at 4 locations in the city and blocks were examined, photographed and re-measured at six-weekly intervals to visually assess the nature and extent of any soiling and possible changes in permeability. At the end of the trial, Scanning Electron Microscopy (SEM) was used to examine sections of the blocks' surfaces in detail.

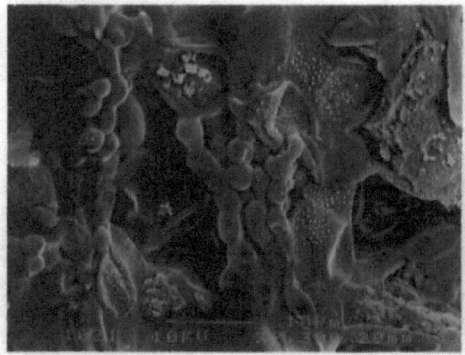

Figure 3. Scanning electron micrograph of a microbial community found on Dunhouse sandstone exposed to rainwash at Belfast Harbour for two years. Scale bar = 10 μm.

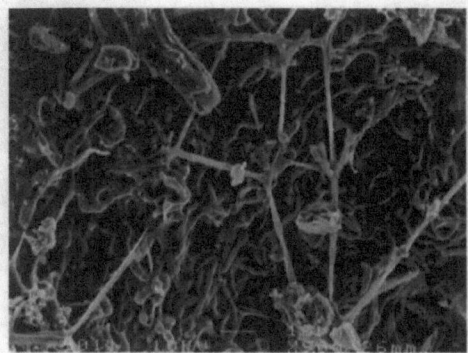

Figure 4. Fungal hyphae and algal cells/filaments found on Dumfries sandstone exposed to rainwash in North Belfast for two years. Scale bar = 10 μm.

Results from the trials revealed the widespread 'greening' of all three of the sandstones that were exposed to rainwash at all of the sites over the two years of exposure. This is particularly evident under the SEM, and Figures 3–5 illustrate the extent of both algal and fungal colonisation and their effectiveness in binding particulate pollutants to the stone surface. The extent and intensity of colonisation did, however, vary with sandstone type. Biological growth was, for example, generally more rapid and extensive on the highly porous Dumfries sandstone. This is thought to reflect the greater moisture holding capacity of this stone type which, when surface permeability is lowered promotes prolonged wetness. The open texture of Dumfries sandstone also promotes rapid colonisation within weeks of exposure. The exposure trial results are in contrast to soiling patterns recorded during previous exposure trials carried out in the early

11

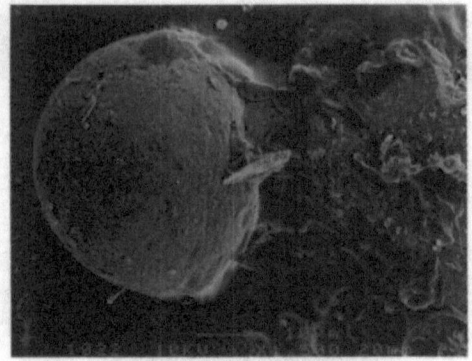

Figure 5. Mucilage binding a coal flyash particle to the surface of a Dunhouse sandstone block exposed to rainwash at Belfast harbour for two years. Scale bar = 10 μm.

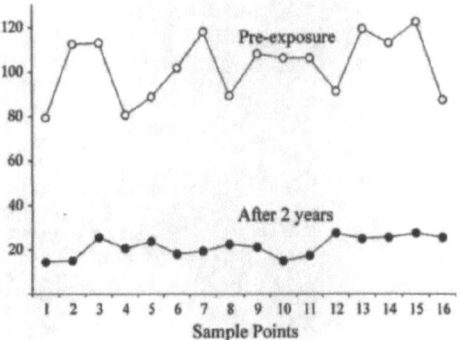

Figure 6. Diagram showing the reduction in surface permeability due to biological colonisation of a Dunhouse sandstone block exposed to rainwash for two years at Belfast Harbour.

1990s in Belfast, in which biological colonisation of rainwashed stone surfaces did not occur to any where near the same extent.

Results from the monitoring of surface porosity/ permeability characteristics show an early increase in permeability followed by a rapid decrease from 5 months onwards. The initial increase could reflect the washing out of pores, together with some possible dissolution of grains and cement. The subsequent permeability decrease indicates occlusion of pores and after <2 years, biological colonisation had caused major reductions on all of the blocks exposed to rainwash (e.g. Fig. 6). The effect of the biological growth is therefore to effectively seal the stone surface in a manner that should have a major impact on moisture absorption and retention – together with any dissolved salts. In particular, it is envisaged that colonisation will restrict moisture loss through evaporation and hence further increase time-of-wetness, establishing a positive feedback encouraging further colonisation. These effects are in addition to the visual damage to structures consequent on greening and enhanced physical and chemical weathering of the stone by a range of organisms.

Increased time-of-wetness is most likely to affect stonework exposed directly to driven rain and rainwash, although increased atmospheric humidity might also reduce evaporation rates from sheltered stonework. However, in the continued presence of atmospheric pollution, results from the sheltered exposure blocks showed both particulate deposition and early evidence of gypsum crust formation. Gas permeameter measurements showed no significant fall in surface porosity/ permeability. However the findings do suggest that buildings could continue to experience black crust formation on sheltered areas that will necessitate ongoing cleaning and re-cleaning.

5.2 Salt penetration into stonework

The possible effects of increased time-of-wetness on patterns of salt-induced decay were investigated through a detailed study of St Matthew's church prior to and during recent conservation. This provided the opportunity to remove and examine complete blocks of weathered stone. Because most studies of building stone decay have been restricted to sampling material that has fallen away or that can be prised from the outer few millimetres of a block, surprisingly little is known of the patterns of salt distribution with depth into weathered stonework. In this study it was therefore possible for the first time to measure salt content throughout complete building blocks c. $30 \times 30 \times 30$ cm. These concentrations could then be related to observed patterns of decay.

The results from this study are given in detail in Warke & Smith (2000), which describes salt and decay relationships in two blocks from the Church, one with a relatively sound outer surface and the other with a surface characterised by active scaling and flaking. Results from both blocks show that calcium and sulphate (gypsum) are present to a depth of 40–50 mm, but are concentrated within the outer 10–20 mm of the more stable block where they are associated with a zone of incipient scaling. Examination by SEM of a polished subsurface profile showed these contour scales to be made up of several thinner flakes held together by crystallized salt. Below the contour scales is a 'pre-weathered' zone in which crystallized salts have opened out the pore structure and reduced the structural integrity of the stone. Once the contour scale falls away, it is possible that this pre-weathered zone will be exploited by salt weathering, leading to the rapid retreat of the block. In contrast to calcium and sulphate

12

measurements, those for sodium, magnesium and chloride show that more mobile chloride salts – possibly derived in part from marine aerosols – had penetrated throughout the blocks.

From these observations, and studies of decay patterns on the church, it appears that the slow accumulation of gypsum progressively weakens the stone by opening out its pore structure. As the porosity is opened out, more mobile chloride salts can exploit resultant fractures. The presence of chlorides can also increase the solubility of gypsum, which could enhance its capacity to disaggregate the stone. This disaggregation eventually manifests itself as a series of surface-parallel fractures. Following this initial delamination, retreat continues through multiple flaking that exploits the combination of a store of 'deep salts' and the 'pre-weathering' that occurred beneath the contour scales (Smith et al. 2002). In a parallel study, other blocks from the church were cut into 20 mm cubes and analysed to produce three-dimensional maps of anion and cation distribution (Turkington and Smith 2000). This study confirmed the presence of chloride and, to a lesser extent, sulphate ions throughout the blocks, but also identified numerous 'hotspots' that could provide potential salt reservoirs to fuel decay as the outer surface retreats. Interestingly, the results also showed inconsistent correlations between, for example calcium and sulphate and sodium and chloride. This could suggest that the widespread distribution of these ions within the blocks may reflect, in part at least, migration by ion diffusion during prolonged periods of saturation.

6 IMPLICATIONS AND CONCLUSIONS

Increased time-of-wetness will require rethinking of not only how stone buildings are managed and conserved, but also how we view the operation of decay processes. For example, how will salt weathering operate on sandstones should there be an increase in wetting and reduction in the effectiveness of drying within the near-surface zone of stonework? Under these conditions it is likely that the outer few millimetres of stone will continue to wet and dry on a regular basis and effects such as granular disaggregation (sanding) could continue. It is also likely that less soluble salts such as gypsum will continue to concentrate just below the surface and ultimately lead to contour scaling. However, it can be envisaged that some gypsum and more soluble salts such as halite will migrate towards the interiors of stone blocks both in solution and by ion diffusion. Consequently, once decay commences through the loss of an outer contour scale, it is more likely to continue as this store of deep salts is exploited. Thus, once decay is triggered it is difficult to switch it off and it becomes increasingly necessary to replace complete blocks of stone during conservation.

Increased time-of-wetness of stone exposed to rainwash is also likely to encourage surface and sub-surface colonisation by algal, fungal and eventually lichen growth. The potential for the 'greening' of building stone is demonstrated on many buildings across Belfast where failure of drainage systems leads to the formation of green watermarks and stains on façades. There is, however, growing anecdotal evidence, now supported by results from exposure trials, that 'greening' has become more widespread and intense in recent years. This is especially true for sandstones that provide acid, open-textured conditions that favour algal colonisation. The implications of 'greening' go beyond simple aesthetics and algae and lichen are effective agents, both directly and indirectly, of stone decay through enhanced dissolution and physical disruption of individual grains (Saiz-Jimenez 1994). Colonisation will also affect moisture absorption and retention characteristics of stone. In particular, it is envisaged that colonisation will restrict moisture loss through evaporation and hence further increase 'time-of-wetness'. In this way a 'positive feedback' could be established whereby further colonisation is encouraged and stonework is kept wet for much of the year.

These changes in soiling and decay patterns also have clear implications for the ways in which stone buildings may have to be managed and conserved in the future. They may, for example, require the more frequent application of biocides, together with a need to investigate the effects of saturation on the penetration and effectiveness of surface treatments. Only in this way will it be possible to determine the optimum quantity, concentration, timing and method of application. The use of water repellents may also have to be re-evaluated in terms of their performance in response to prolonged wetness. As will the currently favoured policy of dressing back stone to remove the outer, weathered layer.

ACKNOWLEDGEMENTS

Financial support was provided by EPSRC grants GR/L99500/01 and GR/L57739/01. Thanks are also due to the staff of Consarc Design Group Ltd and to Gill Alexander of the QUB Geography cartographic unit.

REFERENCES

Betts, N. L. 1997. Climate. In J. G. Cruickshank (ed), *Soils and Environment: Northern Ireland*: 63–84. Department of Agriculture Northern Ireland and Queen's University Belfast, Belfast.
Betts, N. L. 2001. Climate Change in Northern Ireland, In: A. Smith, W. I. Montgomery, D. Favis-Mortlock & S. Allen (eds), *Implications for Climate Change for Northern*

Ireland: Informing Strategy Development: 26–44. London: The Stationery Office.

Carey, P. & Curran, J. M. 2000. High resolution characterisation of permeability in arenaceous building stones. *Zeitschrift für Geomorphologie* 120: 175–185.

Hulme, M. & Jenkins, G. (1998) *Climate Changes Scenarios for the United Kingdom.* UKCIP Technical Note No. 1, Climate Research Unit, Norwich, UK.

McAlister, J. J., Smith, B. J. & Curran, J. M. 2003. The use of sequential extraction to examine iron and trace metal mobilisation and the case hardening of building sandstone: a preliminary investigation. *Microchemical Journal* 74: 5–18.

Moses, C. A. 1996. Methods for investigating stone decay mechanisms in polluted and 'clean' environments, Northern Ireland. In B. J. Smith & P. A. Warke (eds), *Processes of urban Stone Decay*: 212–227. London: Donhead.

Saiz-Jimenez, C. 1994. Biodeterioration of Stone in Historic Buildings and Monuments. In G. C. Llewellyn, W. W. Dashek, & C. E. O'Rear (eds), *Biodeterioration 4*: 587–603. New York: Plenum.

Smith, B. J., Magee, R. W. & Whalley, W. B. 1994. Breakdown patterns of quartz sandstones in a polluted urban environment: Belfast, N. Ireland. In D. A. Robinson and R. B. G. Williams (eds.), *Rock Weathering and Landform Evolution*: 131–150. Chichester: Wiley.

Smith, B. J., Whalley W. B. & Magee, R. W. 1991. Background and local contributions to acidic deposition and their relative impact on building stone decay: A case study of Northern Ireland. In J. W. S. Longhurst (ed), *Acid Deposition: Origins, Impacts and abatement strategies*: 241–266. Berlin: Springer Verlag.

Smith, B. J., Turkington, A. V., Warke, P. A., Basheer, P. A. M., McAlister, J. J., Meneely, J. & Curran, J. M. 2002. Modelling the rapid retreat of building sandstones, A case study from a polluted maritime environment. In S. Siegemund, T. Weiss & A. Vollbrecht (eds), *Natural stone, weathering phenomena, conservation strategies and case studies.* Geological Society Special Publication 205: 347–362.

Turkington, A. V. & Smith, B. J. 2000. Observations of three-dimensional salt distribution in building sandstone. *Earth Surface Processes and Landforms* 25: 1317–1332.

Viles, H. A. 2002. Implications of future climate change for stone deterioration. In S. Siegemund, T. Weiss & A. Vollbrecht (eds), *Natural stone, weathering phenomena, conservation strategies and case studies.* Geological Society Special Publication 205: 407–418.

Warke, P. & Smith, B. J. 2000: Salt distribution in clay-rich weathered sandstone. *Earth Surface Processes and Landforms* 25: 1333–1342.

14

*Geology of dimension stone: from historical
applications to potential resources*

Dimension Stone 2004, Přikryl (ed)
© 2004 Taylor & Francis Group, London, ISBN 90 5809 675 0

The characterization of the Gorgoglione sandstones (Basilicata, South Italy)

A. Calia & M. Masieri
CNR IBAM, Lecce, Italy

A. Lettino
Dipartimento Geomineralogico, Università degli Studi di Bari, Italy

ABSTRACT: Two types of sandstone commonly used as construction material in historical buildings in many places in the Basilicata region (south of Italy) were studied. The two types of material are also widely appreciated in modern construction, to the point that an extension of the quarry is expected. This work represents the first phase of their characterization in order to establish an appropriate and correct use of these stones in new construction, and additionally, to study suitable processes for their improvement and correction. The mineralogical–petrographical, geochemical and physical study revealed that the properties of the two types of stone differ considerably, and are thus likely to behave quite differently in their handling.

1 INTRODUCTION

Two types of sandstones commonly used as construction material in historical constructions in many places in the Basilicata region (south of Italy) were studied. One was used in the *opus incertum* of the masonry, the other in decoration or finishing (frames, thresholds, etc.). Their importance, until recently limited to the local area, is increasing steadily with their export to various Italian regions, mainly in the northern and central areas, due to their superior quality. They are mainly used for paving and cladding, pre-treated to make them water-resistant. The rising demand for these stones requires the enlargement of the extractive site and the establishment of a pool of contractors for the extraction and working of the material. Gorgoglione sandstones are well known and studied from a strictly geological standpoint but there is no study concerning their behaviour in construction works. The stones in buildings undergo varieties of decay processes, whose effects depend both on their intrinsic properties and their exposure (Winkler 1994). This study represents the first phase in an effort to determine their durability. Our purpose is to establish the best and most suitable application of these materials in new constructions, and additionally, to study suitable treatments (either for conservation in historical buildings, or prior to use in new buildings), to correct and improve. With a view to this, in addition to mineralogical–petrographical and geochemical study, special attention was addressed to certain physical characteristics (porosity and behaviour

with respect to water), that are considered as important parameters for characterizing materials, predicting their behaviour under weathering conditions, evaluating the degree of decay and establishing the effectiveness of treatments of conservation or improvement (Borrelli 1999, Torraca 1991). For the physical characterization tests we used the UNI-Normal rules and the Normal recommendations for preservation studies, as they apply to the characterization of stone materials in relation to their degradation and the treatments used (Rec. Normal 20/85, Normal Commission, 1993, Appollonia et al. 1995).

2 QUARRIED MATERIALS

The present extractive site is located mainly along the Gorgoglione river (Fig. 1). There are two different varieties of stone: one generally massive, at times slightly laminated, medium grained, varying from light grey to beige in colour (hereafter Ms) found in layers up to 2–4 m thick; the other is darker, from bluish-grey to ochre-yellow, (hereafter Ls), found in layers up to 0.5 m thick, fine grained, with evident lamination and more cemented. Both types show frequent clear chromatic variations from grey to reddish-ochre. The material is extracted in blocks and slabs, by cutting along the natural joints (fractures and stratifications) in the outcrops. The extracted material is transported to a storage area and then worked in specialized plants. The initial selection of the material is based on its reaction

17

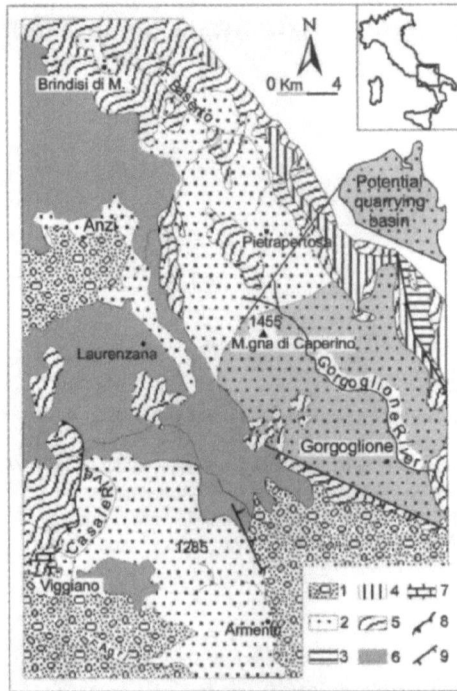

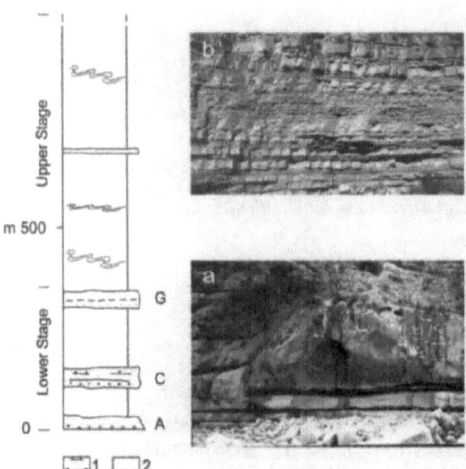

Figure 2. Representative sketch of the most abundant depositional features in the studied stages or systems; 1) Coarse-grained facies (a); 2) Thin beds of mudstone and fine sandstones (b).

Figure 1. Location and geological sketch map of the Gorgoglione Formation. 1) Plio-Pleistocene units; 2) Gorgoglione Formation; 3) Serra Palazzo Formation; 4) Numidian sandstones; 5) Flysch Rosso; 6) Sicilide units; 7) Mesozoic-Tertiary carbonate sequences; 8) Reverse faults; 9) Faults.

to natural freeze-thawing cycles. Due to their textural characteristics, laminated sandstones are not suitable for special stone-working, but are utilized in slabs obtained by exploiting its weakest points, which correspond to the lamination planes.

If we observe the stones inside historical buildings, we note that the Ms sandstone, which is used in decoration or finishing, shows exfoliation and erosion. In contrast, the Ls sandstone, which is used in the *opus incertum* of the masonries, presents a better state of conservation; this could be also due to the fact that the *opus incertum* in the past was protected by the plaster, while at nowadays modern restorations lead to the direct exposition of the surface masonry, without any protective layer.

3 GEOLOGICAL STUDY

The sandstones studied here belong to the Gorgoglione Formation (Fig. 1), a torbiditical silicoclastical succession of the Langhian medium-Tortonian inferior

age (Boenzi & Ciaranfi 1970). The whole succession, in the Gorgoglione area, is considered a turbidite complex (sensu Mutti & Normark 1987) subdivided into two stages or systems (Loiacono 1993) (Fig. 2). In the lower stage, conglomerate and coarse sandstone bodies, up to 100 m thick, are present. The fining upward sequences may be considered fillings of depressions in the underlying deformed units (units A, C and G). Sandy facies, with thick Tb and thin Tc Bouma interval are abundant in this stage. The upper stage is constituted of thin bedded fine grained facies corresponding to a widespread deepening of the basin. The main characteristic of this stage is the presence of middle-thin beds of medium to fine grained sandstones, showing Tb-e and Tc-e Bouma sequences. Within this complex, the two varieties of sandstone described in this study are contained in the Ta-c intervals, which are those that may be exploited. Therefore, based on the geological situation described above, it is possible to identify new areas of interest regarding the extraction of these stones and consequently to suggest points in which a potential enlargement of the extraction area (Fig. 1) may be carried out. Within this area, the layers are near-horizontal; thick massive sandstone layers appear at the lower levels, while thin layers of laminated sandstones prevail at the higher levels.

4 ANALYTICAL METHODS

The two sandstone varieties were characterized from the mineralogical–petrographical, geochemical and

18

physical standpoint. The samples studied come from one of the quarries.

The petrographic study was performed by observing thin sections with an optical microscope, and conducting a modal analysis through point-counting in accordance with the Gazzi-Dickinson method (Ingersoll et al. 1984). The Geochemical study was performed through X-ray fluorescence (XRF) spectrometry; major and trace elements were determined by following the procedures outlined by Franzini et al. (1975) and Leoni and Saitta (1976); the LOI was determined gravimetrically. For the study of the physical characteristics, the following parameters were determined, i.e.:

- specific weight (δr), by measuring the real volume of the samples with a Quantachrome helium picnometer
- volume weight (δa), by measuring apparent volume with a Chandler Engineering mercury picnometer
- compaction degree (C), determined from the ratio between volume weight and specific weight
- total porosity (Pt), measured with the following formula: Pt = ($\delta r - \delta a$)/δr) * 100

In addition, we performed:

- porosimetric analysis, determining open integral porosity (P) and pore distribution in accordance with their diameter (Rec. Normal 4/80. 1980), using a Thermo Finnigan mercury porosimeter, along with a Thermo Quest macropores unit
- tests of water absorption by capillarity (UNI 10859. 2000) on 10 samples, 5 × 5 × 2 cm in size for each type, 5 of which were cut parallel to the lamination planes and 5 perpendicular to them
- tests of water absorption by total immersion (Rec. Normal 7/81. 1981) on 5 cubic samples, 5 × 5 × 5 cm per type
- tests of water evaporation (Rec. Normal 29/88. 1991) on the same samples used in the test of water absorption by total immersion
- tests of permeability to water vapour (Rec. Normal 21/85. 1986) on 10 samples, 5 × 5 × 1 cm in size for each type, 5 of which were cut parallel to the lamination plans and 5 perpendicular to them.

5 RESULTS

5.1 *Petrography, mineralogy and geochemistry*

The massive sandstones (Ms) are grain size ranging between 0.25 mm and 0.5 mm (medium sand). They are light grey to tobacco coloured, medium lithified sandstones with poor to moderate sorting, and are of moderately spherical angular to sub-angular grain (Fig. 3A). The laminated sandstones (Ls) are fine grained, with grain size ranging between 0.125 mm

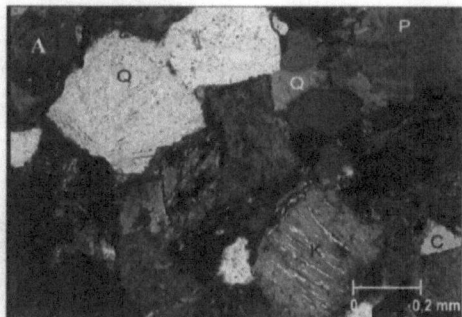

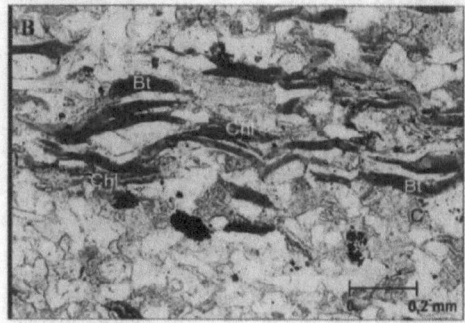

Figure 3. Photomicrographs of sandstone types. (A) Massive sandstone, crossed polarizers. (B) Laminated sandstone; note the alignment of micas (mostly biotite and chlorite) and their concentration on certain levels, plane-polarized light.

and 0.25 mm. In general they are grey-bluish to yellowish coloured, moderately sorted, with constituent grains that are commonly angular to sub-angular with low estimated sphericity (Fig. 3B). The nature of their constituents is the same, although the quantitative ratio between them varies. Major framework components are quartz and feldspar, with lesser quantities of biotite, chlorite, muscovite and granophyre, with granitic, metamorphic and sedimentary rock fragments. Heavy minerals include tourmalina, zircon, apatite and epidote. The average grain composition of the massive type is $Q_{62} F_{32} L_6$. Laminated sandstones have an average composition of $Q_{68} F_{29} L_3$. Therefore, all studied samples can be classified as Arkoses. Quartz (Qt), and specifically monocrystalline quartz (Q_m), is the major detrital phase in the sandstones ($Q_{t(MS)}=49\%$, $Q_{t(LS)}=39\%$). It shows straight to composite extinction, and has zircon, apatite and fluid inclusions to varying degrees. Both plagioclase ($P_{(MS)}=15\%$, $P_{(LS)}=7\%$) and potassium feldspar ($K_{(MS)}=11\%$, $K_{(LS)}=9\%$) are present in all the samples. Plagioclase composition is oligoclase with characteristic polysynthetic twinning.

19

It is commonly altered and replaced by sericite, or more rarely, by calcite. Potassium feldspar is commonly clouded with clay minerals and may also show microperthitic intergrowth with plagioclase (albite dominant). Rock fragments consist of plutonic, metamorphic and sedimentary lithics, in order of decreasing abundance. They are more common in coarse-grained samples. Plutonic rock fragments ($R_{g(MS)=8\%}$, $R_{g(LS)=1\%}$) are clasts, of granitic (acidic) composition and hypabyssal origin; metamorphic lithics ($L_{m(MS)=3\%}$, $L_{m(LS)=1\%}$) come from low-grade metamorphic rocks and vary from quartz-chlorite-sericite to quartz-muscovite in composition.

Sedimentary lithics ($L_{s(MS)=2\%}$, $L_{s(LS)=1\%}$) include carbonate and siltstone. Phyllosilicates ($M_{(MS)=3\%}$, $M_{(LS)=7\%}$) are generally fresh to altered biotite with oxidation and leaching common, apparently the effects of weathering. However, elongate laths and small plates of white mica and chlorite are also present. Matrix is present in the same quantity in both sandstones ($Matr_{(MS)=7\%}$, $Matr_{(LS)=8\%}$) and is generally composed of finely comminuted and altered lithic and feldspathic fragments, quartz, chlorite, micas, opaques and material, too fine to be identified.

From the textural standpoint the two sandstones show considerable differences. In Ms type sandstone, penetrating grains and seams of insoluble residue indicate varying degrees of dissolution pressure. Its cementation is due therefore in the first place to material compaction and secondly to the presence of calcitic cement ($Cem_{(MS)=10\%}$). In Ls type sandstone, the grains are poorly compacted, and the intergranular space is occupied by abundant carbonatic cement, in this case calcite or microspar ($Cem_{(LS)=27\%}$). In addition, Ls type sandstone is characterized by thin laminations due to the presence of levels in which phyllosilicates and iron oxides are concentrated. These laminae also correspond discontinuities within the material, as indicated by the presence of elongated pores.

The macroscopically evident chromatic shifts in Ls from grey to ochre seem to be due to the presence of iron hydroxides, generated in diagenetic environments by the action of fluids that have permeated the material along its discontinuities (see the preferential localization along the stratification structures and the intergranular zones in the matrix).

The major and trace-element geochemistry of the sandstone samples is almost uniform. Ls sandstones have higher CaO ($CaO_{(MS)=6\%}$, $CaO_{(LS)=20\%}$) and lower SiO_2 ($SiO_{2(MS)=69\%}$, $SiO_{2(LS)=56\%}$), reflecting preferential incorporation of calcite cement. Low $TiO2$, Ni and Cr in these samples indicate little or no mafic or ultramafic rocks in the provenance.

These sandstones appear to be derived mostly from granite, granodiorite, and mafic-poor metamorphic rocks, as confirmed by their detrital modes.

Table 1. Summary of physical parameters of the sandstones.

Samples	δr [g/cm³]	δa [g/cm³]	C	Pt [%]	P [%]
Ms	2.65	2.43	0.92	8.3	8
Ls	2.65	2.60	0.98	1.9	1.9

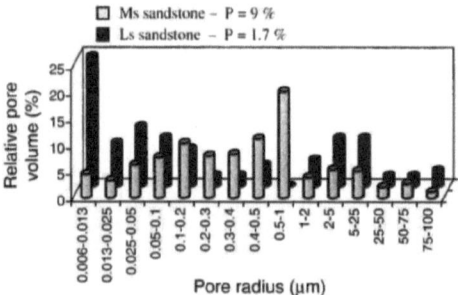

Figure 4. Example of pore size distribution.

XRD analyses confirmed the presence of the main mineralogical phases observed in the thin sections, and showed that the passage from Ms to Ls is characterised by an increase in carbonates and phyllosilicates and a decrease in quartz and feldspars.

5.2 Physical characteristics

5.2.1 Density and porosity
Table 1 shows the mean values of δr, δa, C, Pt, P, measured on 5 samples for each type.

The two sandstones differ considerably in the porosity values (total and open integral porosity) and in porosimetric distribution.

Ms sandstone samples show the majority of pores between 0.1 and 1 μm one micron, with a remarkable peak inside this interval. Pores above 1 μm and under 0.1 μm are scarcely present (about 20%).

In contrast, in the Ls type samples prevail smaller pores (under 0.3 μm); pores having radius above 2 μm are about 30%, while those in the intermediate intervals are very scarcely present. Figure 4 illustrates an example of pore size distribution in the two sandstones.

5.2.2 Behaviour with respect to water
In order to perform a deeper study of the two materials' behaviour in regard to water, and to see its evolution over time, the tests of water absorption by capillarity and by total immersion, and the water evaporation tests were prolonged past the times indicated by the norms, until no more weight variations were noted in the samples.

Figures 5a and b, show the quantities of water absorbed by capillarity over time for samples Ms and Ls respectively, cut parallel and perpendicular to the

20

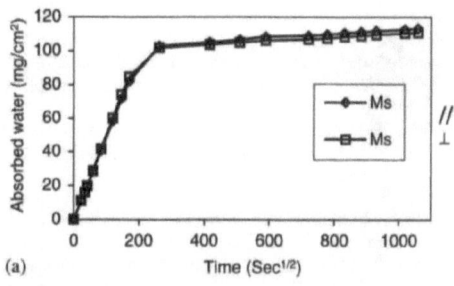

(a)

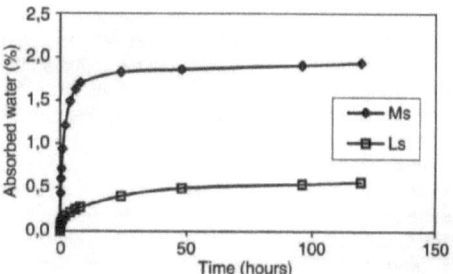

Figure 6. Diagram of water absorption by total immersion.

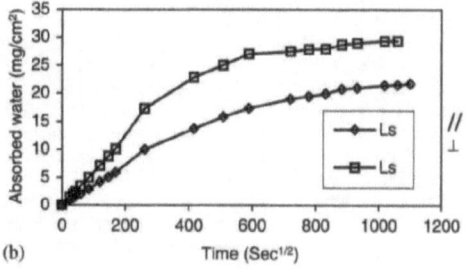

(b)

Figures 5a and b. Diagrams of water absorption by capillarity.

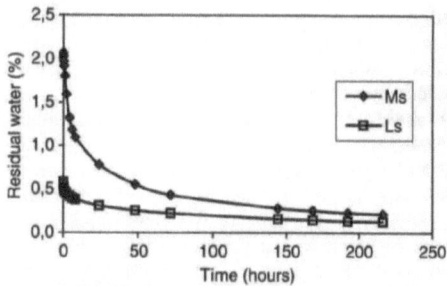

Figure 7. Diagram of evaporation of water.

laminations (// and ⊥ respectively). Note that in Ms sandstone, the influence of the lamination structures' orientation on water absorption is negligible: both series of samples are characterized by fairly fast absorption in the initial phase of the test. The total quantity of water absorbed (calculated after 13 days), is equal to $113.27\,mg/cm^2$ for Ms // samples and $110.64\,mg/cm^2$ for Ms ⊥ samples. The capillary absorption coefficient (CA) for both Ms // and Ms ⊥ samples is equal to $5 \times 10^{-4}\,mg/cm^2\,s^{1/2}$.

Ls samples absorb less water more slowly than Ms samples, with different behaviour in relation to the orientation of the lamination planes. Ls // samples absorb a total quantity of water (calculated at 14 days) equal to $21.17\,mg/cm^2$ and have a CA of $4 \times 10^{-5}\,mg/cm^2\,s^{1/2}$. Ls ⊥ samples absorb in total (after 13 days) $29.42\,mg/cm^2$ of water and have a CA equal to $6 \times 10^{-5}\,mg/cm^2\,s^{1/2}$. As may be noted, water absorption is higher in the samples cut perpendicularly to the lamination planes than those cut parallel to them, with a difference equal to 26%.

Figure 6 shows the percentages in weight of water absorbed during total immersion by Ms and Ls samples over time. Ms samples are characterized by a faster absorption than Ls samples in the initial phase of the test. The imbibition coefficient (calculated at 5 days) is equal to 2.07% for Ms sandstone and 0.59% for Ls sandstone.

Figure 7 shows the percentages of residual water in time-based tests. As may be noted, in Ms type evap-

oration is much faster. In addition, the drying index (calculated at 9 days) is equal to 0.21, while for Ls sandstone it is equal to 0.10. This corresponds for Ms sandstone to 10% of residual water at the end of the test and to 20% for Ls sandstone. Permeability to water vapour in Ms samples is equal to 85.90 and to $92.69\,g/m^2 \times 24$ hours, for the samples perpendicular and parallel to the laminations respectively. In Ls samples it is 3.56 and $4.85\,g/m^2 \times 24$ hours, respectively, clearly less than the other type. Furthermore, while the influence of the orientation of the laminations is minimum on the behaviour of. Ms samples (the difference between the values measured is equal to about 7%), it becomes important in Ls samples, where the permeability is 30% higher perpendicular to the lamination plane.

6 CONCLUSIONS

From the combined geological study and sedimentology analysis resulted that the two Gorgoglione sandstones derive from Ta-c interval of a turbidite complex; this interval is abundant in the Gorgoglione area. Mineralogical–petrographic and geochemical features indicate that source rocks are quartzofeldspathic, mainly acid plutonic to gneissic with associated low-grade metamorphites, and, in minor proportion,

21

sedimentary. The nature of their constituents is the same, although the quantitative ratio between them varies. Both the studied sandstones are compositionally arkoses. Considerable differences are observed at the structural level in the two varieties. In Ms type sandstone, cementation results in the first place from the material compaction and to a lesser degree from the presence of calcitic cement. In Ls type, the grains are not highly compacted and the intergranular space is filled by abundant carbonatic cement. In addition it is characterized by thin lamination due to the presence of levels where phyllosilicates and iron oxides are concentrated. These stratifications also correspond to discontinuities within the material, as indicated by the presence of elongated pores. These differences are reflected in porosity values that, although low, are fairly different for the two materials, and in clearly different porosimetric distributions. Ms sandstones are characterized by pores mostly between 0.1 and 1 μm. From the petrographic study, it may be hypothesised that the porosity is mainly of an intergranular type. In contrast, Ls has very few pores in the same interval and the highest concentration of pores in the very small class (probably porosity of an intercrystallinic type, in the clacitic cement discontinuities), and a fair amount of pores bigger than 2 microns (probably located mainly along the lamination planes observed in the material). With regard to the different porosimetric characteristics, Ms sandstones have the highest and most rapid absorption of water, the quickest evaporation of the same, the highest permeability to vapour and the material behaves isotropically. In Ls, the lower porosity and the prevalence of smaller pores result in slower and smaller water absorption, both by capillarity and total immersion, slower evaporation and lower permeability. Moreover, it behaves anisotropically due to the presence of lamination structures. In conclusion, the mineralogical–petrographical study, the geochemical and the study of certain physical proprieties revealed noticeable differences that lead us to expect different handling properties in the two varieties of material.

These results are part of the first phase of the characterization of these stones in relation to their durability. The study of their degradation on site along with the performance of artificial ageing which reproduces in the laboratory the real deterioration factors, may help us to understand the behaviour in relation to degradation, and the experiments with superficial treatments may suggest possible actions on them.

REFERENCES

Appollonia, L., Fassina, V., Matteoli, U., Mecchi, A.M., Nugari, M.P., Pinna, D., Peruzzi, R., Salvatori, O., Santamaria, U., Scala, A. & Tiano, P. 1995. Methodology for the evaluation of protective products for stone materials. Part II: Experimental tests on treated samples. *International Colluquium, Methods of evaluating products for the Conservation of Porous Building Materials in Monuments*. ICCROM, Roma.

Boenzi, F. & Ciaranfi, N. 1970. Stratigrafia di dettaglio del Flysch di Gorgoglione (Lucania). *Memorie della Società Geologica Italiana* 9: 65–79.

Borrelli, E. 1999. Conservation of architectural heritage, historic structures and materials: porosity. ARC – Laboratory Handbook. ICCROM. Roma.

Catalano, S., Carbone, S. & Lentini, F. 1993. Il Flysch di Gorgoglione nell'ambito dell'evoluzione dell'Appennino lucano. *Giornale di Geologia* 55(1): 165–178.

Franzini, M., Leoni, I. & Saitta, M. 1975. Revisione di una metodologia analitica per fluorescenza X, basata sulla correzione completa degli effetti di matrice. *Società Italiana di Mineralogia e Petrologia* 31: 365–378.

Ingersoll, R.V., Bullard, T.F., Ford, R.L., Grimm, J.P., Pickle, J.D. & Sares, S.W. 1984. The effect of grain size on detrital modes: a test of the Gazzi-Dickinson point-counting method. *Journal of Sedimentology* 54: 103–116.

Leoni, L. & Saitta, M. 1976. X-ray fluorescence analysis of 29 trace elements in rock and mineral standards. *Rend. Società Italiana di Mineralogia e Petrologia* 32(2): 497–510.

Loiacono, F. 1993. Geometrie e caratteri deposizionali dei corpi arenacei nella successione stratigrafica del Flysch di Gorgoglione (Miocene superiore, Appennino meridionale). *Bollettino della Società Geologica Italiana* 112: 909–922.

Mutti, E. & Normark, W.R. 1987. Comparino exaples of modern and ancient turbidite systems: problems and concepts. In J.K. Leggett & G.G. Zuffa (eds), *Marine Clastic Sedimentology*: 1–38. London: Graham and Trotman.

Normal Commission – Protectives Experimentation Subgroup. 1993. Metodologia per la valutazione di prodotti impiegati come protettivi per materiale lapideo. Parte I: Test e trattamento dei campioni. Da *"L'Edilizia"* anno VII.

Rec. Normal 4/80, 1980. Distribuzione del volume dei pori in funzione del loro diametro, C.N.R.-I.C.R., Roma.

Rec. Normal 7/81, 1981. Assorbimento d'acqua per immersione totale – Capacità di imbibizione, C.N.R.-I.C.R., Roma.

Rec. Normal 21/85, 1986. Permeabilità al vapor d'acqua, C.N.R-I.C.R., Roma.

Rec. Normal 20/85, 1986. Interventi conservativi: progettazione, esecuzione e valutazione preventiva, C.N.R.-I.C.R., Roma

Rec. Normal 29/88, 1991. Misure dell'indice di asciugamento (dryng index), C.N.R.-I.C.R., Roma.

Torraca G. 1981.Porous materials buildings. *Materials science for architectural conservation*, ICCROM, Roma.

UNI-Normal 10859. 2000. *Determinazione dell'assorbimento d'acqua per capillarità*. Milano.

Winkler, E. 1994. *Stone in architecture*. 3rd ed., Berlin: Springer-Verlag.

Dimension Stone 2004, Přikryl (ed)
© *2004 Taylor & Francis Group, London, ISBN 90 5809 675 0*

Mineralogy, petrophysical properties and durability of the Udelfangen sandstone (*Muschelsandstein*, Lower Muschelkalk, Germany)

C.W. Dubelaar
Netherlands Institute of Applied Geoscience TNO, Utrecht, Netherlands

ABSTRACT: The Udelfangen sandstone (Lower Muschelkalk, Germany) has been used as a restoration stone in the Netherlands from the second half of the 19th century. Within fifty years after its application part of the Udelfangen sandstone already had to be replaced. To get more understanding of the factors that determine the weathering of the Udelfangen sandstone the mineralogy and petrophysical properties have been studied in detail. The Udelfangen sandstone is dominated by fine-grained quartz, lithic fragments and micas (mainly muscovite) cemented by carbonate (dolomite). A comparison was made between fresh quarry samples and samples from the Remonstrant Church in Rotterdam, built in 1897. Porosity in all samples is high (17–27 vol. %) and permeability varies between 0.75–376 milliDarcy. Differences in carbonate (dolomite) content and the amount of sulphate point to leaching and crystallisation of salts in the weathered samples. It is concluded that the different qualities of the stone, such as, layers rich in loam and layers with high mica and high carbonate content, determine the weathering behaviour of the Udelfangen sandstone.

1 INTRODUCTION

1.1 *Geological setting*

The Udelfangen sandstone is situated at the base of the Lower Muschelkalk serie in the south Eifel, with outcrops near the village of Udelfangen, west of Trier (Fig. 1).

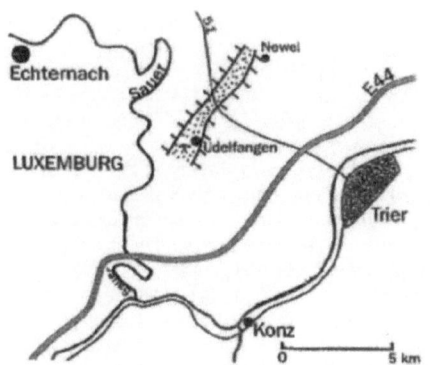

Figure 1. Outcrop of the Udelfangen sandstone in the tectonic high northwest of Trier (Rheinland-Pfalz, Germany). Modified after Negendank (1983).

Deposits of the Lower Muschelkalk in the Trier region are subdivided in the so-called *Orbicularis* layers, a 3–5 m thick sandy dolomite, underlain by a 45 m thick sequence of calcareous (dolomitic) sandstone. Due to the occurrence of a rich mollusc fauna, e.g. *Myophoria Orbicularis*, the sandstone has been referred to as Muschelsandstein in the German literature (Negendank 1983).

1.2 *Historical use of the Udelfangen sandstone*

The first use of the Udelfangen sandstone dates back to the medieval epoch (Liebfrauenkirche, Trier, 13th century). In the village Udelfangen itself the oldest examples are tombstones at the local churchyard which date from the 18th century. In the Netherlands the Udelfangen sandstone became popular in the second half of the nineteenth century. It has been used as a restoration stone instead of calcareous sandstones from the Eocene in Belgium (Fig. 2) and Pleistocene tuffs from the northern volcanic Eifel (Slinger et al. 1980, Dubelaar 1984).

1.3 *Aim and scope of the study*

Time has proven that the Udelfangen sandstone does not belong to the type of stones that are most resistant to weathering. To get more knowledge of the weathering

23

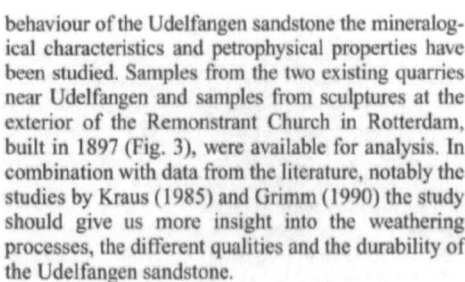

Figure 2. West face of St. Bavo Church in Haarlem. Original, 15th century ashlar masonry of Belgian Eocene calcareous sandstone (centre right) has been replaced by Udelfangen sandstone in 1890. Detail of Udelfangen sandstone blocks, indicated by T, is given in Figure 5.

Figure 3. Façade of the Remonstrant Church, Rotterdam (1897). Sculptures and ornaments are carved from Udelfangen sandstone.

behaviour of the Udelfangen sandstone the mineralogical characteristics and petrophysical properties have been studied. Samples from the two existing quarries near Udelfangen and samples from sculptures at the exterior of the Remonstrant Church in Rotterdam, built in 1897 (Fig. 3), were available for analysis. In combination with data from the literature, notably the studies by Kraus (1985) and Grimm (1990) the study should give us more insight into the weathering processes, the different qualities and the durability of the Udelfangen sandstone.

2 DATA ANALYSIS

2.1 Petrography and sedimentary facies

According to Grimm (1990) the Udelfangen sandstone is dominated by quartz (65%) and lithic fragments (22%) with small amounts of feldspar (about 8%) and mica (2%). Most of the micas are platy muscovite crystals, but biotite and chlorite can be present too.

The sandstone is cemented by carbonate (dolomite) which usually makes up about 3–5% of the mineral content. The colour of the stones varies between yellowish-grey to greenish-grey. A characteristic feature of the Udelfangen sandstone is the occurrence of very small stains of dark coloured manganese oxide.

The quartz grains are fine, with grain sizes between 100 and 150 micron, and with angular to well-rounded forms (Fig. 4). Most of the Udelfangen sandstone layers are very well sorted. In the upper part of the sequence, frequently some thin intercalations and lenses of silty material occur between homogeneous sandy layers. Clay minerals are scarce; probably only some kaolinite, formed by the decomposition of feldspars, is present.

24

Figure 4. Photomicrograph of the Udelfangen sandstone showing angular to rounded quartz grains and carbonate cement. Platy crystals of muscovite can be observed in the upper central and right part of the picture. Base of photograph is about 0.3 cm.

The cementation by dolomite-carbonate is not always through and through. Some levels can be very tough for a stonemason to work with, while other levels are softer and easier to work. Pressure solution features of the quartz grains hardly occur and obviously the formation of new quartz has been subordinate in the diagenesis of the Udelfangen sandstone.

Mineralogy, grain size and content of the marine fossils point to a shallow sea and near coastal environment for the deposition of the Udelfangen sandstone. The high mica content emphasizes the neighborhood of a granitic hinterland which supplied the sandy material.

2.2 Petrophysical properties and geochemistry

Petrophysical properties and geochemical analysis are presented in Table 1 and Table 2. Porosity varies between 17.5 and 27.0 vol. % and permeability ranges from

Table 1. Porosity and permeability data of 10 samples Udelfagen sandstone. G- and Y-codes are fresh quarry samples; R-codes are samples from the Premonstrant Church in Rotterdam.

Sample number	Core depth [m]	Ambient He-porosity [% of Vb]	Gas horizontal permeability [mD]	Emp. Klink. Hor. Perm. [mD]	P mean [g/ml]	Grain density [g/cm³]	Remarks
R1	n.a.	20.4	64.7	57.1	1.62	2.63	
R2	n.a.	24.1	153	142	1.43	2.61	
R3	n.a.	25.7	274	263	1.18	2.62	
R4	n.a.	22.5	65	57.2	1.70	2.64	
Y1	n.a.	23.3	155	144	1.41	2.64	
Y2	n.a.	27.0	376	368	1.18	2.63	
Y3	n.a.	21.5				2.65	No perm. possible (no cylinder from sample)
G4	n.a.	19.5	8.28	6.47	1.52	2.65	
G5	n.a.	17.5	0.75	0.51	1.64	2.66	
G6	n.a.	23.2	183	172	1.33	2.64	

Table 2. XRF analyses of 10 samples Udelfagen sandstone. G- and Y-codes are fresh quarry samples; R-codes are samples from the Premonstrant Church in Rotterdam (Dubelaar 2003).

	G4	G5	G6	R1	R2	R3	R4	Y1	Y2	Y3
SiO_2	70.63	67.39	73.52	70.61	76.14	80.97	71.80	72.18	76.50	62.15
Al_2O_3	8.51	10.11	7.11	8.23	7.82	6.89	9.86	7.12	6.94	6.16
TiO_2	0.363	0.472	0.279	0.290	0.443	0.480	0.527	0.315	0.319	0.296
Fe_2O_3	1.19	1.94	0.82	1.02	0.94	0.77	1.54	1.02	1.00	1.13
MnO	0.054	0.049	0.043	0.038	0.010	0.007	0.038	0.058	0.048	0.055
CaO	3.48	3.66	3.52	3.21	0.36	0.22	2.03	3.62	1.81	10.44
MgO	1.60	2.04	0.96	1.34	0.49	0.20	1.17	1.46	0.72	1.32
Na_2O	0.24	0.26	0.22	0.30	0.47	0.30	0.22	0.22	0.22	0.19
K_2O	4.94	5.25	4.56	4.97	4.88	4.31	5.35	4.46	4.54	3.82
P_2O_5	0.122	0.150	0.122	0.140	0.761	0.144	0.176	0.135	0.141	0.111
S	0.00	0.00	0.00	0.90	0.28	0.15	0.06	0.00	0.00	0.02
Sum	91.14	91.33	91.15	91.01	92.60	94.43	92.76	90.59	92.24	85.68
$CaCO_3$	6.22	6.54	6.28	5.73	0.65	0.40	3.62	6.47	3.23	18.64

25

0.75 to 376 milliDarcy. Lowest values of permeability were obtained from the layers with loamy intercalations (Table 1, G samples). Data from Grimm (1990) reveal that the uptake of water of the Udelfangen sandstone under atmospheric pressure is 6.02 wt. % and under vacuum 9.55 wt. %. Saturation coefficient is 0.63. The amount of micropores (pores < 1 micron) in samples analysed by Kraus (1985) was about 4.4% of the total pore volume. It is well-known that a high percentage of micropores makes many types of dimension stones susceptible to (frost) weathering (BRE, 1997; Dubelaar et al. 1997, Dubelaar et al. 2003). Considering the relative low amount of micropores and the low saturation coefficient it can be expected that the Udelfangen sandstone is resistant to frost weathering.

From both the quarry samples and the samples from the Remonstrant Church in Rotterdam the geochemical compounds were analyzed by XRF (roentgenfluorescence spectroscopy). The most striking differences are the variable amounts of CaO, and thus the calculated percentages of carbonate (Table 2). The values of carbonate in the quarry samples are distinct from the values in the monument samples (samples R2 and R3), probably due to the leaching effect during weathering. Also, the amount of sulphur (and thus sulfate) is distinct in the two sample groups. The fresh samples hardly contain any sulphate, while the monument samples do show a small amount of sulphate. Undoubtedly this is caused by deposition of sulphate salts in the interior of the stones during weathering.

3 WEATHERING AND DURABILITY

Compared to the mature pure quartz arenites, like for instance the Cretaceous sandstones in northwest Germany, the immature Triassic Udelfangen sandstone is more receptive to weathering. Weathering forms are scaling and loss of grains due to leaching of the carbonate cement. Crystallization of salts, like gypsum and halite, near the surface of the stones, results in blistering and flaking and eventually in disruption of the outer layers (Fig. 5). Due to the alternation of levels rich in loam with laminae rich in platy muscovite crystals and layers dominated by fine-grained quartz, the original stratification becomes prominent by weathering in the course of time. In the case of heavy loaded stonefragments in the construction, with the original layering placed in an upright position, the stone may split lenghtwise. Kraus (1985), testing the Udelfangen sandstone in the laboratory, mentioned a remarkable loss in strenght, of almost 40%, between dry and saturated samples.

In Holland, part of the Udelfangen sandstone already had to be replaced within fifty years after its application. However, in other applications, the stones do not show much deterioration.

Figure 5. Detail of the Udelfangen sandstone blocks at the west façade of de St. Bavo church in Haarlem. Dense mortar presumably prevented the water to flow freely from one stone to the other. Crystallization of salts resulted in honeycomb weathering (lower right). Stones are situated just above the bike in Figure 2.

Therefore, given the nearly equal weathering conditions in those cases, the difference in quality of the used stones must have been notable. It is thought that the quality of the stone is largely determined by the carbonate content, the amount of platy micas and the occurrence of loamy laminae. These parameters strongly affect the durability of the Udelfangen sandstone.

4 CONCLUSIONS

Although the number of analyzed samples of the Udelfangen sandstone have been restricted, the study of the chemical compounds and the petrophysical properties have resulted in clear indications about the weathering and durability of the Udelfangen sandstone. Differences in carbonate (dolomite) content and the amount of sulphate point to leaching and crystallisation of salts in the weathered samples. Due to the use of different qualities of the stones from different beds, such as, layers intercalated with loam, beds rich in micas and layers with a high carbonate content, the weathering behaviour, and thus durability, is different in each application. The results of the study are of importance for the Dutch Agency for Monument Care (Rijksdienst voor de Monumentenzorg) and for architects working with natural stone in restoration projects.

ACKNOWLEDGEMENTS

Thanks are due to the Dutch Agency for Monument Care in Zeist (RDMZ) for permission to publish the results of the study. Dr. Karin Kraus (Institute of Stone Conservation, Mainz) gave permission to use data from her thesis. Hendrik Tolboom and Gerard Overeem (RDMZ) are thanked for stimulating discussion. Albert Hebing (Panterra Geoconsultants, Leiden) and Bertil van Os (TNO-NITG) are thanked for the analysis of the samples. Hanneke Hoendervangers assisted with the drawing of Figure 1.

REFERENCES

BRE, 1997. *Selecting natural building stones.* Digest 420. Building Research Establishment.

Dubelaar, C.W. 1984. Steenrijk Amsterdam. Een geologische stadswandeling, KNNV Hoogwoud 88 pp.

Dubelaar, C.W. 2003. Udelfanger zandsteen. Geologisch onderzoek van een aantal monsters uit een tweetal groeven bij Udelfangen (Duitsland) en van monsters afkomstig van het exterieur van de Remonstrantse kerk in Rotterdam. Unpublished TNO-rapport NITG 03-134-B, 28 p.

Dubelaar, C.W., Koch, R. & Lorenz, H. 1997. Weathering of travertine in a temperate humid climate (National Monument, Amsterdam, The Netherlands. In P.G. Marinos, G.C. Koukis, G.C. Tsiambaos & G.C. Stournaras (eds), *Engineering Geology and the Environment; Proc. Intern. Symposium, Athens, 23–27 June 1997*: 3135–3141. Rotterdam. Balkema.

Dubelaar, C.W., Engering, S., Koch, R., Lorenz, H.-G. & Van Hees, R.P.J. 2003. Lithofacies characteristics and physical properties as related to the durability of Portland Whitbed Limestone (St. Johns Cathedral, 's-Hertogenbosch, The Netherlands). In E. Yüzer, H. Ergin & A. Tugrul (eds), *Proc. Intern. Symposium on Industrial Minerals and Building stones, Istanbul, 15–18 September 2003*: 539–544. Istanbul.

Grimm, W.D. 1990. *Bildatlas Wichtiger Denkmalgesteine Bundesrepublik Deutschland.* München.

Kraus, K. 1985. Experimente zur immissionsbedingten Verwitterung der Naturbausteine des Kölner Doms im Vergleich zu deren Verhalten am Bauwerk. Dissertation Universität Köln, 210 pp.

Negendank, J. 1983. Trier und Umgebung. Sammlung Geologische Führer. Band 60. Gebr. Borntraeger Berlin, Stuttgart.

Slinger, A., Janse, H. & Berends, G. 1980. *Natuursteen in Monumenten.* Rijksdienst voor de Monumentenzorg, Zeist.

Dimension Stone 2004, Přikryl (ed)
© *2004 Taylor & Francis Group, London, ISBN 90 5809 675 0*

Historical review of exploitation and utilization of stone in Croatia

S. Dunda & T. Kujundžić
University of Zagreb, Mining, Geology and Petroleum Engineering Faculty, Zagreb, Croatia

ABSTRACT: The utilisation of stone as a natural raw material has been one of the longest lasting economic activities in Croatia. The evidence of this are numerous ancient buildings which are "woven" into our modern towns of Poreč, Pula, Rab, Zadar, Šibenik, Trogir, Hvar, Korčula, Dubrovnik, etc.

Exploitation of stone for construction and stone mason's works in Croatia dates back to the Pre-antique age reaching the top development in the Antique Age, when Greek colonies were formed on the seaside and islands, as well as during the Roman times especially. A number of epigrams are evidence of developed stone mason's trade in the Roman Empire, although it was known during the Greek-Illyrian period too. From that period there are many material remnants of construction work especially the carved stones built in the defence walls of towns. The oldest known stone buildings are Illyrian tumuli made of stone without lime as binder and stone-panelled graves. The used stone was found in the nature and was uncut. The constructionally more developed buildings date back to the Greek-Illyrian period – for example the wall ruins of the Greek colonies on the islands of Vis and Hvar, where local stone was used. That period, 6th century B.C. also comprises the remnants of the Hallstatt culture such as stone sculptures probably from a sanctuary and a part of a sculpture of a horseman.

The Antique period is characterised by intensive utilisation of stone for construction of monumental buildings, aqueducts, bridges, sculptures and sarcophagus. Such exploitation and utilisation of stone is primarily heritage of the Roman times. The ancient quarries on the island of Brač are Plate, Rasohe and Stražišća in the vicinity of the towns of Škrip and Splitska. They presumably date back to the times of the ancient Škrip, named as "civitas" and "oppidum", on the ruins of which the Romans built their fortress of Scripea in 79. B.C. using the stone from these quarries. The stone from these quarries was also used later upon construction of Salona and the Palace of Diocletian. The construction of the Palace, which was the residence of Gaius Aurelius Valerius Diocletianus (284–305) born in the surroundings of Salona, started

in the period of the Late Roman Empire and was completed at the beginning of the 4th century. At that time the quarries of Brač were very active offering extraordinary production opportunities. The quarries had been exploited before the Diocletian's period, which is confirmed in the inscription from the 2nd century A.D. found in the vicinity of Škrip. The inscription mentions a theatre manager (*centrio Quintus Silvius curagens theatdrid*), which is connected to the construction of the theatre in Solin. Besides Salona and the Palace of Diocletian the stone from these quarries, which are even today called the Diocletian's quarries, was used for other construction works in the Roman Province of Dalmatia. A chronicler from the island of Brač, Prodić (1662) cited: "*taglio de sassi nella Brazza per la costruzione del Palazzo, et altre Fabriche di Diocleziano*". Cicarelli (1802) quoted: "The emperor Diocletian got in the year of 286 extracted an excellent stone for his magnificent palace in Split. Under Škrip and above Splitska one can see the quarries two miles in the area which are evidence of how much empire's money and manpower was invested". The extraction of stone in the Antique period is confirmed in the findings of the Roman fragments comprising fibulas, altars, uncompleted sarcophagus, columns and capitels including ancient tools and Roman techniques of block extraction. In the Roman times quarries were state-owned and managed by the army. They were dedicated to the protection by gods. Therefore the quarries on the island of Brač were also state-owned and dedicated to the Hercules and according to some fragmented inscriptions to Mercury. In the quarry of Rasohe there is a rustically carved relief of Hercules in a semi-life size and in the quarry of Plate there is a votive altar erected in honour of Hercules by the quarry supervisor Valerius Valerianus, who was sent from the quarries of Fruška Gora and appointed a chief supervisor of the quarries on the island of Brač. The Roman quarries also bear evidence of Roman technology of block extraction. Especially impressive are hand-made narrow canals up to 5 m deep, the so-called "pašarini" – Tagliatta Romana. They were used to open and prepare rock mass for extraction of large blocks. In the old quarry of Tesišća near Pučišća there are remnants of the undercut in the V-shape, the so-called "kunjere" in which wooden wedges were hammered. After they had been watered, they swelled and separated a stone block from the rock mass by strong force. On the small island of Vrnik near Korčula there are abandoned Roman quarries (untouched for centuries) with underground pits which were made in the Roman times when explosive was unknown and stone was extracted by underground drift.

The Palace of Diocletian was also constructed with the stone extracted in Seget in the surroundings of the town of Trogir. The written records on dimension stone exploitation in the quarry of Seget, which is active today, (Lat. *seco, secui, sectum* = saw, break, cut stone) near Trogir date back to the middle of the 3rd century B.C. Pliny the Elder (23–79) mentioned in his encyclopaedic work "Naturalis historia": "... *Tragurium civium Romanorum marmore notum...*". Exploitation and trade of this stone mostly motivated the construction of "Traguria". The stone of Trogir used to sell well on the world market resisting to the strong competition especially by the marble from Greece and Italy (Paros, Pentelikon, Carrara).

Upon archaeological excavations in Vitlo near Otočac Roman quarry was found in thick limestone layers of Cretaceous Age. The quarry bears clear evidence of Roman technology and processing of blocks of large dimensions by hand-made "pašarini" or narrow canals vertical to layerness, including lifting of blocks according to layer areas in weakly or strongly manifested natural discontinuities.

The most important quarry from the Roman times in the northern part of the Croatian coast in Istria is still the active quarry of Vinkuran, situated south of Pula, the famous Cava Romana. The limestone from the quarry of Vinkuran was used in construction of monumental Antique buildings in Pula. In the first half of the 1st century the Roman emperor Vespasian had the amphitheatre built in Pula, which is today popularly called the Arena. Stone was partially transported from the quarries on the Brijuni islands, the island of St. Jeronimo in the Fažana Channel and from the Bay of Saline south from Rovinj.

The history of stone construction in Istria, after the Illyrian Histra (urban settlements – Istrian castles) and the Romans continued, however not so intensively. The first records of exploitation of today still active quarry of Kirmenjak, the ancient name of which is Orsera (Montraker), date back to the year of 568, when the construction of Venice began. This Istrian stone was built in the most sensitive places of the buildings of Venice, in the areas which were in turns drowned into the sea water and dried, so up to this day the stone has been exposed to the most unfavourable impacts of sea water and urban surroundings. The Kirmenjak stone (pietra di Orsera, Grota from Vrsar) and the stone from the surroundings of Buje was used upon construction of the old Christian Basilica Euphrasiana in Poreč in the 6th century. In this period the blocks of Istrian stone were transported to Ravena for construction of the tomb for the king of the Ostrogoths Theodoric. The monolithic cupola (11 m in diameter and with the approx. weight of 20 t) of this Late Antique building with large stone blocks was made of the Istrian stone from the ancient quarry on St. Nicolaus in Poreč, where stone was extracted in the period of the Antique and Late Antique.

After the Romans had left, the utilisation of stone in architecture, construction and sculpturing was decreased in other parts of Croatia, too. In the Pre-Romanesque and Romanesque periods stone was mainly used in construction of sacral buildings. There are scarcely any records from that period, but they are extremely important as historical documents. The utilised stone was autochthonous and original, partially taken from the Antique buildings, which can be observed on the St. Donat Church in Zadar from the 9th century. In the lower part of the church there are in-built marble pillars from the Antique, probably taken from the Antique buildings in Zadar or the town of Nin nearby. The foundations of the old Croatian churches were also built with ruins of the Roman and old Christian buildings, which this area abounded in. The original shape was preserved of the following churches: St. Cross in Nin (built around the year of 800), St. Peter in Prik near Omiš, St. George in Rovanjska near Obrovac, St. Nicolaus near Selce, St. Michael near Stone, St. Peter and St. Lawrence (11th century) in Zadar, St. Barbara in Trogir, St. Nicolaus in Split (11th century) and many other churches. When around the year of 800 the Croats accepted Christianity they started building churches and church furniture of stone. The baptismal font of Višeslav in Nin was for example made of a whole limestone block with the characteristic motif of the Early Croatian interlacing-ribbon pattern.

In the Middle Ages stone was the basic construction material along the whole Croatian coastline and inland. Besides sacral buildings stone was used in construction of town walls, monumental public buildings and residential areas. Stone houses built in the Romanesque style (13th century) are present in all the towns along the coast, especially in Split and Trogir. The most beautiful building of that period is St. Lawrence's Cathedral in Trogir, which is the most important Romanesque monument in Dalmatia with

30

the monumental portal made in 1240 by the stonemason master Radovan. Radovan's chisel created the top masterpiece of the Romanesque order in Europe. The cathedral was built from top quality limestone, known already in the Antique period, which came from the quarries of Seget. These quarries are even today active. With time the in-built stone acquired a yellowish patina, which is the same as "Porta Aurea" in the golden doors of the Palace of Diocletian. Besides the stone from Seget, the stone from Viniśce presumably from the quarry in Voluja and Jamurine was also used in construction of the cathedral.

The construction of Venice continued with the Istrian stone from the old quarries near Vrsar (Montraker) and Zlatni rat near Rovinj. The role of the Istrian stone (besides the Istrian timber) is especially significant upon construction of Venice. According to some records even 80% of stone in this town came from Istria. The beginning of utilisation of the Istrian stone (Pietra d'Istria) upon construction of Venice was due to economic reasons and easy transportation. After recognising its extraordinary physical and mechanical characteristics, especially resistance to the aggression of complex sea agents, i.e. its durability in difficult conditions of the Lagoon this stone became, according to the decision of the Venetian Senate in 1307, the only stone material built in the facades of churches, palaces, civil, sacral and military buildings, etc. All this contributed to the unique whole of the Venetian buildings, where the Istrian stone plays the main role from the structural and stylistic points of view. There is an additional advantage of this stone. Within the urban context like Venice, where the walls are usually protected by plaster, this looks even better due to chromatic prominence of the white stone making a contrast to the chromatic effect of plaster, surroundings, i.e. water and other materials, which were used in decoration of this town.

The Gothic architecture in the continental part of Croatia mainly used *Lithotamnium* limestone and calcareous sandstone. The plastic art of calcareous sandstone can be seen on the head of St. Stephen's sculpture, whereas the Romanesque and Gothic capitels of the St. Stephen's Cathedral in Zagreb (13th–15th century) are made from *Lithotamnium* limestone. The stone came from the quarries on the southwestern slopes of the Medvednica Mt. near Zagreb. *Lithotamnium* limestone originates from Vrapče and Podsused i.e., Bizek and Bregov near Samobor and calcareous sandstone from Markuševac. Besides this stone there are some small quantities of porous and soft limestone from Vinica built in the Zagreb Cathedral. Garić-grad on the Moslavačka Mt. dates from this period too, where the interior architectural elements are built from *Lithotamnium* limestone, which was probably extracted in the slopes of the Moslavačka Mt. The external walls of the town are built from metamorphic rocks.

Recent works on Medvedgrad near Zagreb have discovered magnificent capitels of *Lithotamnium* limestone in the chapel. The stone elements of a larger size, doorposts, lintels and angle wall blocks are made from *Lithotamnium* limestone, whereas the main walls are made from thick slab limestone of Upper Cretaceous Age near the city.

In the Gothic period small churches were built in the Istrian villages in which the traditional Romanesque order was combined with the Gothic constructive and decorative elements.

The second period of development of urban life in Dalmatia was after 1420 when the political turbulence, after the Venetians had established themselves in that area, calmed down. The centre of artistic and building activities were not more the towns built in the Croatian Romanesque style. Artistically the most active was Dubrovnik and the surrounding area, which was the only politically independent town, followed by Korčula, Hvar and Šibenik. The construction plans were more complex than before comprising not only sacral buildings. The main tasks were: fortification and equipment of the towns (building of streets, squares, waterworks), construction of residential areas, which was enriched by a new type of buildings – villas.

Utilisation of the Dalmatian limestone was flourishing in that period. The ancient Antique quarries were re-opened and some new ones opened. That was the time of cultural and artistic prosperity (Renaissance) when our famous constructors such as Juraj Dalmatinac, Andrija Alešija and Nikola Firentinac used the stone from Brač and Korčula for their buildings throughout Dalmatia and the neighbouring country – Italy: in Ancona, Rimini, Mantova and Tremiti. Besides the quarries of Brač and Korčula the stone was exported to Italy from the quarries on Brijuni islands and Istria.

In that significant period an important role plays the Cathedral in Šibenik, the work of art by Juraj Matejev from Zadar, known as Juraj Dalmatinac, who was a brave entrepreneur and stone expert. He built this cathedral according to his own plans in the period from 1441 till 1475, when he died, including numerous other buildings in Zadar, Šibenik, Split, Dubrovnik and Italy. Juraj Dalmatinac (representative of the so-called "Ornamented Gothic"), a profound sculptor (frieze head and baptismal font in the Šibenik Cathedral) used the stone from the quarry of Veselje on the island of Brač for the construction of the cathedral, into which he himself often came, selected the stone and supervised stone cutting. He also used the stone from the quarries in Korčula and for ornaments he got pink calcareous breccia transported probably from the island of Krk. It was used only for ornaments in the building interiors. The preserved records from 1454/1455 (written partially in Latin and partially in Italian jargon of that time) on the trial held in Split

31

referring to the stone and the quarries on the island of Brač between the artist Juraj Dalmatinac and his former business partner an Italian maistro Zuan Brasora (magister Johannes Brexuola), show that Juraj Dalmatinac exported the stone from Brač into Ancona and Rimini (*perche se io ho portado piere in Ancona e Rimano del isola de la Braza sono state mie et hole fate mi savar ai mie maistri cum lie miei denari a mie spekse per auctorita concesa a mi per lo Conseio de la Braza*). According to the reliable documents he used the exported stone in Ancona, Italy for the construction of Loggie dei Mercanti, Augustine and Franciscan Church. In the records about the portal of the letter is stated that the stone was extracted and processed on the island of Brač (*cavato e lavorato alla Brazza*).

According to some other records a famous artist Andrija Alešija worked in the quarries of Brač in 1452. He worked with Juraj Dalmatinac on construction of the cathedral in Šibenik, for which he, as stated in the records, created pillars, distyles, arches above water and ornamented chapels and sculptured 4 statutes. In an important document of 2nd May 1455 Andrija Alešija got concession from the owners Gašpar and Čeprnje Tomašić from the village of Straževnik to use then the largest quarry on the island of Brač Veselje (*super terreno posito Vesela*) at the price of 10 small libras, whereby he was allowed to cut and process stone on the area from the sea to the church (*Magistri Andree lapicide emptio unius loci vacui in Bratia*). In 1472 Andrija Alešija built the church from the Brač stone in Italy, in Tremiti on the Tremiti islands. He also built the Čipiko Palace in Trogir and some statutes and the baptismal font in the Trogir Cathedral.

The third famous Renaissance artist Nikola Firentinac used the stone from the Brač quarries, when he, after Juraj Dalmatinac had died, continued construction works on the Šibenik Cathedral (1475-1505), which was completed in the Early Tuscan Renaissance style. In 1493 the Church Fathers gave him a special proxy to use the quarry on the island of Brač (*fecit procuratorem magistrum Nicolaum q. Johannis Florentinum specialiter et expresse contra cuiscumque occupationes unius petrariae positae in insula Brachiae*). According to some other documents Nikola Firentinac spent some time on the island of Brač due to the stone needed for construction of the church Valvedere (Italy) (*dove ando a taiar le pere e pagar le maestranze*). He built the Orsini chapel in the Trogir Cathedral (1468). He also chiselled some statutes, the rest were created by Andrija Alešija and Ivan Duknović (Ioannes Dalmata). His Early Renaissance works are present in several buildings in Šibenik, Trogir and Split, the stonewall surfaces and ceilings of which are ornamented by magnificent reliefs, statues and figures.

When working in Trogir Andrija Alešija and Nikola Firentinac used stone from the quarry of Vinišće near Trogir too, which was mentioned in some documents.

Upon construction of the Chapel of St. John of Trogir in the Trogir Cathedral the contract was made on 4th January 1468 between Andrija Alešija, Nikola Firentinac and the representative of the Cathedral, according to which "... *the arched chapel in the above mentioned cathedral is to be made from marble from the quarry of Voluje... all the stones in the chapel interior are to be from Voluje, from the place chosen by the supervisor of the construction works, whereas the stones for the exterior part can be freely chosen by the master craftsmen on the island of Čiovo...*". Then "...*The cover above the grave is to be framed by the stone from Voluje*".

The architecture of Dubrovnik is typical of the 15th and 16th centuries, although some most important construction works were created earlier (regulation of Prijeki in 1296). The town walls, this magnificent set of buildings, which gives Dubrovnik a special architectural charm, were mostly built in the 15th century and completed in the 16th century. Dubrovnik had been fortified before but they got their final looks in the 15th and 16th centuries. Paving of the streets as a form of utility works of the towns in Dalmatia started in Dubrovnik as early as in the first half of the 14th century. The first streets were paved in 1328 and 1335. According to the decision of the Great Senate from 1407 all the streets had to be paved by stone or brick. In 1415 contracted were the paving works of the square in front of the cathedral then.

The quarries of Korčula were flourishing during the construction works in Dubrovnik. The stone from these quarries was also used for the construction of the Šibenik Cathedral and earlier for the Palace of Diocletian. Near the town of Korčula there are still abandoned quarries (some are from the Roman period): Badija, Vrnik, Suprava, Sestrice, Vanjak, Kamenjak. The archipelago of small islands in the vicinity of the town of Korčula and Lubarda has always been known by the top-quality stone, especially the stone from the small island of Vrnik. The traces of this stone can be found on numerous palaces, churches, town walls, pavements, towers and other buildings worldwide. The "pearl" constructed from this stone is the town of Korčula and Dubrovnik. Dubrovnik this real "stone town" presents the most beautiful work of art as whole provided by the mediaeval construction works. In the period from the 14th to the 16th centuries everything in this town was built from white limestone, such as façades of public and private buildings, bell-towers, walls of town towers, etc. This, previously white stone town, acquired with time its present yellowish patina. The stone used in construction of Dubrovnik came from Korčula (Soline and Grbača), Kamenjak and mostly from Vrnik, where a number of stonemasons and sculptors were born. On the small island of Vrnik, which is smaller than a third of a square km, there were once about thirty quarries.

32

In the past centuries even 600 people lived and worked in stonemason's trade there. After the Second World War only about 50 persons dwelled in 30 houses. Today (2004) there are almost no dwellers on the island of Vrnik. The stone from Vrnik was used, for example, in the 15th century for the construction of St. Marcus Church in Korčula by domestic (Hranić, Dragošević, Ratko Ivančić, Marko Andrijić) and Italian (Jakov Correr from the town of Trani in Apulia) stonemason masters. Important is also the Divona Palace (Sponza) in Dubrovnik, built by Paško Mihajlov and N.J. Andrijić in 1520.

During the reign of the Venetians many monumental buildings, town walls, water-works and public fountains were built in small towns along the Dalmatian coast. Stone was used in building of anchorage pillars, tables, benches, vases, fences, etc. In 1606 Poljanica in front of the Holy Ghost Church in Šibenik was paved. In Korčula there are many palaces with balconies and the protective walls are well preserved. The theatre and arsenal with the arched ceiling in Hvar were built in 1612 from recrystallised cryptocrystalline limestone from the quarry of Križna Luka, which has today a bright yellow patina.

The original domestic limestone was used in building of the mediaeval gravestones – standing tombstones or marbles, which were found in Lika and mostly in the area of Imotski and the area of Sinj. They originate from the 14th and 15th centuries and disappeared at the end of the 15th and the beginning of the 16th century when these Croatian parts were controlled by the Turks. Taking into consideration their characteristics, ornaments, symbols, inscriptions, language, script and range they can be defined as tomb characteristics of the Croatian independent culture. Those found in Imotski have specific characteristics such as years presented in numbers, which is different from the tombstones found in Hum (Hercegowina). Being in use for more than three centuries these tombstones are typical of Croatia. Except in the neighbouring country of Hercegowina there are no such tomb-stones in any other land, not even in Bulgaria from which the Bogomil religion spreaded. Their western borders are the areas of Imotski and Sinj. They were chiselled from the whole large blocks of autochthonous limestone extracted nearby. When a stone block was separated from the rock it was shaped and ornamented. Rarely a stone block was brought to the necropolis for final processing. They were in various shapes like slabs, trunks, trunks on pedestals, ridge tiles, and ridge tiles with pedestals. There are simple ones made of one part and those of two parts, whereby one part lies horizontally having another part on it made in the shape of a sarcophagus. The first one actually covers the grave (postament) and the other one in the shape of a ridge tile or trunk has just a decorative role. The most beautiful tombstones – ridge tiles are those chiselled from the live stone in the shape of a house, narrowed on the bottom with a trapezoid roof "on two waters". They are ornamented with relief figures: men, women, horsemen, stars, moons, hinds, horses, dogs, shields, swords and very often crosses. The incraved crosses on the Dalmatian tombstones are one of the proofs that these tombstones were not exclusive tombstones of the Bogomils, because they did not respect the cross. From all the tombstones in Dalmatia only those in the area of Imotski have traces of script, which with time were destroyed and became illegible. The people used to destroy the inscriptions because they believed that the "čivuts" i.e., all foreign travellers, would according to them, find buried treasures in numerous piles near the tombstones. Their belief that the storm would destroy the crops if they touched the tombstones preserved them from further destruction. However, their position near the roads and settlements contributed to their destruction. The large "marbles" were cut and somewhere as a whole built into the walls of churches, graveyards, vineyards, gardens and fountains, but also crushed and used in road constructions.

In the 17th and 18th centuries stonemason's trade in the sacral architecture and sculpturing along the coast came to a standstill. The domestic stonemason masters used to build Dubrovnik in stone in the Baroque order. The town was ruined by earthquakes and re-stored again. In the continental part of Croatia, especially in Hrvatsko Zagorje, soft limestone was used. The Baroque buildings in Varaždin are made from soft and very porous limestone and calcareous sandstone from Vinica near Varaždin. Due to a low degree of hardness this stone is commercially known as "Vinicit". It is easily processed by dressing and sowing and therefore has attracted the constructors since ancient times. The evidence of this is numerous buildings in the Baroque Varaždin and its surroundings. In Vinica there is the "Pillar of shame" from 1643, which is completely preserved although it was exposed to all possible atmospheric agents. In the 18th century the stone processing in this area was well developed, so that the stonemason guild was established in Varaždin gaining all its guild rights from the king Charles III in 1728.

After the exuberant Gothic, Renaissance and the charming Baroque the Croatian architecture of the 19th century saw many buildings and monuments in historic non-styles.

The characteristic of the 20th century buildings is flat surfaces. The modern stone industry adjusted to the new demands of construction works, when the stone as a thin panel acquired a decorative and functional or decorative and protective role. The old quarries were re-opened and the new ones opened. Improvements in exploitation, processing and fitting follow the world trends in the stone industry. In numerous mainly public buildings from the 20th century various stone types

33

regarding their kind, decorative characteristics and properties were used in panelling of exteriors and interiors, vertical and horizontal surfaces, stairways, decoration of environment as well as upon construction of memorial plaques or buildings in order to enrich the environment.

REFERENCES

Crnković, B. & Šarić, Lj. 1992. *Building With Natural Stone.* Series: Stone, book III, pp. 184. Zagreb: Mining, Geology and Petroleum Engineering Faculty.

Didolić, P. 1957. Historical Brač Quarries. *Brački zbornik* No. 3, Split.

Dunda, S. & Kujundžić, T. 2003. Digital textbook: Exploitation of Dimension Stone, http://www.rgn.hr/ Nastava/Download.

Kečkemet, D. 1998. Brač – Guide Through History and Cultural Heritage. *Brački zbornik d.o.o.*, pp. 202, Supetar.

Piplović, S. 1992. Exploitation of Trogir's Quarries in 19th Century. *Klesarstvo i graditeljstvo* 3(1–2): 31–35.

Velić, I. 1990. *Istrian Stone*. Zagreb: Croatian television.

Dimension Stone 2004, Přikryl (ed)
© 2004 Taylor & Francis Group, London, ISBN 90 5809 675 0

Dimension stones: the link of natural and built environment in Friuli 'yellow villages'

A. Frangipane

University of Udine, recently Dipartimento di Ingegneria Edile, University of Naples Federico II, Naples, Italy

ABSTRACT: In vernacular architecture dimension stones constitute the link of continuity between the built and the natural environment, being part of the mountains or hills they are used to build over. It is the case of a quite remote hilly area, between Pordenone and Udine (Italy), with a struggling peculiarity: the traditional use of yellowish stones there quarried, calcareous-dolomitic sandstone and conglomerate, characterized by a warm tonality, in evident contrast with the grey colour of calcareous stones commonly employed in the buildings of the region. The paper, starting from the description of the geological features of the area, documents the presence of some accurate naïf architectural elements still in place in the villages, putting the attention to their strong link with the surrounding natural context. From colour evidence to a general reflection regarding the correct approach of the preservation interventions in cultural heritage, when dealing with dimension stones.

1 PREFACE

1.1 *The topic*

When Joseph Roth was describing the 'white villages' of France, in the homonymous travel book, his attention was addressed merely at the colour of the built environment, white plasters in that case.

Nevertheless, in vernacular architecture, the colour of villages can have a strong link with the surrounding environment, when the dimension stones used are quarried close to the building site, realizing a material continuity between houses and earth.

1.2 *The fact*

Some years ago, trying to find the ancient quarries of the stones used in the main portal of the of the Metropolitan church of Udine (Spadea 1995), the main town of Friuli region, in the eastern part of Italy, close to the no more existing Austrian and Slovenian borders, some *Wandertagen* brought the attention to a quite remote hilly area – between the towns of Pordenone and Udine, crossing the main river of the region, the *Tagliamento* river – with a struggling peculiarity: the traditional use of yellowish stones there quarried, sandstone and conglomerate, with a warm tonality, in evident contrast with the grey calcareous stones commonly employed in the buildings of the region. All the area had been seriously damaged by a

strong earthquake, dating 1976: a fast and effective reconstruction had been carried on, paying a very little attention to the effect that new buildings would have for the environment. The repeated surveys of the area – that the going on research defined as the probable quarrying site looked for (Spadea et al. 1996) – had a sort of enchantment effect: the wild environment, the fact that, after the reconstruction, only few buildings were able to maintain the link with their origin – and, therefore, each new encounter constituted a surprise – the presence of some very accurate naïf architectural elements, all that become a sort of fairy-tale, speaking of the 'yellow villages', castled on their yellow hills, as they were growing out of them.

2 THE SITE

2.1 *Introduction notes*

The studied area, nowadays quite a fringe area, due to modern urbanization and to the distance with the main towns of Udine and Pordenone (more thab 20 km), had a period of relative importance at the end of the Middle Ages, when the Castle of Ragogna, on the right bank, at the top of an emerging headland, and the Castle of Castelnuovo, on the left bank, in the middle of a woody area, were in strict link with the neighbour castles of Spilimbergo and San Daniele (Fig. 1), part of a system, along the piedmont line, defending the

Figure 1. Central area of Friuli region in a late 16th century map (Ortelius 1598): in evidence the quoted towns and the study area.

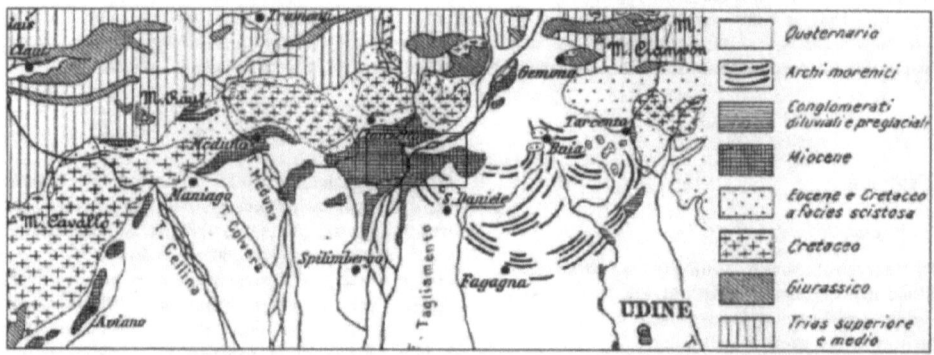

Figure 2. Geological sketch of the area elaborated from Gortani (1926) (*Quaternario* – Quaternary, *Archi morenici* – Moraine arches, *Conglomerati diluviali e preglaciali* – Torrential and preglacial conglomerates, *Miocene* – Miocene, *Eocene e Cretaceo a facies scistosa* – Eocene and schistose facies Cretaceous, *Cretaceo* – Cretaceous, *Giurassico* – Jurassic, *Trias superiore e medio* – Medium and upper trias).

connection roads between Italy, Austria and East Europe countries, that followed the ancient Roman paths.

2.2 Geological outline

The Ragogna headland limits the western part of the moraine arches of the Quaternary Era, deposition of the Tilaventine glacier, the hills today defining the northern boundary of the Friuli plane, mountain mouth of the Tagliamento river (Fig. 2). The weak and friable conglomerate coming from the moraine, is locally named *tof*.

Since oldest geological studies (Taramelli 1881, Feruglio 1925, Gortani 1926) to recent ones (e.g., Frascari & Zanferrari 1977, Martinis 1993) the outcrops

of the area have been recognized to pertain to the Miocene Period (Fig. 3).

In strict contact with the moraine conglomerate, a more consistent dolomitic-calcareous conglomerate is present, in beds 2–5 m thick, imbedded with sandstone 10–20 cm thick. It is locally named *conglomerato del monte* (i.e., mouth conglomerate).

Fronting the river, the hilly area of Castelnuovo presents the alternation of various outcrops in NW-SE direction: molasse (sandstone, calcarenite, silty-clay and marl – locally with lignite, with beds 30–40 cm to some meters thick), conglomerate (the same dolomitic-calcareous conglomerate previously described) and weakly cemented sandstone and conglomerate (calcareous-dolomitic conglomerate and sandstone, weakly cemented with bed of various thickness).

36

Figure 3. Geological map of the area (quotations in legend refers to units of interest; marked little black points indicate localities quoted in the text and in figures. Elaboration from Martinis, ed. 1977).

Both the conglomerate and the sandstone are characterized by a warm yellow tonality. Outcrops are present almost everywhere, showing different stages of cementation and/or alteration.

3 NATURAL AND BUILT ENVIRONMENT

3.1 *Natural environment*

The presence of outcroppings everywhere is, at first, struggling: rude, as conglomerate walls 20–50 m high can be, and gentle, at the same time, for the yellowish colour, when visible far away. Partially hidden by brushes, they follow the roads or they are evident far away, all around, as little clear spots.

Some times very coarse, some times finer, when entering the area, they quickly become a presence and a companion. Around the castle hill of Ragogna (S. Pietro village), they are part of the landscape: the way to the castle passes in fact, through a quite narrow valley bordered by the outcrops. From the headland, following the road down to the river, the conglomerate banks define its limits, until braiding bed of the river, just came out from its mountain course, offers a wide, enchanting view to the plane. The bridge crossed, no more outcrops are visible for a while. The village named Manazzons is reached, where some far yellow spots, preceding the first houses, appear. Continuing toward the hills, direction Celante, through a narrow secondary road, a dreadful infernal view (Fig. 4) materializes. No way of coming back – the road is too narrow – just the possibility of observing the overhanging conglomerate, this time darker, due to mosses and lichens living on the banks. And so on: paths around the area will alternate view of torrent banks, carved in the stone, and wider valley; sometimes the

Figure 4. Outcrops on the road from Manazzons to Celante.

Figure 5. Outcrops on the road to Manazzons.

yellowish emergence disappears, but it is for a short while: outcrops will be soon visible, with their evident embedding, even if not close to the observer (Fig. 5).

37

Figure 6.   Window frame in Paludea.

Figure 7.   Window frame in Franz.

Figure 8.   Window frame with mould windowsill in S. Pietro.

The tour ends: the fact that the area is relatively far from the main connection roads, conceptually isolates it, as a closed world, even if the extension of Miocene formation (Fig. 2), should bring the attention also to the western towns of Meduno, Spilimbergo, Maniago, Aviano, historically known for the capability of Lombard sculptors since 13th century (Someda de Marco 1959), but this is an other story: the relationship between the 'stone-men' and the quarries of the region, still waiting to be written.

3.2 Built environment

The built environment is ugly: after the reconstruction, following the severe 1976 earthquake (nearly 1000 people died in Friuli), everything have changed: the extensive use of reinforced concrete, the will to close yards as soon as possible and the true fear of the weakness of old houses made their course.

Clean, aligned, ready to use constructions seem to have come there from everywhere: no link with tradition, as reported in reference and recent studies (Del Puppo 1907, Scarin 1942, Tentori 1983, Tentori 1984a, Tentori 1984b), apparently no link with the site, they are just something very tidy. Even the preserved existing houses lost their spirit, coated by cement plasters, to tight for them.

So, the only way to understand what really is the essence of the site is to look for feeble traces in ruined houses or where the scarcity of founding stopped the works before their completion. The first elements appear: yellowish window frames show their evidence Figs. 6–9, some hidden dimension stone in the walls texture, as corner or flue blocks (Fig. 10), together with the grey calcareous blocks and cobbles, more frequent in vernacular architecture of Friuli, indicate that there must be something true somewhere around. The warm tonality helps the identification of the material. Even in renewed buildings the architectural elements begin to signify their consistence: both conglomerate and sandstone are present, in good conditions, when sheltered, with evident degradation features, when subject to weathering; for sandstone the weakness orthogonal to the sense of sedimentation is evident in the lateral elements of window frames and in their edges, conglomerate, roughly carved, shows desegregation of its cement.

The survey brings to some surviving relicts of what vernacular architecture should have been there: in close contact with the outcroppings, probably quarried in

38

Figure 9. Window frame with upper wood and brick discharge elements in Paludea.

Figure 11. Traditional house in Paludea.

Figure 10. Corner and flue stones in Paludea.

Figure 12. Traditional house in Paludea.

39

Figure 13.   Traditional house in Paludea.

Figure 15.   Outcrop in Ragogna, in front of the Castle hill.

very close sites, they truly constitute the looked for link between natural and built environment. Dimension stones play the role of protagonist in this relationship: houses seem to emerge from the hills, in a primary idea of sustainable architecture, due to necessity and scarcity of means. As often underlined when dealing with vernacular architecture, but always usefully repeated, the limits imposed by the site bring builders to calibrate each detail, with the consciousness that every effort must go to the goal, with no wastes.

3.3   Quarries

Quarries can be everywhere as spontaneous architecture could never effort far provisioning (Fig. 14). In Ragogna headland, they were found (Fig. 15), in number of five until now, and mapped (Toniutti 2002), located in the northern part of the Castle hill (Fig. 16).

In Castelnuovo area the memory of inhabitants remember the names od Oltrerugo and Costabeorchia as quarrying sites (Fabris, pers. comm.).

Ancient 19th century references, cataloging natural resources, with both a positivist and commercial aim (Marinoni 1881, Pitacco 1884) are not generous, however, with this building material: its use, limited to the area, had no interest elsewhere.

Quarries were, therefore, probably opened by builders in relation with actual needs, in the closest

Figure 14.   Outcrop on the road from Manazzons to Celante.

40

Figure 16. Ragogna Castle. View from the outcrop of Figure 15.

sites presenting a sufficient strength of the stone, roughly reduced in dimension blocks, brought to the yards and there refined, if the case.

Nowadays no way of employment of the stone: quarrying acts exclude the use of local stone, while important quarries of a grey calcareous sandstone, the 'pietra Piasentina' are concentrated in the eastern part of the region: its use, together with that of 'imported' stones, is the only one possible in realizing frames or other architectural elements, often imposed by the planning legislation.

What today visible is the only trace of a passed link with the natural environment: the hurry of reconstruction made demolitions frequent and definitive: frames, steps, dimension stones rest in river beds, buried ruins of a tragedy to forget.

4 CONCLUSIONS

For different reasons in different places of the world the robust link between earth and houses, based on the previous one between men and environment, has disappeared. Modern civilization, in researching its origins, pays attention to multiple aspects: traditional customs, family trees, even preservation of vernacular architecture, but little care is used in the choice of building materials, in an incontrovertible tendency to standardization.

Dimension stones played, on the contrary, an important role in that. The consistence, the colour, the weathering: all these characteristics remembered that buildings were born in a specific place for evident reasons and they were part of it, giving a sense to their being there.

In the case of the area described, the link was substantially broken by severe earthquake, but, reasonably, it would be broken by time, as well. Anyhow, there is something special in the described atmosphere; the relationship between houses and quarries seems to overcome the present: as far as the connection is revealed, the place assume a different aspect: no more a renewed area, but a camouflaged one.

As building components, dimension stones have seen generations passing, things happening, as if they were the true inhabitants of the sites, fronting their natural relatives in quarries.

ACKNOWLEDGMENTS

This research would never start if Professor Piera Spadea, Udine University, would not allow me to join her research of the yellowish stone of the Metropolitan church of Udine. It was the beginning of the renewed interest for ancient and traditional buildings, today full time job. I must not forget it. My thankfulness for the people who helped me in the research: Maria Alberta Bulfon, Emanuele Fabris and Glauco Toniutti, in different ways enthusiastic inhabitants of the land of 'yellow villages'.

REFERENCES

Del Puppo, G. 1907. *La casa in Friuli*. Udine (I): Del Bianco.
Feruglio, E. 1925. *Schizzo geologico della Prealpi Giulie Occidentali e della Prealpi Carniche Orientali*. Udine (I): Società Agraria Friulana.
Frascari, F. & Zanferrari, A. 1977. Geologia della formazioni pre-quaternarie. In Bruno Martiis (ed.), *Geology of the Friuli area primarly involved in the earthquake, 1976*. Milano (I): CNR-National Research Council.
Gortani, M. 1926. *Guida geologica del Friuli*. Included map. Tolmezzo (Udine, I): Stabilimento tipografico Carnia.
Marinoni, C. 1881. Sui minerali del Friuli. In *Annuario statistico*: (III). Udine: Seitz.
Martinis, B. (ed.) 1977. *Geology of the Friuli area primarly involved in the earthquake, 1976*. Included map. Milano (I): CNR-National Research Council.
Martinis, B. 1993. *Storia geologica del Friuli*: 61–63. Udine (I): Editrice La nuova base.
Orteliu, A. 1598. Theatre de l'Univers, contenant les cartes de tout le monde. *Fori Iulii accurata descriptio*.

41

Pitacco, L. 1884. *Descrizione delle pietre e dei marmi naturali che si impiegano nelle costruzioni in Provincia di Udine*. Udine: Tipografia G.B. Doretti e Soci.

Scarin, E. 1942. *La casa rurale in Friuli*. Firenze (I): CNR-National Research Council.

Someda de Marco, C. 1959. Architetti e lapicidi lombardi in Friuli nei secoli XV e XVI. In *Arte e artisiti dei Laghi Lombardi*: 309–342. Como (I): Rivista Archeologica dell'Antica Provincia e Diocesi di Como, Tipografia Editrice Noseda.

Spadea, P. 1995. Studio mineralogico e petrografico dei materiali lapidei e delle malte. in Duomo di Udine. In *Ricerca per il restauro del Portale della Redenzione. Rivista del Centro Regionale di Catalogazione e di Restauro. Villa Manin di Passariano*, (3): 95–116. Trieste (I): Regione Autonoma Friuli Venezia Giulia.

Spadea, P., Perusini, T. & Frangipane, A. 1996. Dolostones used in Middle Age in Friuli (Ne Italy). *Proceedings of the 8th International Congress on Deterioration and Conservation of Stone, Berlin (D)*: 155–157. Berlin (D).

Taramelli, T. 1881. *Carta geologica del Friuli rilevata negli anni 1867–74 e pubblicata nell'anno 1881*. Udine (I).

Tentori, F. 1983. La casa in Friuli. Note per una ricerca. In *Identità*: (4). Tavagnacco (Udine, I): Casamassima.

Tentori, F. 1984a. La casa in Friuli. Note per una ricerca. In *Identità*: (6). Tavagnacco (Udine, I): Casamassima.

Tentori, F. 1984b. La casa in Friuli. Note per una ricerca. In *Identità*: (8). Tavagnacco (Udine, I): Casamassima.

Toniutti G. 2002. Le cave. *Quaderni del Museo civico* (2): 10. Ragogna (Udine, I): Comune di Ragona.

42

Dimension Stone 2004, Přikryl (ed)
© *2004 Taylor & Francis Group, London, ISBN 90 5809 675 0*

Dimension stones in Thailand

A. Hoffmann & S. Siegesmund
Geoscience Center, Georg-August-University, Goettingen, Germany

H. Duerrast
Faculty of Environment and Resource Studies, Mahidol University, Nakhon Pathom, Thailand

K.-J. Stein
Natursteininformationsbuero, Feldberger Seenlandschaft, Germany

ABSTRACT: After the SE-Asian economic crisis in 1997 the Thai dimension stone industry undergoes a significant upturn since 2001. Companies offer a product range with well defined grades, mainly floor tiles that meet quality standards in the processing. Investigations reveal that the country disposes of many varying dimension stone quarries from which the most prominent are quoted in this article. Additionally, data on the tensile strength and thermal dilatation are given for selected dimension stones from Thailand.

1 INTRODUCTION

Natural stones like granite, marble and sandstone are important building materials worldwide and therefore can be seen as valuable natural resources. In Thailand, especially in the Bangkok Metropolitan Area, different rock types from the domestic market were used during times of intensive construction activities. The high variation and large quantities show, that Thailand is a recognized supplier for dimension stones and offers products, which might be interesting not only for SE-Asia, but also for export and the world market.

Against this background, fieldwork and sample collection were carried out from February to August 2003. The research was conducted under the topic "Investigation and assessment of dimension stones in Thailand" and covers the main deposits of northern, north-eastern and central Thailand. Most important element within these investigations will be the study of petrography, texture and petrophysical properties of rocks, which is the base for a reliable assessment of dimension stones.

Finally, in combination with already existing data of current dimension stone deposits, the results of fieldwork and laboratory work can be used to establish a catalogue for governmental and commercial institutions, which might contain the documentation, assessment and quality control of the domestic dimension stone deposits as well as their supply, demand and prices.

2 ECONOMIC SITUATION

In the Kingdom of Thailand the production and processing of dimension stones are well established, mainly to satisfy the demand of the national market.

Based on information from the Thai Marble & Granite Association Thailand has considerable capacities for the processing of dimension stones in the region, after China. As a result of the Asian economic crises in 1997 several dimension stone quarries and processing sites were closed. For some sectors the production decline was more than 90% for several years. Not until the beginning of 2001 the production of dimension stones in Thailand restarted, especially in the central and northern part of the country, whereas in the South this positive sign is still to come. A clear signal was the recommencement of granite quarries in the Tak region in 2001, associated with an increase in the production capacities. Products from there are mainly used in bigger construction projects in Thailand, like the new underground stations in Bangkok. The expansion process for granite and marble sites is mainly carried by companies defined as large-scale enterprises, whereas new quarries in the

43

sandstone sector often started by small and medium enterprises.

Import of dimension stones to Thailand was in terms of quantity and value significantly high during the period of prosperity, mainly from China, Brazil, Vietnam, Norway and other countries. In the last years the import of dimension stones was relatively low due to the governmental import restrictions. Some of these restrictions are lifted until now, so that for example since March 2003 marble can be imported again, but only in form of unprocessed blocks. Still in terms of dimension stones Thailand can be seen as a relatively closed market.

The majority of the enterprises dealing with dimension stones is organized in the Thai Marble & Granite Association, TMGA, with its head office in Bangkok. Currently the association has about 144 members.

3 PROCESSING DATA

3.1 Technical equipment

Depending on the rock type, the technical equipment of quarries and of facilities for further processing is different. In the granite sector the processing is mainly in the hand of enterprises with bigger capacities. Predominately they are using gang saws and polishing equipment from Italian manufactures. But gang saws, block saws and other smaller machines are also coming from China or South Korea. One of the biggest processing factories, the Silamanee Factory Cutting in the Tak Province, District Ban Tak, has more than ten gang saws and two polishing lines. In the quarries the production of blocks is done by carefully applying explosives. After that the blocks were finally made using wedges. The oblique attachment of the wedges to the net block results in an additional material loss of 15% to 20% for the final slabs after cutting and surface treatment.

For the production of limestone and marble blocks in the quarries modern blade- and wire-saws are often used. The further processing is done with block-saws (one to four blade saws). Some factories are using wire-saws for preparing the blocks for final cutting. In the sandstone sector smaller production and processing units are dominating, mainly with the use of block saws. The production of the blocks there is carried out with simple breaking methods, rarely using explosives.

3.2 Products and quality

Part of the investigation was to measure and check the size of polished tiles, similar to a quality control. Companies produce mainly polished floor tiles with standard size of $80 \times 40 \times 2$ cm for granite and $60 \times 30 \times 2$ cm for marble. The final products are generally completely polished. The size, the angles and the cutting borders of the tiles usually meet quality standards. But the tiles often show considerable variations in the thickness. In the companies every tile undergoes a last check. At this stage marble and limestone tiles will be sorted by colour and structures, like joints or veins, defining the different grades/varieties of each product group. Therefore a relatively constant quality of each grade can be expected in a certain frame which has to be defined (Fig. 1).

4 GEOLOGICAL OVERVIEW

From the geological point of view Thailand can be divided in three major areas: (1) The Northeast is a Precambrian terrain, the Khorat platform. Later it was covered by a sequence of sandstones, in which the Tertiary and Quaternary basic and acid vulcanites intruded. (2) The western part of the country from the North to the South is made out of the Shan-Thai Para Platform. A variety of differentiated metamorphic rocks can be found there. (3) Between these platforms is a mobile zone, also with metamorphic rocks.

From Perm to Cretaceous granites with different chemical composition intruded into the metamorphic sequences of the mobile zone. As a result the granites show different grain sizes and various colours. These intrusions are often associated with ore mineralization. In the past the exploitation of these minerals was

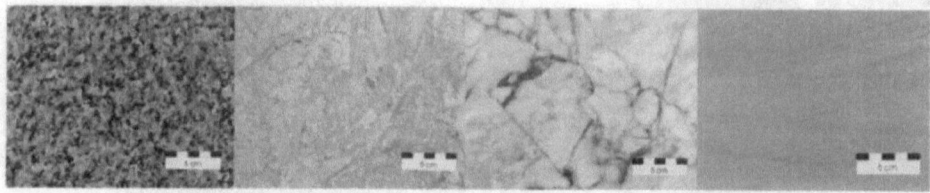

Figure 1. Selected dimension stones from Thailand. From left to right: granite (Tak), limestone (Tak), marble (Nakhon Ratchasima), sandstone (Nakhon Ratchasima). Scale bar for each rock type is 5 cm.

a major economic factor for Thailand. Today the exploration and production is still continuing, but on a smaller scale and focusing on certain minerals. For the dimension stone industry these granites are a promising potential for further development.

5 ROCK MATERIAL AND LOCATIONS

5.1 Marble

Since more than 20 years the fine crystalline limestone in the Kamphaeng Phet region is produced from two quarries. The reddish to yellowish, light to medium gray or dark gray micritic limestone is mainly re-crystallised, but sedimentary structures and fossil parts are still visible. This limestone is very dense showing nearly no open pores. A characteristic feature is the intensive brecciating, in which the clasts are good healed with carbonate matrix, often in the same shade. The overall colour of the stone is often blurred, but sometimes clearly separated. Small veins in white, gray, black and violet colour, and often developed as fine plumages, give the stone an intensive décor. In another quarry the limestone is nearly completely white and fine crystalline, in the appearance similar to the Carrara Marble. Unfortunately the reserves of this material are nearly depleted.

A changing décor can be found within the marbles from the District Pakchong, in Nakhon Ratchasima Province (Fig. 2). In several quarries the main décor changes from white and light gray, over pink and reddish, to white bluish colours over relatively short distances. Layers with intensive breccialike structures can be found, but these are usually good healed. Joints in different colours give the stone a varying décor.

The fine crystalline marbles of the Thung Saliam District in Sukhothai Province are also heavily brecciated. The fragments are good healed with a matrix in the same colour, but still many open pores remain

Figure 2. Marble quarry (Nakhon Ratchasima).

visible. The colours are intensive and ranging from light gray, over beige, pink to many gray variations. They can be both clearly separated or merge into another. The marble here is greatly affected by tectonic events resulting in the appearance of many small joints and veins in different colours, from white and gray, over yellowish to brown, violet and red. The intensive tectonics in this area is also reflected in the small sizes of the blocks, which are limited by the occurrence of joints.

Marble with an exceptional quality can be found in Uttaradit Province, Ban Tamdin District, showing colours from white to light gray. The very fine grained types are often completely white and their quality might be compared with marbles from Carrara. Besides a white to light gray variety appears, which shows crystal sizes up to 5 cm. Unfortunately this quarry was closed during the Asian economic crisis.

Further deposits of light gray, gray and reddish limestone, dolomite and marble can be found in the southern part of the country, in the Provinces Prachop Khiri Khan, Nakhon Si Thammarat and Yala.

5.2 Granite

The Tak Province offers a high variety of granitoide rocks, different in texture, grain size and colour. The deposits are very big with favorable conditions for quarrying and production. The Saha Heng-Mining Company there has the largest single quarry site. The granite from there is in general grayish-orange with a relatively homogeneous texture. This quarry can produce blocks with the maximal size for the processing. Most of the quarries in Tak Province are producing from near surface boulder material. It remains to be seen what textures and therefore what décor will be found when they will reach deeper quarrying levels.

Nearly at the border to Myanmar in the Province Chiang Rai, a white to slightly yellowish granite crops out. There was no production in recent years but efforts were made to prove an excavation of the material for a short time. From this amount, a limited quantity of rocks was once available on the domestic market. The rock exhibits a well developed foliation pattern by the alignment of biotite and feldspar. Mega-clasts of feldspar with sizes up to 10 cm are irregularly distributed leading to an inhomogeneous appearance of the rock in general. It remains to be seen either the final product will include these parts of the rock or not.

In Nakhon Sawan, about 200 km north of the capital, deposits of pink coloured, fine- to medium-grained granite can be found, but they are relatively limited in size. The shade can vary from light red to intensive flesh-coloured with a characteristic spot colouring. The intensive red coloured, cm to dm large areas of potassium feldspars represent the spots, which are

45

surrounded by lighter coloured, alterated feldspar, biotite and epidotite minerals in netlike structure. The biotite is optically less visible which makes the overall colour of this granite lighter. Quartz takes the reddish shade of the feldspar minerals resulting in the continuously red colouring of the stone, only broken by the lighter alterated areas. Noticeable are the mm large open areas in the polished surface, which are usually filled in the factory. Certain dm^2 to m^2 sized areas of the deposit can show a higher content of green, cm-large aggregates of epidote. It is possible that with further production and going deeper these alterations will be much less than now.

The Hornblendite from the Pak Thong Chai District of Nakhon Ratchasima is an exception to the frequently used dimension stones. The rock consists of nearly 90% Hornblende and exhibits a high variation in grain size from <1 to several 10 cm. The feldspar content varies up to 10% and aggregates are spot-like situated between the euhedral hornblende. Pyrite is frequent in some areas but irregularly distributed in general. While in former times a conserving excavation with wire-saws was common, the method of exploitation changed to carefully blasting.

5.3 Gneiss

The Province Prachuap Khiri Khan, situated between the Andaman Sea and the Gulf of Thailand on the southern peninsular, disposes of a strongly foliated garnet-gneiss. The rock shows alternating layers of mafic and felsic minerals in mm-scale, while the lighter parts of the material are double in size. Biotite and feldspar build up the foliation, quartz is partly elongated or in aggregates parallel to the foliation pattern. The garnet content is low with around 2–3%.

5.4 Sandstone

The sandstone deposits in the western part of the Province Nakhon Ratchasima reveal good colour intensity. Colours might change very abruptly so the production has to be careful and selective. The white, fine grained sandstone might be polished at its surface. Decorative rainbow colours can be found in the transition to the yellowish and reddish varieties. The green sandstones can be easily split along the layers with light mica minerals (Fig. 3), but a production of larger blocks is also possible. The sandstones from Nakhon Ratchasima have the potential to get an acceptable place on the international market, due to their technical quality, colour intensity and the fact that their deposits are comparable large.

5.5 Limestone

About 150–200 km ESE of Bangkok in the Province Sakaeo a deposit of Permian limestone occurs. The

Figure 3. Hand-made plates of green sandstone inside the quarry (Nakhon Ratchasima).

fossil-bearing rock shows both red and gray colours, while the reddish variety is dominant in the deposit and gray colours are of minor importance for further processing. The material exhibits a conglomerate-like structure with micritic clasts from 5 to 40 mm that slightly show a different shade. Frequent elements are white, calcitic veins that usually range from 1–10 mm in thickness. These are mostly parallel to each other and very linear in their distribution. Parts of the rock where no conglomerate-character is developed are mixed with red colours and milky white calcite, while transitions from one colour to another are gradual.

Another limestone-deposit is located in the Mae Phrik District of Tak Province. The Mae Phrik limestone is frequent on the domestic market and obtainable in around 20 varieties. Colours range from dark brown to gray and yellow, while a few of them tend to be quite similar from the macroscopic point of view.

6 SAMPLE DESCRIPTION

For the investigations of petrophysical parameters rocks with different developed fabrics from isotropic to anisotropic, varying grain sizes and mineral assemblages from monocrystalline to polycrystalline were chosen. At present, one diorite, two sandstones and one marble were analysed. Investigations on the directional dependent parameters are made in three directions (x, y, z) with perpendicular orientation to each other. The orientations are chosen by macroscopic visible fabric elements on the rock, like foliation and lineation.

6.1 Diorite, Tak Province

The diorite (sample no. TH03/06AH) chosen for the investigation is a fine grained mafic rock with the main constituents of hornblende and feldspar. Accessory minerals are biotite, quartz, sphene and pyrite. The rock displays no preferred orientation of the minerals.

6.2 Brown sandstone, Nakhon Ratchasima Province

The brown sandstone (sample no. TH03/23AH) is a fine-grained rock with slightly visible horizontal layering.

6.3 White sandstone, Nakhon Ratchasima Province

The white sandstone (sample no. TH03/27AH) is a fine- to medium-grained sandstone with different, reddish to violet colour shades resulting from the migration of fluids. These precipitations are irregular in their distribution and most intensive in the adjacent area of joints. In many cases, the structures lie with high angles up to perpendicular to the bedding of the rock. Additionally, single layers of the rock can also show reddish colours. The sandstone is characterized by an explicit amount of open porosity that is macroscopically noticeable.

6.4 Marble, Nakhon Ratchasima Province

The marble (sample no. TH03/29AH) is a fine- to medium-grained rock with grain sizes up to 2 mm. The deposit shows many varieties of marble depending on the grade of tectonic stress, as a clearly defined fracture-zone strikes the quarry leading to an intensively structured breccia-marble. The rock chosen for this study is a white marble characterized by reddish to brownish veins that are usually in the scale of 3–5 mm thickness. No preferred orientation or any sedimentary structures are visible.

7 TECHNICAL DATA

7.1 Density, porosity

Density and porosity of all samples were determined by buoyancy weighting of cubes with 100 mm in length. Samples were weighed both dry and water saturated at air and water saturated under water. Figure 4 gives an overview of the results.

7.2 Tensile strength

Tensile strength was determined by means of the Brazilian test, using disc-shaped samples in compression along the diameter. The specimen size was 40 mm in diameter and 20 mm in height. For each test a strain rate of $30\,N\,s^{-1}$ has been applied.

Sample No.	Density [g/cm³]	Porosity [Vol. %]
Diorite	2.88	0.35
Sandstone (brown)	2.39	10.12
Sandstone (white)	2.34	11.35
Marble	2.70	0.31

Figure 4. Density and porosity for selected dimension stones from Thailand. Further explanations are given in the text.

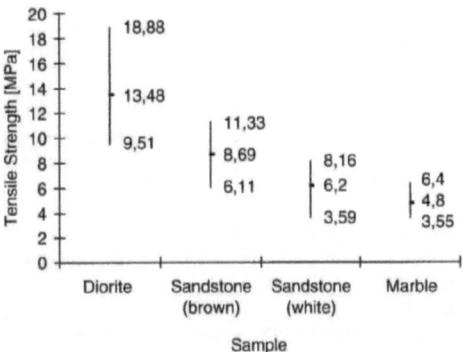

Figure 5. Tensile strength for selected dimension stones from Thailand, including maximum, minimum and mean values for each sample. Mean values were calculated by the average data from x-, y- and z-direction. Further explanations are given in the text.

Although the data for the diorite from Tak Province vary obviously from 9.51 (x-direction) to 18.88 MPa (y-direction) there is no significant correlation between the strength and the investigated orientation, as mean values for single directions are nearly similar with 12.48 MPa for x-direction, 14.66 MPa for y-direction and 14.38 for z-direction. Mean value for the analysed diorite is 13.48 MPa (Fig. 5).

Many sedimentary rocks show the smallest tensile strength perpendicular to the bedding (z-direction, loading parallel to the foliation). However, the investigated brown sandstone behaves different, as mean values for the z-direction are highest with 9.12 MPa, followed by nearly identical values for x- and y-direction (8.42 and 8.52 MPa). Single values range from 6.11 MPa to 11.33 MPa, mean value for all collected data is 8.69 MPa (Fig. 5).

The white sandstone is obviously lower in tensile strength than the brown one and displays the lowest values perpendicular to the foliation (z-direction, mean value 4.96 MPa). The z-direction yields also the minimum value of all tested specimens with 3.59 MPa. The highest tensile strength with 8.16 MPa is still below the mean value for the brown sandstone (Fig. 5). Additionally, the mean value of the white

47

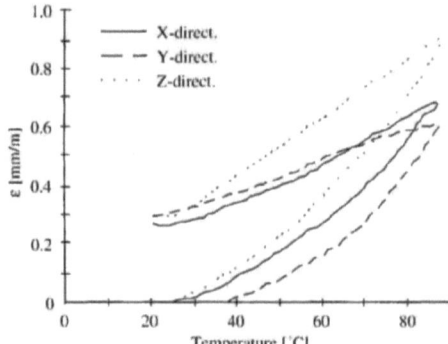

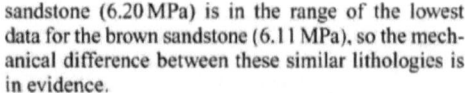

Figure 6. Thermal expansion behaviour of a marble from Nakhon Ratchasima.

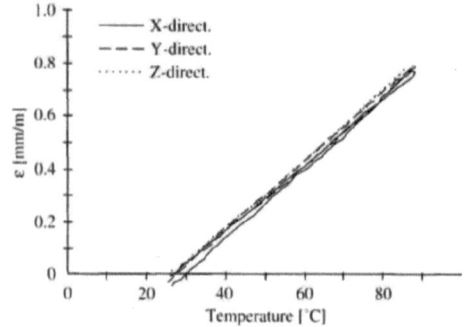

Figure 7. Thermal expansion behavior of a sandstone from Nakhon Ratchasima.

sandstone (6.20 MPa) is in the range of the lowest data for the brown sandstone (6.11 MPa), so the mechanical difference between these similar lithologies is in evidence.

The tensile strength for the marble with a mean value of 4.80 MPa is generally low compared to other rocks of the same composition (Peschel 1983). Mean values for the different directions lie between 4.26 and 5.25 MPa. Single values start from 3.55 MPa up to 6.40 MPa and show the smallest range within the four tested samples (Fig. 5).

7.3 Thermal expansion

The thermal expansion behavior was investigated by cylindrical samples (15 mm diameter, 50 mm length) which were prepared with respect to the reference system (x, y, z). The coefficient α expresses the volume change of the cylinders parallel to their long axis as a function of temperature. The coefficient α for calcite is anisotropic with $\alpha_{11} = 26 \times 10^{-6} K^{-1}$ parallel and $\alpha_{22} = \alpha_{33} = -6 \times 10^{-6} K^{-1}$ perpendicular to the crystallographic c-axis (Kleber 1959). In the result, calcite contracts normal to the c-axis and expands parallel to the c-axis during heating.

The coefficient α for the marble from Nakhon Ratchasima indicates a weak directional dependency, as the values for x- and y- directions are close to each other $(10.88 \times 10^{-6} K^{-1}, 12.30 \times 10^{-6} K^{-1})$ and only slightly higher values are obtained from the z-direction $(13.88 \times 10^{-6} K^{-1})$. Although there are differences in the maximum dilatation, all directions display an almost identical residual strain of around 0.3 mm/m, which is significantly high (Fig. 6).

With a linear increase and decrease of the thermal dilatation the brown sandstone from Nakhon Ratchasima reveals a common behavior for sandstones.

Values for α range from $11.45 \times 10^{-6} K^{-1}$ to $12.74 \times 10^{-6} K^{-1}$ and show no difference of any direction (Fig. 7).

8 PREVIOUS STUDIES

Explicit studies on the quality of Thai rocks as dimension stones have been conducted for a hornblendite and a diorite from Nakhon Ratchasima Province (Wang 1994) as well as a granite from Tak Province (Kyi 1991). Both authors published data from Brazilian Test, but a much higher stress rate was applied and the investigations were carried out without any consideration of different directions.

The description of the mineral composition for the hornblendite coincides with the already given information for this rock type, although the sample seems to be more homogeneous in the crystal size. The rock shows an average tensile strength of 10.08 MPa with the lowest and highest value of 6.73 MPa and 14.24 MPa respectively.

For the diorite an average tensile strength of 24.59 MPa (min. 18.80 MPa, max. 28.61 MPa) was obtained which is significantly higher than data for the diorite from Tak Province. The tested diorite TH03/06AH from Tak reveals maximum data that are in the range of the minimum for the diorite from Nakhon Ratchasima.

As the Tak area exhibits many different varieties of granitoide rocks, the data for tensile strength determined by Kyi (1991) cannot be seen as general characteristics for granites from this region. The rock chosen for the study is coarse-grained with an average grain size of >2 mm and of pink colour. Main constituents are K-feldspar, plagioclase and quartz with accessories of biotite and opaque minerals. Tensile

48

strength for the investigated sample lies between 6.30 MPa and 11.86 MPa with an average of 9.42 MPa.

9 CONCLUDING REMARKS

The results presented in this article show just a section of the envisaged testing program. Altogether 31 different samples have been collected that are now in preparation for further investigations. Detailed studies of their mechanical and thermal behaviour will provide a much deeper insight into the petrophysical properties of Thai rocks.

REFERENCES

Kleber, W. 1959. *Einführung in die Kristallographie.* Berlin: VEB Verlag Technik.
Peschel, A. 1983. *Natursteine.* Leipzig: VEB Deutscher Verlag für Grundstoffindustrie.
Kyi, A. 1991. *Engineering properties of the Tak Granite, northern Thailand.* Bangkok: unpubl. master thesis.
Wang, D. 1994. *Dimension stone quality of hornblendite and diorite from northeastern Thailand.* Bangkok: unpubl. master thesis.

Dimension Stone 2004, Přikryl (ed)
© 2004 Taylor & Francis Group, London, ISBN 90 5809 675 0

Ancient building stone sources of Bratislava's monuments

R. Holzer, M. Laho & T. Durmeková
*Comenius University Bratislava, Department of Engineering Geology, Faculty of Natural Sciences,
Mlynská dolina, Bratislava, Slovak Republic*

ABSTRACT: The paper deals with characteristics and properties of the dimension stone excavated in ancient quarries and used on Bratislava's cultural monuments. The building stones are represented prevailingly by weak sedimentary rocks that during their long existence have been undergone various exogenous factors like weathering, diverse pollutants, temperature and humidity changes, etc. Knowledge and assessment of the building stone composition, properties and durability are important for proposed restoration stages.

1   INTRODUCTION

Bratislava is distinguished as the centre of political, societal and cultural life during several historic periods by many significant monumental buildings having been constructed prevailingly in Gothic, Baroque and Renaissance style. Similarity of architectural styles in the broader Bratislava's space (Bratislava-Vienna-Sopron) has been displayed in the often use of alike, even the same building material – natural dimension stone, which was at that time exploited with certainty nearby to places of its application. It means that source areas of the quarrying had to be undoubtedly in the geological surroundings of the SW points of Small Carpathians Mts., Vienna Basin and Leitha Mts., respectively.

The research presented in paper was oriented on the monuments of a high cultural value, first of all on the St. Martin's Cathedral and Bratislava Castle, and partly on another sacral structures, plague columns, and objects like portals and outer walls of historic habitable objects, etc., in Bratislava – parts of which are in bad technical state and the casual restoration is planned.

2   RESEARCH METHODOLOGY

The following basic steps have been realized:

- study of the technical state of the building stone on chosen objects (outer walls and dimension stone foundation);
- sampling from building blocks of objects by a portable drilling machine;
- detailed description and identification of rocks (mineral composition, paleontologic and structural analyses, deteriorative changes);
- findings of rock sources in surrounding quarries based on the comparison with the building stone;
- sampling and quarry stone lithological characterization;
- laboratory determination of basic physico-mechanical properties (real and apparent density, porosity, water absorption, unconfined compressive strength on dried, saturated and after freezing-thawing cycles of specimens, coefficients of softening and freezing, point load strength, resistance of the rock to salt crystallization in Na_2SO_4 solution, slake durability index and resistance to wear by micro-Deval test);
- special laboratory research concerning the porosity of rock (mercury porosimetry and nitrogen adsorption – BET method);
- environmental assessment of possible re-use of abandoned quarries or by setting up of new quarries into operation.

The main goal of this research was to locate the main alternative resources of the building stone for unavoidable restoration works of ancient valuable monuments. By the reconstruction works in the recent past has been used not always the most appropriate kind of substitute dimension stone, neither from the view point of its quality, nor concerning its colour and appearance.

The most feasible and complex research methodology depends on the type, access possibilities and preservation necessity of the structure exposed to environmental influence, as well as on sampling and laboratory testing possibilities.

51

3 ANCIENT SOURCES OF THE EXCAVATED BUILDING

The limited transport possibilities in the past had marked out the present retrieval of historic quarries in the vicinity of Bratislava. According to the microscopic-petrographic and paleontologic composition of the building stone of Bratislava's monuments, the probable source areas of it had to be on peripheral slopes of mentioned mountainous units built of detritic littoral sediments of the Vienna Neogene Basin. No quarries with similar rocks were found in the vicinity of Bratislava (Durmeková et al. 2003). Findings of old quarries with the similar building stone in Austria directed our research towards Hundsheim Mts. and Leitha Mts. But it could not be excluded that some historic excavations had been situated on the SW edge of Small Carpathians on the territory of Slovakia.

According to the proper documentation, mineralogical and petrographical analyses and archive studies it was confirmed that the most popular building and ornamental stone in Bratislava (and Vienna and Sopron region, too) were the Neogene's rocks due to their suitable properties – colour, structure, easy workability and relatively sufficient durability. The Neogene strata provided a sufficient amount of relatively soft, generally weakly cemented sedimentary rocks of Badenian and Sarmatian age (Rohatsch 1997). Practically all studied rocks belong to the group of different limestones with sandy or organic admixture, or locally to calcareous sandstones and carbonate conglomerates.

The use of the rocks from the northeastern Austrian quarries as a building stone can be dated back to the first centuries of the so-called Roman period (Czijzek 1852). Many of historical monuments from the mentioned epoch were destroyed and the building stone was also often reused in the 13th–14th centuries.

Basic types of dimension stone were excavated in the following recently known quarries (some of them abandoned) in Wolfsthal, Hundsheim, Oggau, Mannersdorf, St. Margarethen – all on the territory of Austria (Fig. 1).

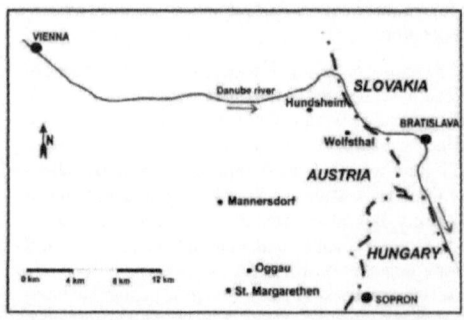

Figure 1. Area of interest and localisation of quarries.

The building stone of Bratislava's monuments was transported with high probability over the Danube River from the mentioned quarries in the Hundsheim Mts. and Leitha Mts. in Austria.

4 PETROLOGICAL CHARACTERISTICS AND QUARRY ROCK PROPERTIES

Detail petrological, paleontological and laboratory studies have been carried out on following types of rocks samples:

1. Oolitic limestone, quarry Wolfsthal: Light-coloured rock of Sarmatian age is built by calcareous ooids with kernels of foraminifers fragments or of quartz clasts, muscovite, granite and sandstone. Contact sparite-calcite cement and bivalves embody them. The rock was often used in the construction epochs of the Cathedral St. Martin's and the Bratislava Castle. The rock was used for its very easy workability, suitable colour and short transportation distance.
2. Leitha limestone, (abandoned quarry) in Oggau and in Mannersdorf (operating quarry): The rock is light white coloured, slightly porous and fine-grained. The main amount of organic substance belongs to red coral algae – genera *Lithophyllum sp.* and *Lithothamnium sp.*, bryozoans and foraminifers. This lithological type is present almost in the whole plinth part of the St. Martin's Cathedral walls, further on many of Bratislava's monuments (plague columns, historic habitable buildings, churches, Baroque palaces, etc.). The rock was used mainly for its very good workability and white colour of the material. The building stone is sufficiently resistant.
3. Sandy limestone/calcareous sandstone, quarry Wolfsthal: The rock is light coloured, porous and fine-grained.

Components are built by ooids, foraminifers *(Miliolina, Nubecularia sp., Quinqueloculina div. sp., Triloculina sp., Elphidium sp.)* and bivalves. The lithoclasts are formed by sandy grains, fragments of granite, quartz, biotite and plagioclase. The material is cemented by calcareous substance. The rock was used in oldest constructional epochs and is very often present in the Bratislava castle frontage. The second phase of its use was connected with years about 1760, during the so-called Rococo rebuilding phase (as window fringes).

4. Sandy limestone, quarry St. Margarethen: The Neogene sandy limestone is light-yellowish coloured, organogenic rock. It originated in the shallow-water environment rich in foraminifers of species *Textularia sp., Triloculina sp.* and *Elphidium sp.* There occur frequent bivalves and ostracods, which

52

aragonite tests were leached-out and replaced by calcite druses. Fossils and clasts are weakly cemented with calcite cement. The important historic quarry is in St. Margarethen (dated back to the Roman period), a part of it still operating. The rock had been very often used for its easy extractability and good workability, also in various building phases of the St. Stephen Cathedral in Vienna (Müller et al. 1993).

5. Lumachelle limestone, quarry Wolfsthal: The rock is white-coloured, with specific rough structure, which is responsible for its high porosity. It is completely formed by of foraminifers clasts *(Miliolina, Nubecularia sp., Quinqueloculina div. sp., Triloculina sp., Elphidium sp.)* and bivalves cemented by the calcareous phase. This lithological type belongs to special rock type and has been found as dimension stone very seldomly. It could not be excluded that it was applied in some wall parts of historic objects covered by the mortar.

6. Fine-grained and coarse-grained carbonate conglomerates, quarry Hundsheim: In this rock group, the light-grey organogenic conglomerates and well-sorted, fine-grained carbonate conglomerates (beach sediments) or even limestones of slightly pink

colour (sublittoral zone) were confirmed. The main amount of organic substance belongs to red coral algae – genera *Lithophyllum sp.* and *Lithothamnium sp.*, tubes of worms, bryozoans and foraminifers. Very variegated association represent foraminifers of genera *Elphidium sp., Textularia sp., Spiroloculina sp., Heterolepa sp.* and *Triloculina sp.* Terrigeneous admixture consists of quartz clasts, granites and carbonates fragments. Bioclasts and lithoclasts are bounded by calcite cement coatings. The conglomerates were found in St. Martin's Cathedral external and basement walls. The same rock material was identified also during remedial works on the NW Bratislava Castle tower. Conglomerates were used also during the so-called Pállfy reconstruction (1616–1635). The rock is relatively sound, however harder workable.

Some assessed physico-mechanical properties of mentioned individual rock types are presented in the Table 1. However, not all rock types were completely tested. From the Table 1 is clearly visible the great difference in the estimated properties, which reflects the different resistance against weathering factors – temperature, moisture and frost cycles, winds, industrial

Table 1. Results of physical and mechanical properties.

Locality/ Lithological type	Apparent density ρ_d [g/cm³]	Absorption capacity N [%]	Uniaxial compressive strength (MPa)			Coefficients		Loss of weight in Na₂SO₄ M [%]	Slake durability test I_d [%]
			σ_c	σ_c'	σ_c''	k_1	k_2		
1. Wolfsthal/ lumachelle limestone	2.09	4.12	5.4	4.4	4.6	0.814	0.846	0.70	95.96
2. Wolfsthal/ oolitic limestone with quartz	2.52	1.07	68.2	46.4	44.2	0.680	0.648	7.24	98.03
3. Wolfsthal/ oolitic limestone	2.17	5.12	15.4	15.0	15.0	0.976	0.975	5.23	97.23
4. Wolfsthal/ calcareous sandstone and sandy limestone	2.05	7.81	9.8	8.7	6.6	0.888	0.679	11.07	96.58
5. Hundsheim/ medium-grained carbonate conglomerate	2.40	2.87	11.9	11.8	11.5	0.985	0.963	60.62	98.28
6. Hundsheim coarse-grained carbonate conglomerate	2.51	1.36	36.3	34.9	33.9	0.960	0.934	6.02	98.05
7. Mannersdorf Leitha limestone	2.68	9.24	49.2	38.2	—	0.776	—	—	—
8. Oggau Leitha limestone	2.69	1.14	65.0	59.1	—	0.909	—	—	—
9. St. Margarethen sandy limestone	2.20	7.25	15.8	13.6	—	0.860	—	—	—

Note: σ_c – on dried samples, σ_c' – on saturated samples, σ_c'' – on samples after 25 cycles freezing and thawing, k_1 – coefficient of softening, k_2 – coefficient of freezing.

emissions, acid rains, etc. Suitable test for the evaluation of resistance against external influences is the determination of rock resistance to salt crystallisation in Na_2SO_4 solution (Fig. 2). The deteriorative processes are manifested in various types of failure on the stone surface – presented in dissolution and recrystallization of calcite, oxidation of iron-bearing phases, gypsum crusts, dusts and decomposition of stone even to the sandy particles, changes in colour, etc.

Figure 2. Different behaviour of two various lithological types during the test in Na_2SO_4 solution: a) Lumachelle limestone from Wolfsthal quarry, loss of weight – 0.7%, left specimen – stage before the test, right specimen – stage after the test; b) Fine-grained carbonate conglomerate from Hundsheim quarry, loss of weight – 60.6%, left specimen – stage before the test, right specimen – stage after the test.

5 SPECIAL LABORATORY RESEARCH

Beside tests results that are mentioned in Table 1, several particular tests on rock specimens have been provided.

Pore surface area was determined by nitrogen adsorption BET method (NOVA 2200 Quantachrome) for temperature 60°C and 110°C. Results are presented in Table 2. Results of the *total porosity and pore radius* are based on the joint evaluation of data obtained by Hg porosimetry and nitrogen adsorption (BET method) and are presented in Table 3. The results reveal remarkable differences between the samples regarding their porosity, characteristics such as total porosity, pore size distribution, pore radius and pore surface area. Results of pore size distribution of study rock determined by Hg porosimetry are in the Table 4.

6 DISCUSSION AND CONCLUSIONS

The complex analysis of Neogene building stones aiming on identification of their favourable exploitation sites and the workability processing was evoked by:

- the need of the reconstruction and the dimension stone applicability;
- the reflection of the negative acting external phenomena in various kinds of the rock decay;
- the need to obtain the information on mineralogical, physico-mechanical and special properties, which are decisive for the behaviour of ancient building stone.

Data obtained have confirmed that combined influence of intrinsic rock properties and external, mostly

Table 2. Results of pore surface area determined by nitrogen adsorption (BET) and Hg porosimetry.

No.	Rock	Locality	60°C, BET [m²/g¹]	110°C, BET [m²/g¹]	Hg porosimetry [m²/g¹]
1.	Oolitic limestone	Wolfsthal	0.731	0.531	0.355
2.	Leitha limestone	Oggau	1.031	1.157	0.792
3.	Sandy limestone	Wolfsthal	0.308	0.378	0.396
4.	Oolitic limestone (slightly weathered)	Wolfsthal	0.214	0.185	0.208
5.	Sandy limestone	St. Margarethen	0.924	1.179	1.970
6.	Lumachelle limestone	Wolfsthal	0.343	0.313	0.198
7.	Leitha limestone	Mannersdorf	0.623	1.106	1.822
8.	Fine-grained conglomerate	Hundsheim	0.556	0.674	0.262
9.	Conglomerate	Hundsheim	0.289	0.356	0.276

atmospheric conditions is responsible for the stone decomposition. Therefore different Neogene rocks used as a building stone on Bratislava's monuments show severe forms of decay. The high amounts of atmospheric pollutants (SO_2 and soots in dust together with temperature and humidity changes) are responsible for the accelerated weathering. If rock material containing small and very small pores is placed in the environment with high concentrations of soluble salts, it cannot resist the periodic cycles of solution and recrystallization during decades of years.

Results of several analyses have proved a great differentiation in rock properties e.g. strength and compactness in comparison to the weak resistance in the Na_2SO_4 solution of the same rock from the conglomerate quarry in Hundsheim. The Leitha limestone, on the other hand, is significant with high percentage of porous area surface (BET test) caused by the great number of micropores but macroscopically they raise the impression of a sufficient compactness and resistance.

From the viewpoint of appearance and durability the most suitable building stones are the Leitha limestones from the Mannersdorf quarry and oolitic limestone with quartz from the Wolfsthal quarry, possibly coarse-grained conglomerates from the Hundsheim quarry (none of them is not operating).

In Slovakia there are some alternative sources in:

- the abandoned sandstone quarry in Sokolovce;
- the abandoned quarry in Cretaceous conglomerates in Chtelnica.

One of the next important tasks in the methodology has to tackle to the chemical composition oriented on:

- the clay and carbonate minerals content in rock types. In the near future we cannot avoid the
- procedure of testing as for these mineral components are one of the most decisive for the progress of rock degradation, as well
- the crusts and encrustations investigation by electron microscopy, ion chromatography, atomic absorption, atomic emission spectrometry, X-ray diffractometry, thermal analysis, elemental analyses by combustion.

Effective protection, conservation and reconstruction strategy of historic monuments of Bratislava has to be based on the multidisciplinary approach of

Table 3. Results of porous properties determined by Hg porosimetry.

No.	Rock	Locality	1st run		2nd run	
			Pore radius [μm]	Total porosity [%]	Pore radius [μm]	Total porosity [%]
1.	Oolitic limestone	Wolfsthal	44.083	14.4	15.674	3.5
2.	Leitha limestone	Oggau	0.005	2.2	10.077	0.7
3.	Sandy limestone	Wolfsthal	27.119	16.7	32.934	9.2
4.	Oolitic limestone (slightly weathered)	Wolfsthal	41.754	11.1	27.499	5.5
5.	Sandy limestone	St. Margarethen	44.656	18.5	51.598	8.7
6.	Lumachelle limestone	Wolfsthal	42.314	11.1	18.139	3.6
7.	Leitha limestone	Mannersdorf	43.966	16.9	45.391	13.4
8.	Fine-grained conglomerate	Hundsheim	10.032	12.2	0.986	6.5
9.	Conglomerate	Hundsheim	0.215	5.0	0.214	2.3

Table 4. Results of pore size distribution determined by mercury porosimetry (1st run).

No.	Rock	Locality	100–10 [μm]	10–1 [μm]	1–0.1 [μm]	0.1–0.01 [μm]	0.01–0.001 [μm]
1.	Oolitic limestone	Wolfsthal	50.4	22.1	19.7	7.8	0.0
2.	Leitha limestone	Oggau	33.9	19.0	9.2	13.7	24.1
3.	Sandy limestone	Wolfsthal	71.8	15.5	5.9	6.6	0.3
4.	Oolitic limestone (slightly weathered)	Wolfsthal	64.4	21.7	7.5	6.3	0.1
5.	Sandy limestone	St. Margarethen	34.0	30.9	1.0	20.2	2.0
6.	Lumachelle limestone	Wolfsthal	57.2	20.2	5.9	6.7	0.0
7.	Leitha limestone	Mannersdorf	67.9	13.9	3.0	10.7	4.5
8.	Fine-grained conglomerate	Hundsheim	26.4	38.2	29.0	6.5	0.0
9.	Conglomerate	Hundsheim	9.0	13.2	59.3	18.5	0.1

55

geologists, architects, restorers and cultural monument management.

ACKNOWLEDGEMENTS

The paper was worked out in the frame of the Agency VEGA Grant project No. 1/0117/03.

REFERENCES

Czijzek, J. 1852. Geologische Verhältnisse der Umgebungen von Hainburg, des Leithagebirges und der Ruster Berge. *Jb.k.k. Geol. Reichsanstalt* 35–55.

Durmeková, T., Holzer, R. & Wagner, P. 2003. Weak rocks in engineering practice. In *Geotechnical Measurements and Modelling*. Lisse: Swets and Zeitlinger.

Müller, H.W., Rohatsch, A., Schweighofer, B., Ottner, F. & Thinschmidt, A. 1993. Gesteinsbestand in der Bausubstanz der Westfassade und des Albertinischen Chores von St. Stephan. *Österr. Zeitschr. für Kunst und Denkmalpflege, Sonderdruck* 106–116.

Rohatsch, A. 1997. Gesteinskunde in der Denkmalpflege unter besonderer Berücksichtigung der jungtertiären Naturwerksteine von Wien, Niederösterreich und dem Burgenland, Habilitationsschrift, Universität für Bodenkultur Wien: 284 p.

Dimension Stone 2004, Přikryl (ed)
© 2004 Taylor & Francis Group, London, ISBN 90 5809 675 0

The relationship between spatial distribution of hydrothermal silicification in Keuper sandstone, primary facies and early diagenesis (The Wendelstein-Quartzit near Nuremberg, Southern Germany)

R. Koch
Institut für Paläontologie, Angewandte Faziesforschung-Bausteinforschung, Erlangen, Germany

U. Zinkernagel
Consulting Laboratory, Bochum, Germany

ABSTRACT: The so-called "Wendelstein-Quarzit", a local variety of Upper Burgsandstein in the South of Nuremberg (Franconia) has been intensively silicified by silica-rich hydrothermal solutions along a system of subvertical fractures trending between 100–120°. It is the most resistant sandstone to weathering of the region. The outcrops form a chain of monadnock hills in an area of 1 to 4 km in size running parallel to the fracture system of Upper Cretaceous age. Several cross sections trending north-east were studied, i.e. vertically to the fault lines. The sandstone bodies display large progradational cross-bedding of up to 5 m to 50 m in size and are mainly composed of medium and coarse-grained detritus. The sandstones close to the fault display hydrothermal quartz precipitates in secondaryly porous fabrics forming syntaxial, often euhedral dust rims on the surface of detrital quartz grains. There are repeating phases of quartz cementation. With increasing distance from the main fault line the sandstones display diagenetic features as they are encountered in Keuper sandstones of the broader environs. Feldspar is often partly and/or completely dissolved and large authigenic kaolinite vermicules form characteristic clay cement. Illitic clay coatings occasionally outline the boundary of leached feldspar grains. Such sandstones commonly display features of heavy compaction of sutured grain contacts. The size and spatial distribution of specific types of pores, i.e. intergranular, moldic, or secondary pores and neoformations, had influence on the formation of pathways for hydrothermal fluids. In sandstone fabrics heavily hydrothermally silicified succeeding diagenetic effects were less effective. This interaction of primary facies, early diagenesis and migration of hydrothermal silica-rich solutions on pre-existing pathways has to be taken into account when quarrying for sandstones as building stone of constant quality.

1 INTRODUCTION

The region of Nuremberg-Fuerth-Erlangen in Southern Germany is famous for numerous siliciclastic sediments that are sources of natural building stones. Sandstones of the Keuper Formation, as well as sandstones of the Lias and Dogger periods in particular were quarried in the vicinity of these towns. Furthermore, the famous "Treuchtlingen Marmor" ("marble") of Upper Jurassic Age occurs in outcrops within the Southern Franconian Alb, about 40 km south of Nuremberg.

"Burgsandstein" is one of the sandstones used as building stone. Its name is derived from all the castles ("Burgen") built of Keuper sandstones topping the hills e.g. in Nuremberg, Fuerth and Cadolzburg. Thus, Nuremberg castle is situated on a hill composed of Blasensandstein, Lower, and Middle Burgsandstein.

Clay stone horizons (so-called "Letten") intercalate in sequences of sandstones forming local or even more spacious aquifers. Two to three thousand wells are known to have been dug in medieval Nuremberg reaching water tables at depths of between 6 and 15 metres.

The reddish, cross-bedded rectangles of medium to coarse-grained sandstones distinguish the townscape of Nuremberg. The varying resistance to weathering of the Keuper sandstones can often be observed in one building. This is especially true of the Burgsandstein commonly used in Nuremberg. These variations in the utilisation of the sandstone for buildings reflect the sedimentological variations caused by various depositional environments and diagenesis.

The so-called "Wendelstein-Quarzit", a sandstone of the Upper Burgsandstein age which has been intensively silicified by hydrothermal solutions along a

57

tectonic fault line is the most stable sandstone of the broader environs.

The topic of this study is the spatial distribution of this unusual diagenesis created as result of the pre-existing facies conditions, determined by the depositional system and early diagenesis.

2 GEOLOGY AND SEDIMENTOLOGY OF THE KEUPER

Keuper sediments were mainly deposited in non marine environments of the Germanic Basin on the even surface of the preceeding Muschelkalk-deposits (Freudenberger & Schwerd 1996, Geyer & Gwinner 1991).

Keuper sediments are to be found in outcrops at the western border of the crystalline basement close to Amberg and Regensburg, in the North and West around Bayreuth, Wuerzburg, Crailsheim and in the South of Nuremberg. Stratigraphic subdivision, including lithostratigraphic nomenclature and lateral interfingering facies have been published most recently by Beutler et al. (1999) and Glaser et al. (2001).

Detritus supply of the Lower Keuper (Werksandstein) and the Lower Middle Keuper (Schilfsandstein) occurred from the north along so-called "sandstrings" (Wurster 1964) within a brackish to limnic facies (Nordic Keuper). A study by Dittrich (1989) revealed a complex of sedimentary/tectonic interactions which caused the differentiated distribution of various facies in the Keuper Basin.

Sediment transport from the Vindelician-Bohemian Land (so-called Vindelician Keuper) in the south-east that provided the detritus of the deposits of the Middle Keuper, i.e. "Blasensandstein", "Coburger Sandstein", and "Burgsandstein", occured in the area around Nuremberg and resulted in sediments of mainly coarse-grained sandstones enriched in feldspar (Freyberg 1936, Freudenberger & Schwerd 1996, Beutler et al. 1999).

This study is focused on the Upper Middle Keuper sandstones of the "Burgsandstein" (thickness about 80 m) that were and are still of great importance to the supply of natural building stone.

Interbedded claystone (so-called Basisletten) of varying thickness help to subdivide the sequences of sandstone (Tab. 1, Fig. 1 and 4) into Lower (kmBU), Middle (kmBm) and Upper Burgandstein (kmBo) (Haunschild 1985, Gudden & Haunschild 1993). Additionally intercalations of "Zwischenletten" result in a more complex composition of the sandstone sequence.

North of the line Ansbach-Erlangen Lower burgsandstein marine environmental conditions dominate. In the South of that line deposits accumulated of terrestrial environment starting with the deposition of the Blasensandstein. Bed flows and large alluvial fans

resulted in the deposition of non-homogeneous (composition, sorting, thickness), poorly-sorted sandstone of varying feldspar content.

In the Middle Burgsandstein dolomitic arcoses are encountered composed of poorly-sorted, coarse-grained sandstone of a Northern playa facies (Coburg, Kulmbach). In the area of Nuremberg the Keuper

Table 1. Stratigraphy of the Keuper sediments in Bavaria (from Geyer & Gwinner 1986).

Rhaetian	Upper Keuper	Mudstones, lower and upper sandstone
Coburg –	**Middle**	Feuerletten
Formation	**Keuper**	**Burgsandstein**
Ansbach –		**UpperBurgsandstein**
Formation		**Middle Burgsandstein**
Stuttgart –		**Lower Burgsandstein**
Formation		Obere Bunte Mergel
		Coburger Sandstein
		Blasensandstein
		Untere Bunte Mergel
		Schilfsandstein

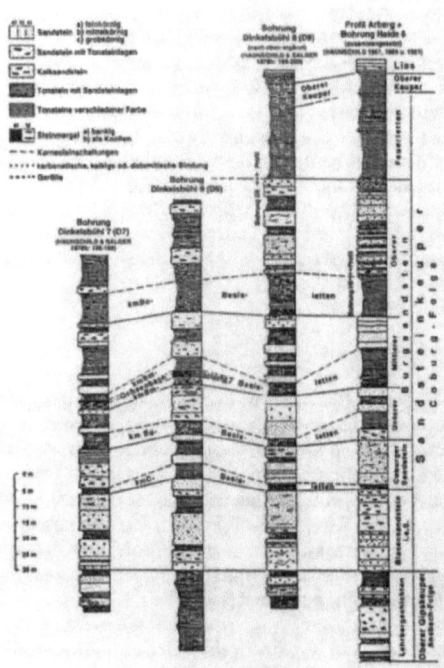

Figure 1. Profiles through the "Sandsteinkeuper" between Dinkelsbühl and Arberg (from Haunschild 1987 in Schmidt-Kaler 2003).

58

sediments are characterized as sandy terrestrial border facies also present in the Upper Burgandstein deposits. In the area of Erlangen-Nuremberg such sandstones are composed of middle, coarse, and fine-grained detritus and are generally thick-bedded and rich in feldspar (Freudenberger & Schwerd 1996, Emmert 1964). Sandstones of marine environments only occur North of the river Main.

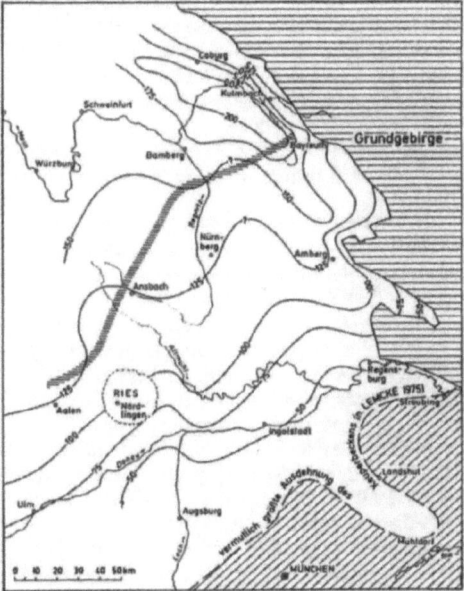

Figure 2. Distribution and thickness of the sediments of the "Sandsteinkeuper" (without Feuerletten) with a possible boundary to the Vindelician land (from Haunschild 1985). Note the SE boundary of the basin facies, the so-called Heldburgfacies running from the area of Bayreuth to Aalen and the location of the outcrop area of Wendelstein Quarzit (South of Nuremberg) (from Schmidt-Kaler 2003).

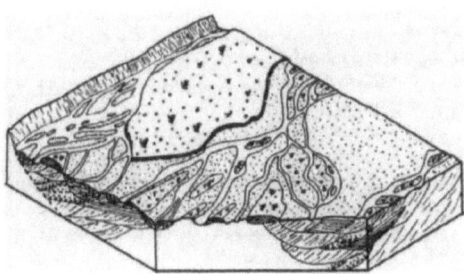

Figure 3. Schematic drawing of the depositional environment of Blasensandstein of Lower "Sandsteinkeuper" (from Frings 1983).

Feuerletten top the Burgsandstein. They form the uppermost member of the Middle Keuper and represent a sequence of red to pink-coloured claystones of 40–60 m thickness, deposited in a brackish to lacustrine environment. In the Upper Keuper (Rhät, ko), marine influenced sediments predominate and are encountered in the area of Northern Bavaria. These clayey-sandy sediments were deposited during periods of rapidly changing regressive and transgressive phases.

2.1 Building stones in the Nuremberg area

The Keuper sandstones, predominantly quarried in and around Nuremberg reveal great differences in thickness and in the type of beds. This is caused by changing depositional systems of meandering rivers, of local alluvial fans, of channel systems and of sand sheets. Generally, rapid lateral changes are observed even in the same quarry where medium-grained sandstones of cross-bedded bar and channel systems change laterally into siltstones or even claystones (Fig. 3).

3 THE WENDELSTEIN-QUARTZIT

"Wendelstein-Quarzit" is a local variety of Upper Burgsandstein (Haarländer 1955) that is distinguished by its hardness and resistance to weathering. It is found in outcrops south of Nuremberg close to the motorway interchange "Nürnberg-Süd" in an area where sandstone has been quarried for centuries (Fig. 4). Due to its resistance to weathering "Wendelstein-Quarzit" forms a chain of monadnock hills in an area of 1 to 4 km in size trending NW-SE.

The sandstones are composed of fine to middle, occasionally coarse-grained sandstones which were deposited under terrestrial arid conditions (v.Gehlen 1956). The extraordinary hardness is caused by hydrothermal impregnation of secondary quartz.

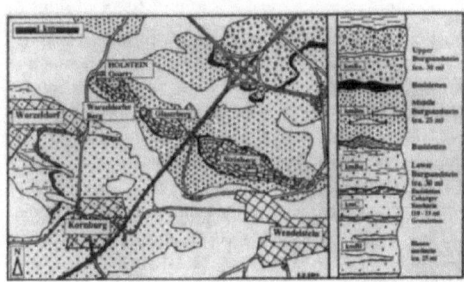

Figure 4. The geology of the Wendelstein-hills including the location of the most important closed-down quarries of Upper Burgsandstein. The Holstein quarry at the NW edge of the chain of hills has recently been reactivated (modified from Berger 1977).

59

Hydrothermally silicified "Wendelstein-Quarzit" (Dorn 1926, Gehlen 1956, Spöcker 1964, Grimm 1990) has been used since medieval times for millstones, gravestones and building stones. The range of commercial distribution of such products can be traced along the River Danube down to the Balkans. In more recent times "Wendelstein Quarzit" has been used to restore the Opera House and the Justice Palace in Nuremberg because of its excellent suitability as building stone (Spöcker 1964).

The rocks of the Wendelstein hills are heavily faulted by a system of subvertical fractures trending in a direction of 100–120°. Silicification and neoformations of other minerals is almost exclusively related to these fractures (v. Gehlen 1956). The hydrothermal source of silicification is documented by the occurrence of other hydrothermal neoformations, such as barite, pseudomorophs after quartz to barite, fluorite, fluorite negative crystals in quartz, needles of apatite, galena, and sphalerite (Dorn 1926). The mineralisation occurred subsequent to the formation of the herzynian fractures and is assigned to the Upper Cretaceous (Senonian; v. Gehlen 1956).

The intensity of silicification decreases with increasing lateral distance from the fault system (Knetsch 1929). Furthermore, Gehlen (1956) mentioned lateral silicification along pathways which were active at distinctive phases. Dorn (1926) described feldspar to be less likely disintegrated by weathering in rocks being heavily affected by silicification. Feldspar, commonly, converts into kaolinite. Furthermore Dorn (1926) mentioned that the presence of early formed clay suppresses succeeding silicification process. This effect has been repeatedly described also from other sandstone formations (Gaupp 1996, Gaupp et al. 1993, Horn 1965, Drong et al. 1982) and is related to the decrease in the nucleation on the surface of quartz grains to precipitate quartz from dissolved silica.

3.1 *Recent sedimentological and diagenetic studies*

In the course of sedimentological studies, first facies analyses were carried out at the outcrops on the tops of the Wendelstein hills and in the recently reactived Holstein Quarry (Fig. 4). Several NE trending cross sections were studied, i.e. vertically to the fault lines. In this way the lateral decreasing influence of silicification on the sandstones can be traced best.

Large progradational cross-bedded sandstone bodies mainly composed of medium and coarse-grained detritus from 5 to 50 m in size can be observed as well as channel fills of planar cross-bedded sandstones cutting into variegated claystones.

In the NE range of the Holstein Quarry, these sandstones are friable and of reddish and yellowish colour and commonly contain green clay chips. Little pans filled with greenish clay form clay lenses. Towards SE,

in direction of the main fracture line within a lateral distance of about 50 m the colour changes from red to yellow to light yellow to a darker colour, giving a glassy look to the rocks. Probably due to increased rock density the flanks of the hill are more steep, indicating an increased resistance to weathering of the silicified sandstone. A vertical wall of about 15 m in height displays some smaller lenses only moderately cemented, whereas others are very dense and entirely cemented.

On the SW flank the same features are encountered. With increasing distance from the top of the hill, i.e. the source of the hydrothermal silica, the sandstone become more friable and colour becomes more red and contains various lens-like sandstone-bodies of varying intensity of cementation.

Similar features are encountered at other quarries along the main hill line, which runs parallel to the fault line.

The sandstones close to the fault display hydrothermal quartz precipitates in a secondary porous fabric forming syntaxial, often euhedral dust rims on the surfaces of detrital quartz grains. There are repeated phases of quartz cementation. At the present stage of the study it is not yet possible to enumerate diagenetic cementation phases or to distinguish these from those created by hydrothermal supply. As is demonstrated in Figure 5c their presence cannot be doubted.

Normal diagenetic features of Keuper sandstones can be observed in sandstones at greater distance from the main fault line. Feldspar is often partly and/or entirely dissolved (Fig. 5a) and large uthigenic kaolinite vermicules commonly occur as neoformations (Fig. 5b), forming characteristic clay cement within the sandstones. Illitic clay coatings occasionally outline the boundary of leached feldspar grains indicating that leaching postdates the formation of clay.

In these areas the sandstone commonly reveals intensive compaction features with sutured grain contacts (Fig. 5e) the formation of which has been enhanced by the early clay coatings.

Diagenetic processes within the Keuper sandstones continued over geological periods if not interrupted by overprinting events such as hydrothermal impregnation. Hydrothermal silicification of the Wendelstein area preserved the diagenetic stage of the Cretaceous period in those sandstones being heavily impacted by hydrothermal quartz.

Closer to the main fault line, the increasing degree of density coincides with an increasing degree of silica cementation as documented by abundant idiomorphic silica overgrowth on detrital grains (Fig. 5d).

Along the fault line, which is composed of varying smaller faults running parallel like those of an auxillary fault silica cementation within the sandstones is obvious.

One striking observation looking at thin sections is not fully understood up to now. The secondary pore

60

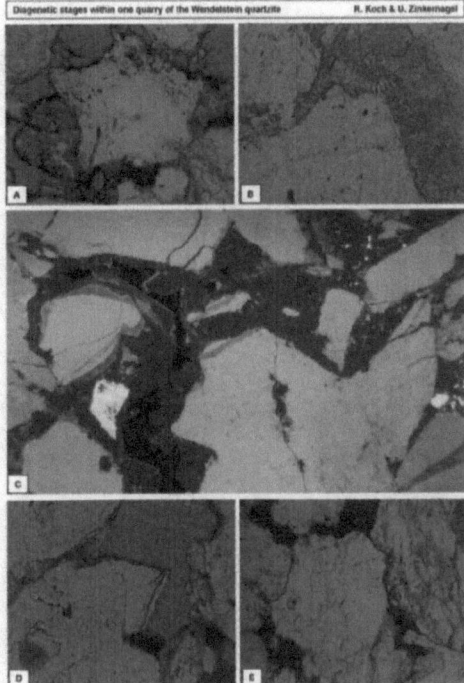

Figure 5. Microscopic evidence of the diagenesis in studied rocks. For explanation see text.

spaces of sandstones enhanced by leaching of diagenetically unstable detritals are never entirely closed by cement. The pores in those sandstones that are heavily impregnated with hydrothermal quartz are free of any remnants of dissolved grains. With the abating influence of hydrothermal impregnation, the amount of remains of leached detritals increases.

4 RESULTS

The study shows that primary facies and early diagenesis formed characteristic pore spaces in Keuper sandstones of the Wendelstein hills as it is also the case in other Keuper sandstones of the broader environs. The size and spatial distribution of porosity types (intergranular, secondary due to dissolution; neoformations) triggered the pathways for hydrothermal fluids originating from a fault line trending NW-SE and intersecting the Keuper sediments over a distance of about 10 km.

The size and extension of the sandstone bodies, their porosity and permeability, and their early diagenetic history permitted hydrothermal fluids to penetrate into Keuper sandstones and to impregnate the fabrics to varying extent.

Where hydrothermal silicification occurred, later diagenetic effects such as the dissolution of feldspar, neoformation of kaolinite or illite coatings did not take place.

This interaction of primary facies, early diagenesis and the migration of hydrothermal silica-rich solutions on pre-existing pathways have to be taken into account when quarrying for sandstones of constant quality. This is even true if only planning the quarrying for a very restricted area as it is that of the Wendelstein-Quarzit.

REFERENCES

Berger, A. 1977. *Geologische Karte von Bayern 1 : 50 000, Blatt Nürnberg-FürthErlangen mit Erläuterungen.* München.

Beutler, G., Hauschke, N. & Nitsch, E. 1999. Faziesentwicklung des keupers im Germanischen becken. In N. Hauschke & V. Wilde (eds), *Trias – Eine ganz andere Welt*: 129–174. München: Pfeil.

Dittrich, D. 1989. Der Schilfsandstein als synsedimentär-tektonisch geprägtes Sediment – eine Umdeutung bisheriger Befunde. *Z. dt. geol. Ges.* 140: 295–310.

Dorn, P. 1926. Geologie des Wendelsteiner Höhenzuges bei Nürnberg. *Z. dt. geol. Ges.* 78: 522–564.

Drong, H.J., Plein, E., Sannemann, D., Schuepbach, D. & Zimdars, J. 1982. Der Schneverdinger Sandstein des Rotliegenden – eine äolische Sedimentfüllung alter Grabenstrukturen. *Z. dt. geol. Ges.* 133: 699–725.

Emmert, U. 1964. *Erläuterungen Geologische Karte von Bayern 1 : 50 000*, 91–120. München.

Freyberg, B. v. 1936. Die Randfazies des mittleren keupers in Mittelfranken. *S. Ber. phys.-med. Soz. Ges. Erlangen* 67: 167–246.

Frings, U. 1983. Sedimentologische Untersuchungen im oberen Blasensandstein in Nürnberg. *Geol. Bl. NO-Bayern* 32: 13–34.

Freudenberger, W. & Schwerd, K. 1996. Erläuterungen zur Geologsichen karte von Bayern 1 : 500 000, 329 pp. München: Beil.

Gaupp, R. 1996. Diagenesis types and their application in diagenesis mapping. *Zbl. geol. Paläontol.* 11: 1183–1199.

Gaupp, R., Matter, A., Platt, J., Ramseyer, K. & Walzebuck, J. 1993. Diagenesis and fluid evolution of deeply buried Permian (Rotliegendes) Gas reservoirs, Northwest Germany. *Am. Ass. Petrol. Geol. Bull.* 77: 1111–1128.

Gehlen, K.v. 1956. Sekundär-hydrothermale Mineralisation im Burgsandstein des Wendelsteiner Höhenzuges bei Nürnberg. *Geol. Bl. NO-Bayern* 6: 12–21.

Geyer, O.F. & Gwinner, M.P. 1991. *Die Geologie von Baden-Württemberg.* 4th ed., 482 pp. Stuttgart: Schwizerbart.

Glaser, S., Lagally, U., Schenk, P., Eichhorn, R. & Brandt, S. 2001. Geotope in Mittelfranken. *Erdwissenschaftliche Beiträge zum Naturschutz* 3: 127 pp. München: Bayer. Geol. Landesamt.

Grimm, W.-D. 1990. Bildatlas wichtiger Denkmalgesteine der Bundesrepublik Deutschland. *Bayerisches Landesamt für Denkmalpflege, Arbeitshefte* 50: 255 pp.

Gudden, H. & Haunschild, H. 1993. Die Trias in der Forschungsbohrung Abendberg 1001. *Geol. Bavarica* 97: 47–66.

Haarländer, W. 1955. Geologie des Blattes Röttenbach. *Erlanger geol. Abh.*, 13: 16 pp.

Haunschild, H. 1985. Der Keuper in der Forschungsbohrung Obernsee. *Geol. Bavarica* 88: 103–130.

Haunschild, H. 1987. In Schmidt-Kaler, H. 2003. Wanderungen in die Erdgeschichte (14) – Von der Frankenhöhe zu Fränkischen Seenland. 58 pp. München: Pfeil.

Horn, D. 1965. Dagenese und Porosität des Dogger-beta-Hauoptsandsteins in den Ölfeldern Plön-Ost und Preetz. *Erdöl und Kohle* 18: 249–255.

Knetsch, G. 1929. Der Keuper in der bayerischen Oberpfalz. *N. Jb. Mineral., Beil.- Bd.* 61: 83–150.

Spöcker, R.G. 1964. Die geologischen und hydrologischen Verhältnisse im Untergrund von Nürnberg. *Abh. Naturhist. Ges. Nürnberg* 33: 136 pp.

Wurster, P. 1964. Geologie des Schilfsandsteins. *Mitt. Geol. Staatsinst., Hamburg* 33: 140 pp.

Dimension Stone 2004, Přikryl (ed)
© 2004 Taylor & Francis Group, London, ISBN 90 5809 675 0

Origin and processing of precious stones in the interior of St. Wenceslas chapel, St. Vitus cathedral (Prague, Czech Republic)

I. Kolaříková & R. Hanus
Charles University, Faculty of Science, Institute of Mineralogy, Geochemistry and Mineral Resources, Prague, Czech Republic

ABSTRACT: The origin and processing of precious stones decorating the St. Wenceslas chapel in Prague (Czech Republic) is described. Amethyst-jasper material is related to the quartz veins located in Podkrušnohoří area, Ciboušov locality (western Bohemia). A number of buried shafts document an extensive mining activity in this region during the 14th century. The spoil was cut using multiplex stringed saws and ground by metal blocks covered with abradant. Final polishing consists in mulling of diatomaceous earth and annealed pyrite on the stone surface.

1 INTRODUCTION

The history of mankind has been accompanied by the use of natural precious stones for churches and dome interior decorations. One of the oldest and most valuable decorations can be found in St. Wencelas chapel in Prague (Czech Republic).

Many legends are told about the origin of precious stones decorating the famous St. Wenceslas chapel. The memoirs (Krabice 1366) contain many speculations about the places, where quartz precious stones could be found. Strange virtues were also ascribed to these stones (Krabice 1366).

This paper is focused on historical data and explains the real origin of precious stones used in incrustations of St. Wenceslas chapel.

2 ST. WENCESLAS CHAPEL (PRAGUE)

The St. Wenceslas chapel (Fig. 1) is a part of the metropolitan St. Vitus cathedral situated in the area of Prague Castle. The St. Wenceslas chapel with ground 10×10 m and inwrought walls dates back to the 14th century.

The decorations consist of 1484 polished plates up to 3.6 m high (Fig. 2). More than 1479 minerals and 5 rock types can be found in these incrustations, mainly different varieties of quartz – e.g. rock-crystal, amethyst and cryptocrystalline varieties such as jasper, chalcedony, rare agate and labradorite porphyry – *porfido verde antico* and *porfido rosso antico* (Krabice 1366, Žežulka 1985).

Labradorite decorative stones were probably brought to Bohemia by Charles IV., Czech king and Roman

Figure 1. St. Wenceslas chapel in Prague.

emperor in the 14th century. The original rocks were mined in the eastern Egypt (Wadi aber Maamal, Jabal Dokhan) and brought to Italy. Famous Italian churches (e.g. St. Eufrasius basilica) were decorated by Egyptian labradorite rocks.

Figure 2. Part of the interior (St. Wenceslas chapel).

3 ORIGIN OF QUARTZ-TYPE PRECIOUS STONES

Restoration works in 1995 have risen a demand for quartz precious stone material similar to that which was used in the 14th century.

Several misleading hypothesis about the origin of quartz-type precious stones used for decorations were published during the 1970s. Urban (1973) states, that the precious stones were mined in Podkrkonoší area. It was also thought that amethyst-jasper material came from Schlotwitz locality (Germany) because of nearly identic mineralogical composition of these stones (Kašpar 1980). However, the structure and texture of Ciboušov and Schlotwitz quartz material is different (see Fig. 3).

Figure 3B shows lace amethyst cockades, whereas Figure 3A demonstrates the mural type of amethyst. In addition, samples from Schlotwitz contain less jasper mass than material coming from Ciboušov.

According to typical paragenesis and structure of quartz precious stones (jasper-amethyst), relation to the quartz veins located in Podkrušnohoří area near Ciboušov can be declared.

4 GEOLOGY AND MINERALOGY OF THE CIBOUŠOV AREA

Ciboušov is located on the southern slopes of Krušné hory Mts., approximately 3 km SSE from Klášterec nad Ohří (Fig. 4).

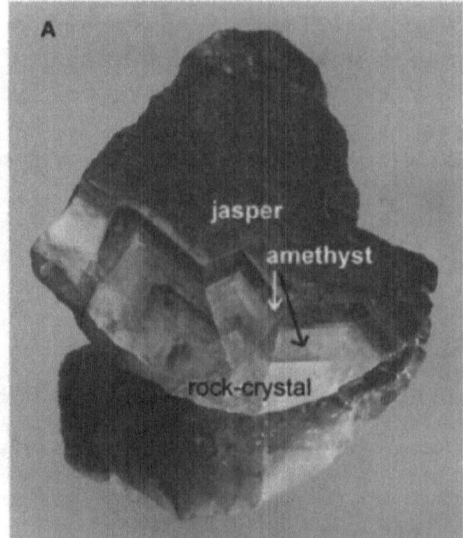

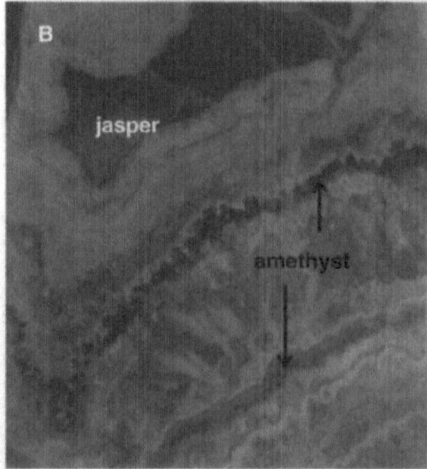

Figure 3. Comparison of precious stones coming from Ciboušov, Czech Republic (A) and Schlotwitz, Germany (B).

The region is underlined by a complex of crystalline basement represented by muscovite orthogneiss often affected by migmatization (Žežulka 1985). In general, the gneisses consist mostly of quartz, orthoclase, muscovite, biotite and accessory minerals (garnet, calcite and chlorite).

The whole region is bound by several faults, including major northern fault zone (Podnikla dislocation district). The width of the most important faults and fault zones is variable, up to 50 m in case of Podnikla

64

Figure 4. Geographical location of Ciboušov locality.

Figure 5. Podnikla valley in 2002.

fault. Cracked gneiss debris lined with cockades of quartz, rock-crystal and amethyst and cemented by several generations of red jasper were formed as a result of frequent movements connected with Podnikla fault zone (Žežulka 1985).

Quartz veins located in the close proximity of fault zones are typical of strong zoning (Tuček 1983):

(1) Zone of brittle discontinuites with feather joints 2 m thick and approximately 100 m far from the main Podnikla fault are filled with debris lined with grey-white milky quartz. Small cavities formed between blocks of quartz can be filled with crystals of rock-crystal or amethyst (maximal size about 20 cm). This material was secondarily broken and subsequently cemented by younger generation of milky quartz. In addition, thin brittle discontinuities were partly filled with red jasper.
(2) Farther zone of faults (200–500 m far from the main dislocation) was markedly affected by younger movements. However, the cementation processes were analogical to that described for the nearer zone. Red jasper cemented thin cracks (several mm) containing zonal amethyst crystals (see Fig. 3).

Amethyst-jasper material can be found at exposures near Ciboušov up to the present day. This rare material was used for redesigning of St. Wenceslas chapel as well as St. Cross chapel interior in Karlštejn castle.

5 MINING HISTORY

Mining history of Podnikla district was described by Urban (1973) and Švenek (1979). The successful mining of quartz veins resulted during the medieval times in production of high quality decorative stones.

Shafts were sunk in the rock and carried 7 or 8 m below the estimated depth of the bottom of quartz

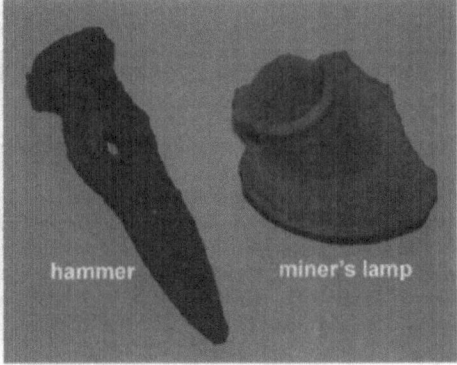

Figure 6. Artefacts from the 14th century (Ciboušov locality).

veins. After a main cross-cut has been driven, from a shaft bottom to a point directly under the centre of the amethyst-jasper layer, two main galleries were opened up to follow the precious layer. The crosscuts, in turn, were connected by tramming drivers 3 to 4 m apart, depending on the adopted system of mining: (1) blocking, where the spoil contained large lumps (up to 1 m) or (2) panelling, where the width of each block was limited to 30 cm (Švenek 1979). The spoil was transported using sophisticated system of narrow paths and tunnels following contour lines.

The decline of Podnikla region was as quick as was its expansion. The richest veins were exhausted and mines closed in the late 14th century (Švenek 1978).

Many old buried shafts and dumps document mining activities in Podnikla valley and its close surrounding (see Fig. 5).

Several artefacts coming from 14th century and documenting the gallery driving were found in Podnikla area in 2002 (Fig. 6).

65

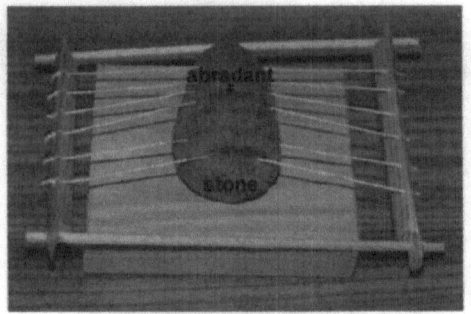

Figure 7.   Model of the multiplex stringed saw.

6   PROCESSING

Material was cut using multiplex stringed saw (Fig. 7). Each string was powdered with abradant (milled garnet of topaz) and subsequently damp-cured with hot water-oil emulsion (Krabice 1366).

Rough-grinding of cut plates was done using metal blocks and abradant. As a result of hand grounding, plenty of striae can be found on the plate suface.

Final polishing represented the most important step of this precious stone processing. Stonecutters usually used old techniques brought from Middle Asia which consist in grinding of diatomaceous earth and annealed pyrite on the plate surface using wooden lamina (Belda 1979).

7   CONCLUSIONS

Precious stones used for decoration of St. Wenceslas and St. Cross chapel in the 14th century come from quartz veins of Podkrušnohoří area.

Specific techniques of cutting, grinding and polishing used in the 14th century came from Middle Asia and gave a unique appearance to the precious decorative stones.

REFERENCES

Belda, J. 1979. Czech stonecutters (In Czech). Čas. NM 5: 1–9.
Kašpar, J. 1980. Gemstones cutting in the 14th century (In Czech). Geol. průzk. 7: 203–205.
Krabice, B. 1366. Memoirs (In Czech). Manuscript.
Švenek, J. 1979. Precious stones of quartz veins (In Czech). Čas. NM 2: 2–20.
Tuček, K. 1983. History of The First Czech lapidarium (In Czech). Čas. NM, řada přírodovědná. 125/1: 37–60.
Urban, J. 1973. The origin of precious stones decorating the St. Wenceslas chapel (In Czech). Geol. průzk. 4: 107–109.
Žežulka, J. 1985. Precious stones around us (In Czech). Interní zpráva 45881/3, Geofond.

Dimension Stone 2004, Přikryl (ed)
© 2004 Taylor & Francis Group, London, ISBN 90 5809 675 0

Greek marble through the ages: an overview of geology and the today stone sector

K. Laskaridis

Institute of Geology and Mineral Exploration, Department of Economic Geology, LITHOS Laboratory, Athens, Greece

ABSTRACT: The article overviews the Greek marble sector and provides information on the historic development of the Greek marble sector and its geological background. Reference is made to the major marble-production areas. The great wealth of Greece with respect to deposits of high quality marble, mainly white types, in combination with a very long tradition in the art of marble, the roots of which reach the depths of centuries back ago, have much contributed to the development of the modern and dynamic Greek marble industry, that is classified among the top world producers of decorative natural stones, as concerns both sizes of production and exports.

1 INTRODUCTION

In ancient Greece the use of marble had been very wide. Marble and stone were the materials that deeply touched the human sensitivity and drove man to the world of aesthetics and symmetry.

Marble has been used in Greece for the construction of sacred buildings since early recorded times. The ancient Greeks were the first among many ancient civilisations to notice the unique properties of this remarkable stone that lasted so long, remained so beautiful and shaped to their needs so easily. Marble quarrying in Greece started several centuries ago. Since the Medium Neolithic era (about 5.000 BC) we have marble female idols, whilst later the series of the famous Cycladic idols followed. The first monuments of Greek sculpture (marble was used in combination with porous-materials), appeared as early as 630 BC. Representative samples of such monuments were the temple of Zeus in Olympia, as well as the temple of Apollo in Delphi, with marble of Paros in the façade and porous-stone for the rest part of the construction. The peak of the Greek classical period is represented by such outstanding structures as the Athens Acropolis with Parthenon and Erechtheum, the Aphrodite of Milos, the Hermes of Praxiteles etc.

The history of the Modern Greek marble industry started in the 1960s when building activities and standards of living rose remarkably. The number of marble quarries has been continuously increasing since the 1960s.

2 TERMINOLOGY – GEOLOGY

The commercial use of the term 'marble' ('marmaro' in Greek) does not involve only the metamorphic limestone deposits, but also any ornamental stone that can be cut to standard dimensions can be cleaned to a mirror finish – polished and that can be used in the decorative marble art. In general, the name marble is given to various types of rock such as metamorphic rocks – e.g. crystalline limestones and/or sedimentary rocks – e.g. calcareous alabasters, etc. Currently, although there is still a habit of calling ornamental stones by their traditional names – some of them given in antiquity – norms are being developed (e.g. by CEN TC 246, EN 12440) to characterise a rock by its traditional name, by its petrographical name and its place of origin.

The geological history of Greece has been influenced by conditions of intense orogenesis, magmatism and metamorphosis that led to the creation of extended areas of deposits of ornamental stones. Today, the following ornamental stones are widely used in Greece:

• metamorphic stones: calcitic marble, dolomitic marble, cipollins and ophicalcites;

- sedimentary stones: limestones, travertines, brace, onyxes and alabasters;
- magmatic rocks: granites, granodiorites and gneiss.

3 GREEK MARBLE PRODUCING AREAS (DEPOSITS) IN THE ANCIENT AND TODAY

Greece, which is extremely privileged with regard to marble deposits, is one of the major marble producing countries. It provides the world market with rare varieties of marble, which can scarcely be found elsewhere and which have greatly contributed to the history of civilisation.

Greek white marble is widely known because of its use for the construction of historic monuments of art. Marble was widely used by the ancient Greeks to create masterpieces such Parthenon, Erechtheum, Propylaea of the Acropolis of Athens, the temple of Olympius Zeus and the famous statues Aphrodite of Milos, Niki of Samothraki, Hermes of Praxiteles and many others.

Marble quarrying in Greece started from ancient times. During the 6th and mainly the 5th century intensive quarrying is reported in the following exploitation centers: Penteli and Ag. Marina of Attica area, the islands of Naxos, Paros (the famous marble of Paros, better known as 'lychnitis' has been quarried since antiquity in the island) and Thassos (white marble quarried at the Cape of Alyki and Cape of Fanari and Vathi) and Philippi Kavala for white marbles. Tinos island, Styra – Karystos in Eubea (famous in the ancient times for the marble quarried here with the name 'Karystia lithos' or 'Cipollino of Karystos'. Karystía Líthos (= Stone of Karystos) is a widely used structural material since archaic times. The term 'Karystía Líthos', as it is mentioned by Strabo, Pliny et al., refers to cipollino marble, schist etc. that are still quarried in the southern part of Euboea Island, under the commercial name 'Karystos Schists'), Hassabali Larissa and Krokees Peloponnese (famous for the greenish 'krokeatis lithos') for green marbles. Eretria Eubea for red marbles and finally Skyros Island ('marble of Skyros', this name was given to the conglomerates that in the ancient times were quarried) for multi coloured marbles.

Marble quarrying continued in Greece with rare breaks such as in Byzantine times and the Turkish occupation. In the early 20th century, extensive exploitation started and marble has been exported in W. Europe since then. Thus, the Greek marble became well known abroad. The Greek marble industry starts in the sixties, supported by the abrupt development of construction in urban centres and the higher standard of living. Marble became a widely consumed industrial product with increasing demand particularly in big constructions and specific uses.

To meet these demands, further increase of marble production is necessary and is achieved through the exploitation of new reserves all over the country. In parallel new modern cutting and processing units are created.

The following paragraphs of this article provide a brief description and the current status of the major marble producing areas in Greece.

3.1 Drama, Kavala-Thassos regions (eastern Macedonia)

In the marble-bearing beds of Kavala-Drama units (Rhodope massif), the metamorphic carbonate rocks (calcite and dolomite) are widely distributed. Thassos is one of the most known ancient marble quarrying centres and one can see quarries of all times. In the century from 540–440 B.C., dolomite marble from Cape Vathy occasionally reached markets well beyond Macedonia. At the end of the Hellenistic period, Thassian dolomite marble began to reach Italy. In the Roman Imperial period, dolomite marble from Cape Vathy was exported to distant parts of the Mediterranean in steadily increasing volume, reaching a peak of intensive in the 2nd century A.D.

This region is the most important marble quarrying centre of the country with modern units of primary production and processing. The total marble production (blocks and shapeless rough blocks) exceeds ca. 200,000 m^3 annually, deriving from 90 active quarries (Data of the year 2001). The most important processing plants are over 25 in number, while the employed personnel exceed 2500 persons (quarries and factories). It should be pointed out that 80% of the totally exported Greek marbles derive from this region.

The snow-white dolomitic marble of Thassos, well known from the ancient Greek times, covers an extremely high proportion of these exports.

The most important marble producing locations in this area are: Thassos (white of Limenas Thassos 'PRINOS', white of Saliara Thassos, Crystallina of Thassos) Nestos, Limnia, Nikissiani, Piges, Volakas (white of Volakas, the most exported Greek marble in China), Elaphohorio, Dysvato, Stenopos, Vathilakos, Palia Kavala and Nikisiani.

3.2 Kozani, Veria regions (western Macedonia)

From this area white and white-whitish and coloured marble of superior quality is quarried (Koumaria, Veria and Tranovaltos, Zoodochos Pigi, Zidani, Roditis and Servia of Kozani).

The marble of Veria is considered to be one of the most famous pure marbles of Greece, since ancient times.

The white-whitish marble of Tranovaltos of Kozani started to be quarried, since 1950, due to its beauty and the excellent qualitative characteristics. Today more than 20 quarries operate in this area with

68

estimated total production of 20,000 m³/year. Note worthy exploitation of the green marble of Veria has also been developed.

3.3 Ioannina region (Ipiros)

In the area of Ioannina the characteristic 'beige' limestone is quarried. It used to be the most used ornamental stone in Greek construction due to its nice colour and low price.

The production is of the order of 15,000 to 20,000 m³/year. A great number of cutting – processing plants of small to medium sized enterprises exists in this area.

3.4 Larissa-Volos region (Thessalia)

The area of Larissa and Volos offers a great variety of white, whitish, pink, and coloured marbles.

More specifically, the Volos area has become an important centre of intensive quarrying activity and total production of marble blocks exceeding 30,000 m³/year. The marble of Tissaion mountain (pink of Lafkos), in the south end of Magnesia peninsula, is of great economic value. Intensive exploitation took place during the ancient times.

3.5 Penteli (Attika region)

In the region of Attika, the marble quarries started their operation mainly after the Persian Wars. This area produced the world famous Pentelikon white marble, by the names 'Bianco di Pendeli' or 'Marmo Greco Fino'. This marble has been widely exploited during antiquity and the Hellenistic period and has been used in the construction of the famous immortal masterpieces of sculpture and architecture of Greece.

Penteli marble quarrying started in the ancient times and continued systematically up to 1976, when the quarrying activity in the south- western slopes of Penteli has ceased due to measures taken for environmental protection. The white marble quarrying has been restricted ever since in the north part of the Penteli Mountain. In the areas of Dionyssos and Agia Marina about 15,000 m³ are produced per year. Today systematic exploitation (mainly underground exploitation) of the Dionyssos marble continues to take place in the area of Penteli.

3.6 Levadia-Domvrena (Sterea Hellas)

The area of Levadia-Domvrena offers a great variety of pink-white, whitish, black and coloured marbles. The Levadia and Hellikonas areas have become an important centre of intensive quarrying activity with total production of marble blocks exceeding 20,000 m³/year. The most important marble types in the area are: Black

of Levadia, Pink of Levadia, and the Whitish of Helikona.

3.7 Other regions

Apart from the above mentioned important production and processing centres, note worthy marble exploitation occurs also in many other districts such as the Aegean islands: Naxos, Tinos, Paros, Evia, Crete, and Argolis regions (Peloponnese), where white, whitish, beige and coloured marbles types are quarried. The production is of the order of 20,000 to 25,000 m³/year.

4 EXTRACTION

4.1 Marble quarrying in antiquity

Ancient evidence about marble quarrying methods does not exist; but by findings and studies about and around the remaining ancient quarries, it seems that ancient quarrying procedures did not differ much from those applied by quarrymen till a few years ago, before the extensive use of the modern quarrying machinery (drilling machine, diamond wire cutting, etc).

The ancient quarries were classified into open and underground, as that in Paros where the 'Lychnitis lithos' was produced.

At the open quarries, the extraction of blocks was made by vertical and horizontal channels by saw and sand. Next, openings were made in order to insert iron or wooden wedges. After excavating the block, the quarrymen had to hew the stone in order to get rid of the undesirable burden and to make transport handling easier.

The transport of blocks from the quarry to the workshop was called 'lithagogia'. When they had to cover small distance and the blocks were not very big, they used wooden handling cylinders.

Transport by sea, a much cheaper means, was made by cargo boats upon which the smaller blocks were put, whilst the bigger ones – in order to become less weighty – were hung in the sea water with wooden beams, which were supported by the two other cargo boats.

4.2 Marble quarrying today

Greek marble deposits usually occur in layers of small, medium or big thickness with horizontal or nearly horizontal inclination. Criteria for the selection of the quarrying method are:

• the relation between waste and exploitable blocks; it is directly related to the thickness of the overlaying waste, the thickness of the marble layers, the layer inclination, its direction and trend;

69

- the in depth extension of the deposit;
- the tectonic shape of the deposit;
- the topography of the deposit.

The thickness of the overlaying wastes, in the currently exploited Greek marble deposits, is usually small. Consequently, opencast quarries are the most common in Greece. Nevertheless, underground quarrying activities have started in the recent years in order to minimise environmental impact.

During opencast quarrying, the stone is removed from its original position by means of the so-called upstream cut, creating blocks that can vary in size from a dozen to several hundred cubic meters. Subsequent operations generally involve overturning the blocks in the quarry yard and squaring them into marketable sizes. Commonly used techniques and equipment are:

- autonomous electricity generator;
- compressed air techniques;
- pneumatic block-cutter, with a jackhammer for vertical and horizontal drilling;
- drilling machine for diamond-wire insertion in primary cuts;
- diamond wire cutting;
- continuous cutting with chain saws, mainly in gallery activities;
- set of jackhammer of medium weight;
- truck/dumper and loader;
- derrick crane etc.

The diamond wire has led to significant progress in quarrying crystalline metamorphic rocks and siliceous metamorphic rocks. The diamond wire consists of a three-stranded steel cable set with beads at different intervals that have an electroplated or sintered synthetic-diamond coating. The material is cut by the action of the diamond wire, which is laid in a closed loop around the section to be cut and is water-cooled. The advantages of the technique are:

- improved productivity;
- increase of the percentage of the deposit that can be exploited;
- reduction of necessary squaring activities;
- better overview of the deposit;
- quality control of the extracted blocks;
- better working conditions for personnel;
- improved security – no use of explosives;
- reduced cost in comparison to other traditional techniques;
- no need for specially trained personnel that become more and more rare.

Greek marble producers have performed major investments relating to mechanical and technical quarrying equipment. Now a day, 90% of traditional quarrying techniques (such as explosives) have been substituted by modern methods.

5 MARBLE TRANSFORMATION – PRODUCTS

The marble blocks cut from the mountain are transformed through successive work cycles, which take place in cutting and transformation factories that are usually geographically located in the vicinity of the end-user. The work cycles can be grouped as follows:

- creation of massive pieces for use in artistic or monumental works (columns, blocks, tombstones, and gravestones)
- sawing of blocks successively from large-sized, medium to thin slabs (from 1.5 to 6 cm) for use in manufacturing a wide range of building products (for flooring, facing, cladding);
- sawing to produce standard sized pieces starting directly from the block and without going through the big-slab phase;
- Working curved-surface pieces used to create facings and manufactured articles in complex shapes based on specific designs.

6 STONE SECTOR IN GREECE – CURRENT STATUS

The modern and dynamic Greek marble industry is classified among the top world producers of decorative natural stones, as it concerns both sizes of production and exports.

Now-a-days, quarrying companies are scattered all over Greece since there are marble deposits almost in the whole Greek area.

The number of the companies engaged in the marble sector is estimated to be about 4000 (6.7% from the Dimension Stones Sector in Europe) and includes small, medium-sized and also, several big units that they have realised important investments and ranks among the best industrial units in Europe.

The Greek marble sector employs more than 60,000 persons (12% from the European employing in the sector); a great number of them have a high level of speciality in the fields of quarrying, processing and installation of marble. The Greek marble companies are mainly engaged in one of the following fields:

- quarrying;
- cutting or/and processing;
- manufacture of art works, ecclesiastical elements and memorials;
- trade of marble blocks and products in home and foreign market;
- installation and applications.

However, there are several companies that have achieved vertical organisation and thus they can be engaged almost in all the above fields of production.

70

These companies usually exploit more than one marble quarries, own very modern cutting and process factories and also dynamic trade departments to assist their exporting activities.

The marble quarry production has impressively increased during the last years. The total mine production has risen sharply. In year 1966, the quarry production of marble blocks was 141,000 tons. In 2002, the annual production was ca 2,100,000 tons.

According to 1996 data, collected from statistical Institutes of E.U. Countries, Greece occupied the fifth position in the world quarry production and exports of ornamental stones, after Italy, China, Spain and India. However, due to strong competition Greece is gradually loosing this position. It may be that today Greece occupies only the seventh position among the top ornamental stone producing countries in the world.

Exporting activity has also been heavy with total unprocessed and processed marble exported amounting to around 377,840 tones in 2002 from 206,770 tones in 1991. The balance of trade of processed and unprocessed marbles is positive, but with a negative tendency (2001–2002), marking a drop of around 1.33% compared to 2001, due mainly to the increasing imports. The problems with the environmental legislation and the growing bureaucracy for new quarrying licenses have led the marble sector companies to increase imports of raw material. Importing activity has increased from 74,757 tones in 1998 to 233,020 tones in 2002. The import value has increased too, from 17,083,000€ in 1998 to 41,384,000€ in 2002.

The marble sector, being as it is export oriented, constitutes one of the few sectors in the Greek economy, which is in a position to compete in the international market. Total exports value (both of processed and of unprocessed marble) is of 107,213,000€ for 2002, marking a drop of around 14.3% compared to the 2001 value (125,115,000€). However, when compared to 1991 export value (€ 57,872,000) an increase of 185.3% is recorded.

There is also a constant increase of the share, the processed marble exports have in comparison to the raw material exports, as a result of the investments on modern technology and the development of new products.

The first 10 export-markets occupy a share of 71.6% in value of the total exports, with USA having by far the biggest share in Greek exports.

The main markets for Greek marble in 2002, according the value, were USA, China, Spain, Cyprus, Hong-Kong, Germany, Japan, Saudi Arabia, Italy, and Brazil.

The first 10 export-markets occupy a share of 55.88% in quantity of the total exports, with China having by far the biggest share of Greek exports.

The main markets for Greek marble in 2002, according the quantity, were China, Spain, Saudi Arabia, Cyprus, USA, Hong-Kong, Italy, Germany, Japan, and Brazil.

The first 10 import-markets occupy a share of 82.1% in quantity of the total imports, with Turkey having by far the biggest share of Greek imports.

The main imports markets for ornamental stones in 2002, according to quantity, were Turkey, FYROM, Italy, China, India, Bulgaria, Morocco, South Africa, Egypt, and Brazil.

7 CONCLUSIONS

Marble has been used in Greece for the construction of sacred buildings since early recorded times.

Greece offers a wide variety of marbles, of different aesthetic and technical characteristics appropriate for all uses. Modernised quarrying takes place today all over Greece. Commercial companies are active all over the world and promote Greek marble exports.

REFERENCES

Giannaros, I. 1998. The Greek marble sector, "Foundation for Economic and Industrial Research" (IOBE), Sectorial Study 162.
Athens Chamber of Commerce & Industry.
National Statistics Service of Greece (NSSG), 2003.
Ornamental Stone from Greece 2003. Edition Hellenic Marble, 20.
Marmaro. 2001. Edition Hellenic Marble, 3.
Laskaridis, K. & Patronis, M. 2003. "Karystía Líthos": a diachronic structural material, 4th Symposium of Archeometry, Athens.
Hermann, J. 1995. The exportation of dolomitic marble from Thasos. Actes du Colloque International: 57–74, Thassos, Limenaria.

Dimension Stone 2004, Přikryl (ed)
© 2004 Taylor & Francis Group, London, ISBN 90 5809 675 0

Types of dimension stones in the Teplá monastery in Western Bohemia

G. Lehrberger & S. Gillhuber

Technische Universität München, Lehrstuhl für Ingenieurgeologie, Munich, Germany

ABSTRACT: The Teplá monastery is one of the landmarks in the district of Karlovy Vary in Western Bohemia and focussed cultural activities in the whole area over 800 years. It will be probably become a first category national monument of the Czech Republic in the next future. A first study of the dimension stones yields a great variety of rock types. Trachyte and sandstone are the main construction materials. A case study within a Czech-German research and conservation project shall yield an inventory of the dimension stones and their decay mechanisms. The impact of the extremely acid environmental conditions during the 1970s and 1980s is one of the focuses of the investigations.

1 BUILDING HISTORY OF THE MONASTERY COMPLEX

The oldest parts of the buildings, mainly the central church, date back to 1193, when the Bohemian noble man Hroznata founded a monastery near the spring of the Teplá river. The monastery was named after the nearby town of Teplá.

Over the centuries the monastery developed to be one of the most important institutions of this type in this area in the heart of Europe. Affected by rebuilding, extension, many fires and subsequent renovations and restorations the monastery buildings reflect the different phases of the construction. The final layout of the monastery goes back to Baroque architects, who reconstructed major parts after disastrous fires in the 17th century. In that time such famous architects as Christoph Dientzenhofer were involved in the planning and construction of the convent and prelature buildings, which still exist.

The monastery had a very prosperous time after the foundation of Marianské Lázně (Marienbad) in the late 18th century on ground which belonged to the monastery of Teplá. The whole development of the spa facilities including medical services and accommodation stayed in the hands of the monastery and yielded to an enormous wealth during the second half of the 19th century and early 20th century.

Not affected by deconsecration as most of other central European monasteries, the Teplá monastery was continuously renovated and extended. A major phase of reconstruction and extension is represented by the new buildings of the neo-Baroque library and

Figure 1. View of the calvary group (sandstone) and the western facade of the Church of monastery (mainly trachyte).

the museum buildings, which were erected around the turn of the 19th to the 20th century.

The today's appearance of the monastery mirrors mainly the effects of the last major renovation of

Figure 2. Aerial view of the vast monastery complex of Teplá at its high times at the beginning of the 20th century. Many of the buildings were destroyed after 1950 (after Grassl 1910).

the western facade of the church and of the towers. Many blocks, damaged by fire or weathering were changed in 1930/1931. These blocks can be easily identified by their slightly red colour and a smooth surface. The red colour of the surface is caused by an oxidation process, which is in detail not yet fully understood.

From 1950 to 1990 many of the building were (ab)used by troops of the Czechoslovakian army, which are responsible for vast damages. The whole monastery was given back to the Premonstratensian order during the restitution in 1990. From this year on many activities of restoration were undertaken. The renovation is partly financed from the Czech monument conservation state office, private donations and entrance fees of the many visitors of the monastery.

2 GEOLOGICAL SITUATION AND DIMENSION STONES SOURCES

The Teplá monastery is situated almost on the boarder between the metamorphic Teplá crystalline complex and the meta-mafic to ultramafic Marianské Lazně complex. The diversity of rocks in this geotectonic units and late to posttectonic magmatism, post orogenic basin fillings and volcanic events in the Tertiary are responsible for the great variety of rocks found in the walls of the monastery.

Over the centuries and in different construction periods different raw materials were obviously available. A walk along the walls of the cloister building is similar to a geological excursion to the whole area around Teplá, since almost all rocks were used for different purposes at different times.

A mapping of selected pilot planes shows that conglomerate, sandstone and trachyte are the main dimension stones in facades and constructive elements.

Figure 3. The large Špičák quarry east of the Teplá monastery. The quarry is run by a Czech-German company.

3 DIMENSION STONES AND QUARRIES

3.1 Trachyte

Today most of the dimension stones used in the buildings consist of trachyte, a bright volcanic rock, which consists mainly of alcaline feldspar. The occurrence of this rock is fairly rare in Europe. Famous deposits are located in the Rhine province and the Westerwald in Germany, in the Euganean Hills near Padua in northern Italy and in the Auvergne region of Central France.

The local source in Teplá for trachyte is a single quarry approx. 3 km east on the former ground of the Teplá monastery. The massive trachyte was quarried probably from the 12th century onwards. The volcanism occurred some 12.5 million years ago in the context of the faulting along the so called Ohře rift (Egergraben) and the related volcanic activity. These

74

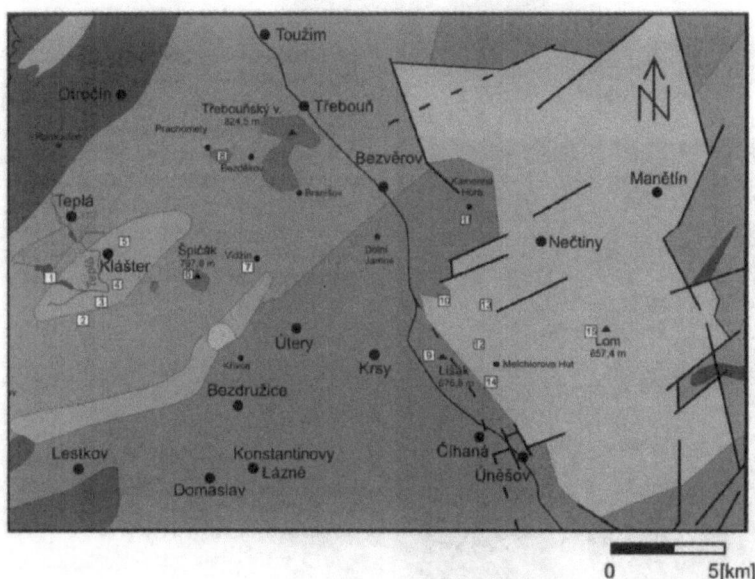

0   5[km]

Figure 4.   Location of the Teplá monastery and possible sources of the dimension stones. 1. Quarry Betlémský lake (ortho- and paragneiss, gabbro known from literature); 2. Quarry Křepkovice (paragneiss, abandoned); 3. Quarry Křepkovice (paragneiss, abandoned); 4 Quarry Smrčí Dvůr (orthogneiss, abandoned); 5 Quarry Ovčí rybnice (granite, granodiorite, abandoned); 6. Quarry Špičák/Štenská (trachyte, active); 7. Quarry Vidžin (paragneiss, known from literature); 8. Quarry Prachomety (trachyte, abandoned); 9. Quarry Lišák (conglomerate, abandoned); 10. Quarry Skelná Hut' (conglomerate, abandoned); 11. Quarry Kamenná Hora (conglomerate, abandoned); 12. Quarry Vojtěšín (conglomerate, abandoned); 13. Quarry Zlatý Lom (conglomerate, abandoned); 14. Quarry Melchiorova hut' (conglomerate, abandoned); 15. Quarry "Lom" (conglomerate, abandoned).

processes produced large amounts of basaltic lavas and only minor intermediate to acid volcanic rocks.

Thus, trachyte is a very rare material, even though two other dome shaped subvolcanic intrusions are known at Prachomety and Třebouň northeast of Teplá. However, these occurrences were only locally used because minor jointing distances allow only small block sizes.

3.2   Permocarboniferous sandstone and conglomerate

Grey and yellow-reddish sandstones and conglomerates are found as dimension stones frequently in the area east of Teplá. In the monastery itself they were used mainly for the walls of the Gothic parts of the church, as bases for the Baroque buildings and for monument sculptures.

These sediments occur in the Permocarboniferous basin of Manětín. They are part of the "Upper Grey" member which is located between coal seems. These were mined in many localities around Plzeň and had a major economic importance. The sandstones and

conglomerates resemble a kind of molasse sediments of the Variscan orogen. Samples of these rocks were already collected by the famous German poet and part-time geologist Johann Wolfgang von Goethe, who visited the Teplá monastery around 1820. He mentioned in his texts the making of mill stones from whitish sandstones. The evaluation of topographic maps, study of local history books and field prospection yielded a number of quarries, which can be regarded as the source of the dimension stones used in Teplá. Remains of the millstone production can be found in the quarries, too (Fig. 6).

3.3   Rocks of the Marianské Lázně complex

The Marianské Lázně complex occurs west of Teplá. Amphibolites and serpentinites are typical rocks of this unit. Irregular blocks were used for the construction of the outer walls of the monastery property. Different types of amphibolites including garnet bearing amphibolites can be found. There are no special quarries known as material sources. They can be easily identified by their dark appearance and the

75

intensively red garnet grains, mainly pyropes. Inside the church several altar tables were made of dark green serpentinite from the Mnichov area north of Marianské Lazně ("Rauschenbacher Serpentin").

Figure 5. Several meters thick sandstone layers in a quarry near Trhomné (Trahona), east of the Teplá monastery. The sandstones were widely used as dimension stone.

Figure 6. Half-ready millstone near a quarry in the Zlatý Lom forest near Skelná Hut'.

3.4 Rocks of the Teplá crystalline complex

3.4.1 Mica schists and paragneisses
The less favourable shape of the mica-rich metamorphic rocks made them only usable for lime mortar bound wall constructions and walls, which were covered by plaster. The loss of the plaster shows, that most of the walls consist of such stones. These rocks occur as field stones everywhere around the monastery. Even small sized and irregularly shaped pieces were obviously collected by the farmers. A few quarries are known in the surroundings of Teplá.

3.4.2 Teplá orthogneiss
The only massive metamorphic rock used as construction material is the socalled Teplá orthogneiss, which occurs in a northeast-southwest trending zone east of the monastery. Abandoned quarries can be found there. The orthogneiss is characterized by several centimetre sized phenocrysts of the former porphyric granite, which were slightly deformed during Variscan metamorphism (Fig. 7). The orthogneiss was used mainly for the construction of the outer property walls.

3.4.3 Granite
Granite was used for the reinforcement of the wall of the north aisle of the Church. It was worked in the nearby quarry of "Ovči Dvůr", which is temporarily used until today. Regularly shaped granite blocks were used at the beginning of the 20th century, when the northern walls of the main church building of the Teplá monastery had to be stabilized by pillars and additional walls at the bottom. Irregular granite stones are found as well in the walls around the property.

3.5 Bright fine-grained sandstone

A bright, fine grained and cross bedded sandstone was used for the neo-Romanesque main portal of the

Figure 7. Teplá orthogneiss with large phenocrysts in the western wall of the monastry court.

76

church, which was added to the building in the years 1893–1895. The provenance of the sandstone is still unclear. On the surface, thick salt crusts can be observed, which consist of Ca- and K-sulfates and can be the reaction products of the interaction of acid rainwater with cement and the rock itself. On the surface, gypsum crystals grow along old scratches and trace these. Probably, fluids migrate through the scratches in the crusts of the patinated surface and precipitate gypsum crystals at the surface (Fig. 8).

3.6 "Exotic" rocks in the interior

In the church, the convent rooms and the buildings of the library and the museum dimension stones from regions far from Teplá can be found.

Even though the complete inventory of rocks is not yet ready, some interesting materials have been identified.

In the church limestones from Kelheim and Solnhofen in Germany have been used for the pavement. A greyish nodular limestone of so far unknown provenience and a red limestone (probably "Deutsch-Rot" from Northern Bavaria were combined to mosaic pavements in the altar area of the church. In the stairways of the library and museum red columns of red limestone on bases and with capitels of white marble (Carrara) are found. The limestone is a typical "Adnet marble" in a homogeneous and a patchy variety. The balustrades of the stairs are made of "Unterbergs marble" from Austria.

A masterpiece in stone can be seen in the Kapitelsaal, where an altar mensa is made of a famous Silurian orthoceras limestone of the Kopanina formation. The dark greyish limestone comes from the Jirasův quarry near Zadní Kopanina southwest of Prague.

4 STUDY OF THE DECAY OF DIMENSION STONES AND CONSERVATION METHODS

Within a project funded by the Deutsche Bundesstiftung Umwelt (DBU, German Federal Environmental Foundation), the association of the "Freunde von Stift Tepl e.V.", the Czech State monument preservation authorities, the German-Czech Future Fonds and private donations, the state of decay and the development of restoration concepts and methods for massive acid to intermediate volcanic rocks are being developed. One of the most prominent objects of the Teplá monastery is the mask portal probably from the late 16th century. It will be studied and restored as a pilot object (Fig. 9).

A parallel project of the DBU deals with similar trachyte rocks from Drachenfels near Bonn (Germany) which were used for the construction of the Cologne and Xanten cathedrals. The cooperation between the research teams including workshops with on-site exchange of experience supports the development of generally applicable methods for restoration of trachyte.

4.1 Decay typology of the main rock types

Only a minor part of the dimension stones in the walls of the church can be regarded to represent original material, based on a study of the history of the building. This is the reason, why dimension stones of very different state of decay can be found close together.

In general the Špičák trachyte is a relatively stable, particularly when used in thick blocks for statues or constructive elements. Only where changing humidity and strong insulation play an important role, stronger decay was observed.

The dominant type of damage and decay is the formation of thin flakes. Scales were mainly identified

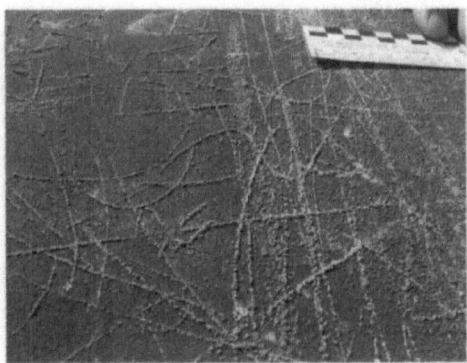

Figure 8. Positive relief of former scratches on sandstone surface of the main portal of the church.

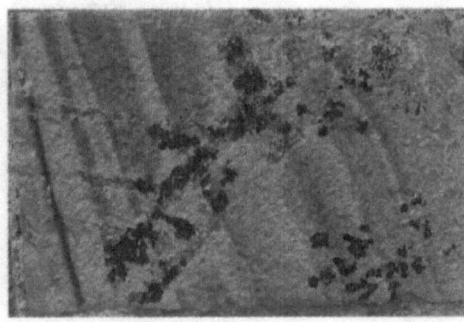

Figure 9. Manganese-oxide dendrites in red banded trachyte in the Teplá monastery. The darker bands are hematite coloured zones probably intensified by the heating during a fire.

77

Figure 10. The mask portal of the Teplá monastery, an artistic masterpiece of the 16–17th century, trachyte from Špičák.

on blocks, which were exchanged in the early 20th century. Probable causes for the damage are improper mortars or surface tooling techniques, which caused deep microcracks on the surface and a subsequent influence of frost. Microcracks reaching some millimetres depth even on freshly worked trachyte surfaces support this possibility. They are made visible by embedding of samples in coloured resin under vacuum.

The alteration of trachyte is favoured by natural inhomogeneities, which are related to the percolation of oxidizing fluids. These caused a transformation of magnetite to ferric iron oxides and hydroxides. Such oxidation fronts can be observed even in trachytes of other provenience and seems to be a characteristic feature of this type of massive volcanic rock.

Green alteration zones are typical for the Špičák trachyte used in Teplá. The colour is obviously related with finely dispersed ferrous iron, but so far no mineralogical changes were observed in the alteration zones. The green coloured parts of the rock are less resistent to weathering and material loss is typical.

In contrast, manganese oxide dendrites which are very characteristic for the trachyte from the Špičák hill are more resistant than the matrix and yield to elevated relief on the surface (Fig. 9).

4.2 Documentation of decay and damage

The developed classification of decay types based on the proposed scheme of Fitzner et al. (1995) was applied to the dimension stones in damage mappings. Detailed studies of pilot areas were conducted to document the exact distribution of decay structures for the restoration or comparisons of changes in future. The maps were prepared with the software metigoMAP by the Focus Company in Leipzig. This program allows an exact quantitative evaluation of areas of different decay patterns.

From the detailed mappings of the quality of the dimension stones a classification of damage for the practical planning and calculation of restorations is derived. The damage classes refer to the urgency and the quantity of restoration works.

REFERENCES

Fitzner, B., Heinrichs, K. & Kownatzki, R. 1995. Weathering forms – classification and mapping. In R. Snethlage (ed.), Denkmalpflege und Naturwissenschaft. Natursteinkonservierung I: 41–88. Berlin.

Grassl, B.F. 1910. Geschichte und Beschreibung des Stiftes Tepl. Selbstverlag: Pilsen.

Gnirs, A. 1932. Topographie der historischen und kunstgeschichtlichen Denkmäler in den Bezirken Tepl und Marienbad: 368–471. Dr. Benno Filser: Augsburg.

Ulrych, J., Novák, J., Lloyd, F., Balogh, K. & Buda, G. 2002. Rock-forming minerals of alkaline volcanic series associated with the Cheb-Domažlice Graben, West Bohemia. Acta Mineralogica-Petrographica 43: 1–18, Szeged.

Svoboda, J., ed. 1966. Regional geology of Czechoslovakia. Part I The Bohemian Massif. Prague: Geological Survey of Czechoslovakia.

Dimension Stone 2004, Přikryl (ed)
© *2004 Taylor & Francis Group, London, ISBN 90 5809 675 0*

Roman quarries in the northern part of Noricum – Austria

H.W. Müller
*University of Natural Resources and Applied Life Sciences, Department of Structural Engineering and
Natural Hazards, Institute for Geotechnical Engineering, Vienna, Austria*

C.F. Uhlir & W. Vetters
*University of Salzburg, Department Geography Geology Mineralogy, Division Regional and
Applied Geology, Salzburg, Austria*

ABSTRACT: Within the area of the Roman Province Northern Noricum (Salzburg, Styria, Upper – and
Lower Austria/Austria) about 170 Roman stone monuments and marble quarries in Lower Austria were sampled
(core drilling with a diameter of 10 mm). The following methods were used for material estimation and its correlation to the quarries: macroscopic description, thin section analysis, determination of the stable ^{18}O and ^{13}C
isotopes and chemical analysis of 25 elements were performed with ICP-MS. The following non-marble materials where used: Untersberg "marble", a fine grained carbonatic breccia (Salzburg), Mönchsberg conglomerate (Salzburg); Kremsmünsterer Nagelfluh, a tufaceous conglomerate (Upper Austria); Enns conglomerate and
conglomeratic sandstone (Upper Austria); Flysch sandstone (Upper and Lower Austria); Hollenburg sandstone
and sandy conglomerate (Lower Austria); Mauthausen granite and material of aplitic and pegmatitic veins
(Upper Austria). The local marbles are from the Bohemian Massive, from the Dunkelstein forest and Häusling-,
Hiesberg- and Lunzen-marble from south of Melk (Lower Austria): The range of the material's utilization was
dependent on its quality, the area of distribution was dependent on the possibility of river transportation. The
use of Mauthausen granite introduced Roman mining activity north of the Limes along the Danube.

1 INTRODUCTION

Recently, investigations on the origin of Roman
marble monuments have been conducted in the northern Roman provinces Dacia, Pannonia, and Noricum
(Benea et al. 1998, Moens et al. 1992, Müller et al.
1996, Müller & Schwaighofer 1999 and Müller 2002,
Müller et al. 1995, 1997, 1998). Rock samples have
been taken from the quarries in question. They were
investigated by means of varifous analytical methods
(Herz 1985, 1988a, b). Samples from the archeological objects were analyzed with the same methods,
which made a correlation between monuments and
quarries possible. The present project, funded by the
Austrian Science Foundation (FWF, grant number
15669), aims at the research on archeological finds and
quarries in northern Noricum. The questions, which
quarries had been used by the Romans, at which time
and to what extent, should be answered during this
project.

2 METHODOLOGY

For the analyses about 170 Roman stone monuments or
fragments have been sampled by core drilling (10 mm
diameter). From each investigated quarry several samples have been used for analyses.

The methodology on the petrographic and geochemical material estimation of roman monuments and
its material's quarries is depending on the material:

For conglomerates and sandstones and granites
mainly macroscopic analysis and thin section analysis
was used. For each material the main and accessory
constituents and the texture were identified.

For marbles the stable ^{18}O and ^{13}C isotopes were
determined in accordance with Craig, H. (1957) using
a conventional standard (PDB). From each sample
50 mg marble powder was treated with H_3PO_4 100% at
25°C for 24 hours. The resulted CO_2 was collected in
glass tubes, frozen at −70°C using liquid nitrogen and
than analysed. The chemical analyses of 25 elements

were performed with ICP-MS. The grain size of the calcite granoblasts in marbles was measured in thin sections with the aid of a polarising microscope. For each sample the main and accessory constituents and the texture were identified.

3 THE QUARRIES

3.1 Breccias, conglomerates and sandstones

3.1.1 The Untersberg "marble" a carbonatic breccia

The Upper Cretaceous Untersberg "marble" (brand name) is a very dense and fine carbonatic breccia, of Dachstein- and Plassen-limestone, cemented with calcite having typical red colored clasts of bauxite, the colour is light grey to yellow and light red (Kieslinger 1964). It was used locally as construction material and has been exported to the northern Noricum as ash-chests, columns, milestones, inscriptions stones and gravestones.

3.1.2 The Mönchsberg conglomerate

The Quaternary Mönchsberg conglomerate of the city mountains of Salzburg (Roman town IUVAVUM) is a coarse to fine grained, porous, partially sandy conglomerate mainly composted of different limestone having minor contents of quartz, green schists and geniss. The colour is greyish brown. It was used locally as construction material and for gravestones.

3.1.3 The Kremsmünsterer Nagelfluh (tufaceous conglomerate)

The Quaternary white Kremsmünsterer Nagelfluh (conglomerate) mined at the river terraces near Kremsmünser (Upper Austria) is a mainly coarse grained, porous, tufaceous conglomerate composed of various limestone and minor contents of brownish grains of flysch sandstone and white quartz (Angerer 1919). The colour is light to dark grey. It was used locally as construction material and in the surrounding

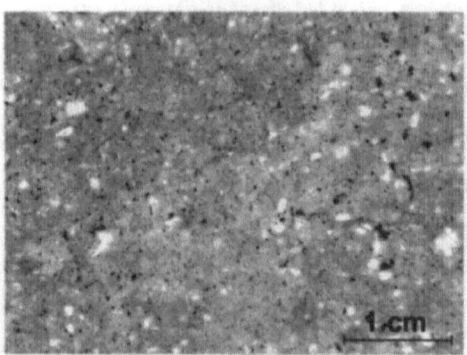

Figure 1. Untersberg "marble" fine breccia with its typical bauxite clasts.

Figure 3. Section of a gravestone made of tufaceous conglomerate (Kremsmünsterer Nagelfluh).

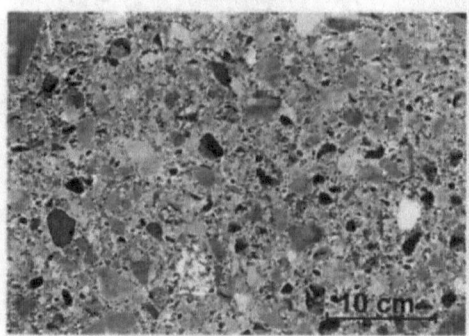

Figure 2. Mönchsberg conglomerate mainly coarse grained with sandy matrix.

Figure 4. Section of a grave stone made of sandy Enns conglomerate.

80

of about 50 km for ash-chests, sacrophages and grave stones.

3.1.4 The Enns conglomerate and conglomeratic sandstone

The Quaternary Enns conglomerate mined along the river Enns near LAURIACUM (Enns, Upper Austria) is a sandy coarse to fine grained, dense conglomerate, cemented with calcite. Partially fine and very uniformed grained variations occur. It is composed of various limestone and minor contents of brownish grains of Flysch sandstone and white quartz. It was used locally as construction material and in the surrounding of about 50 km for inscription stones, sacrophages and gravestones.

3.1.5 The Flysch sandstone

The Middle Cretaceous Flysch sandstone, occurres with frequent outcrops between Salzburg and Vienna, is a fine grained, dense quartzitic sandstone cemented with calcite. It is composed mainly of quartz having

Figure 5. Section of a gravestone made of Flysch sandstone.

Figure 6. Section of a gravestone made of sandy Hollenburg conglomerate.

minor contents of green shists and sideritic grains. The colour is greyish yellow to greyish brown having frequently brownish streaks. It was used locally as construction material and in the surrounding of 50 km of the outcrops as gravestones.

3.1.6 The Hollenburg sandstone and sandy conglomerate

The Tertiary (Palaeocene) Hollenburg sandstone and sandy conglomerate mined near Krems (Lower Austria) is a sandy coarse to fine grained, porous conglomerate and sandstone cemented with calcite. It is composed mainly of various carbonates and minor quartz, feldspar and grains of Flysch sandstone. The colour is light grey to yellowish–brownish grey. It was used locally as construction material and in the surrounding of 50 km of the outcrops as gravestones and inscription stones.

3.2 Marbles of the Bohemian Massif from the Dunkelstein forest and south of Melk (Lower Austria)

The marbles of Häusling, Hiesberg and Lunzen (Lower Austria) have well defined and only marginally overlapping fields of the ^{18}O and ^{13}C isotopes and have characteristically high contents of strontium (between 700 and 2000 ppm).

3.2.1 The Häusling marble

The Häusling marble mined on several outcrops in the Northwest of Melk, Dunkelstein forest (Lower Austria) is a coarse grained very uniform marble having high contents of pyrite and minor mica. The colour is light grey. The weathering resistively is low because of the high content of pyrite. It was used in the surrounding of about 50 km as construction material, for inscriptions stones and gravestones.

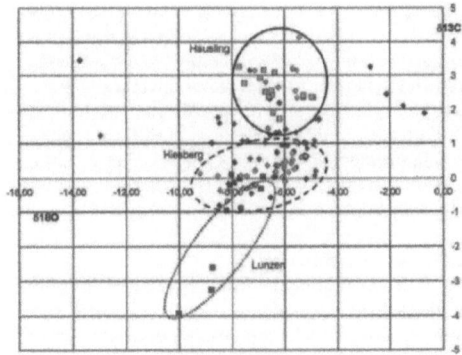

Figure 7. Fields of the ^{18}O and ^{13}C isotopes from the marbles of Häusling, Hiesberg and Lunzen.

81

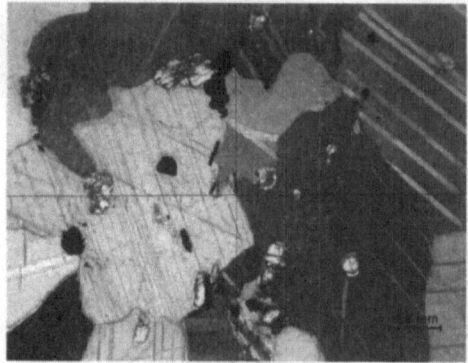

Figure 8. Thin section of Häusling marble, shows well-distributed and partially regular shaped pyrite and muscovite.

Figure 10. Thin section of Lunzen marble, shows typical curved twinning of calcite and mylonitic zones.

Figure 9. Thin section of Hiesberg marble, similar to Häusling marble but shows 3–5 mm large grains of altered hornblende and contains less pyrite.

3.2.2 The Hiesberg marble

The Hiesberg marble mined in the Southwest of Melk (Lower Austria) is a medium to coarse-grained partially layered and partially dolomitic marble having along streaks higher contents of hornblende and mica. The colour is light grey with typical yellowish to light brownish streaks. It was the main used local marble in Roman settlements between Linz (Upper Austria) and St. Pölten (Lower Austria). The main use was for inscriptions stones, grave monuments and gravestones.

3.2.3 The Lunzen marble

The Lunzen marble mined in the Southwest of Melk (Lower Austria) is a medium- to coarse-grained marble having along streaks and layers higher contents of hornblende and mica and has typical fine-grained mylonitic zones. The colour is medium grey with light brownish streaks and layers. It was used in the

surrounding of about 50 km as construction material and for inscriptions stones and gravestones.

3.3 Granites of the Bohemian Massif near Enns and Linz

The Romans rarely used granite for gravestones and inscription stones but there are a view examples of medium to fine Mauthausen granite (identified by typically zoned plagioclase), a dark colored fine grained applite of granitic composition and coarse grained pegmatites mainly consists of feldspar and quartz with minor mica.

As they occur near Lauriacum (Enns, Upper Austria) at the northern side of the Danube, this could be a first hint that Romans did mining north of the Limes.

4 CONCLUSIONS AND PERSPECTIVES

The main material the Romans mined in northern Noricum where conglomerates, tufaceous conglomerates, sandstones, carbonaceous breccias and marbles. The use and transportation was definitely depending on the technical quality and colour of the material. Conglomerates, tufaceous conglomerates and sandstones where used rather local and in the surrounding of 50 km. Carbonaceous breccias and marbles where transported mainly along rivers up to 150 km and had a wide range of utilisation. The use of granite which occur north of the Limes border along the Danube is spectacular and hints on the possibility that there was also near Spitz (Lower Austria) also Roman mining activity for a not jet identified white marble with high contents of Strontium which is typical for that region.

82

ACKNOWLEDGEMENTS

We have to thank the Austrian research community on Roman monuments for a close cooperation and discussions on the monument's material. Special thanks to several specialists on ancient marble and granite quarries in Upper and Lower Austria.

We thank for financial support the Austrian Science Foundation (FWF grant no. P15669) is gratefully acknowledged.

REFERENCES

Angerer, L. 1910. Geologie und Prähistorie von Kremsmünster. *Jahresb.d. Stiftsgymn. Kremsmünster:* 10–17.

Benea, M., Müller, H. W. & Schwaighofer, B. 1998. The single Roman marble quarry in Romania. *Proc. 31st Intern. Symp. Archaeometry, Budapest.*

Craig, H. & Craig, V. 1972. Greek marbles: determination of provenance by isotopic analysis. *Science* 176: 401–403.

Hemmers, C., Traxler, S., Uhlir, C.F. & Wohlmayr, W. in press. "Stein – Relief-Inschrift". Konturen eines Forschung-sprojektes. *Proc. of the VIII. Internationales Colloquium über Probleme des provinzialrömischen Kunstschaffens. Zagreb 2003*: in press.

Herz, N. 1985. Isotopic analysis of marble. In G.J. Rapp & A. Gifford (eds), *Archaeological geology:* 331–351.

Herz, N. 1988a. Carbon and oxygen isotopic ratios: a data base for classical Greek and Roman marble. *Archaeometry* 29(1): 35–43.

Herz, N. 1988b. Geology of Greece and Turkey: potential marble source regions. In N. Herz & M. Waelkens (eds), *Classical Marble: Geochemistry, Technology, Trade (Nato ASI Series, Ser.E: Applied Sciences* 153: 7–10.

Kieslinger, A. 1964. *Die nutzbaren Gesteine Salzburgs.* Salzburg, Stuttgart: Das Berglandbuch.

Moens, L., Paepe, P. & Waelkens, M. 1992. Multidisciplinary research and cooperation: keys to a successful provenance determination of white marble. *Acta Archaeologica Lovaniensia, Monographiae:* 4: 247–252.

Müller, H.W., Schwaighofer, B., Benea, M., Piso, I. & Diaconescu, A, 1996. Greek marbles in the Roman Province of Dacia. *Proc. 3rd Symposium for Archaeometry, 6–9. Nov. 1996:* 22–23, Athens, Greece.

Müller, H.W. & Schwaighofer, B. 1999. Die römischen Marmorsteinbrüche in Kärnten. *Carinthia II:* 549–572. Klagenfurt.

Müller, H.W. 2002. Provenance determination of marble sculptures from Pannonia. *Proc. 31st Archaeometry Symposium Budapest 1998:* 767–775.

Müller, H.W., Schwaighofer, B. & Benea, M. 1997. Die Gesteine des Forums von Sarmizegetusa. *Acta Musei Napocensis:* 34, 1, 837–848. Muzeul national de istorie a Transilvaniei Cluj-Napoca.

Müller, H.W., Schwaighofer, B., Benea, M., Piso, I. & Diaconescu, A. 1998. Provenance of marble objects from the roman province of Dacia. *Jahresh. d. Österr. Archäol. Inst.:* 66: 430–454.

und seine Zeit – Ptuj 1999, Archaeologia Poetovionensis: 93–97, Ptuj.

Müller, H.W. 2001. Herkunftsbestimmung von römischen Marmorobjekten aus der Gegend des Balaton, Ungarn, *Balácai Közlemények VI:* 245–254, Vesprém.

Müller, H.W. 2002. Provenance determination of marble sculptures from Pannonia. *Proc. 31st Archaeometry Symposium Budapest 1998:* 767–775.

83

Dimension Stone 2004, Přikryl (ed)
© 2004 Taylor & Francis Group, London, ISBN 90 5809 675 0

Serpentinite: a potential natural stone in Spain

D. Pereira, M. Peinado, J.A. Blanco, M. Yenes
Departamento de Geología, Universidad de Salamanca, Spain

A. Fallick
Scottish Universities Environmental Research Centre, Glasgow, Scotland

B. Upton
Department of Geology, Edinburgh University, Scotland

ABSTRACT: Serpentinization is a universal process, present in ultramafic massifs. This alteration procedure involves a change in chemical and physical properties of the affected rocks. Geochemical study of major, trace and volatile elements, together with stable isotope values can give information on the serpentinization degree, and the origin of this. Structural and textural characteristics can vary in serpentinites, this being a positive factor regarding aesthetics, although fractures can be detrimental features during mining. Their mechanical behaviour is dependent on this, having low hardness and being grouped under the marble section of building stones. But this could lead to mistakes in their use, and we propose a correct characterization of serpentinites, using traditional methods in geology such as geochemistry, petrography and geo-mechanical measurements, to define properly these rocks.

1   INTRODUCTION

In Spain there exist three important serpentinite areas: the Galician ultramafic massifs (NW Spain), the Ronda massif in Malaga and the small massifs included within the metamorphic material in Sierra Nevada (SE Spain). In this work we have studied serpentinites from Cabo Ortegal, located at the NW Iberia peninsula (Fig. 1).

2   STUDIED AREA

Cabo Ortegal ultramafic complex is one of several that were emplaced during the Variscan orogeny at the NW part of the Iberian massif. This complex presents a variable serpentinization degree, and the studied samples show a different composition related to this evolution, being more impoverished regarding major and trace elements when the serpentinization is more evident.

3   RESULTS

Cabo Ortegal complex is made up of three different units, showing a different serpentinization degree. We have analyzed major and trace elements, including some volatiles, as well as oxygen isotopes, on samples from the three of them. Some of the elements are below detection limit when the serpentinization is considerable. Total serpentinized samples show the most radical geochemistry of the set, having the highest content of boron and the lowest content of chlorine (Fig. 2). The content of volatiles is dependent on the composition of the fluid that produced the serpentinization; therefore these values are indicative of the environment where the alteration was produced.

Oxygen isotopes are very low for some of the samples (lower than 5.5‰) indicating the existence of surface-derived fluids exchange. $\delta^{18}O$ is dependent on the mineralogy: the main phases of these rocks are olivine, pyroxene and spinel. Most of the studied samples are crossed by shears filled up with a variety of minerals from the serpentine group. In general it is possible to distinguish the original mineralogy, except on the most intensely serpentinized. Oxygen isotope values are lowered by high temperature interaction with surface-derived fluids (Fig. 3); this seems common at Cabo Ortegal. Some authors have demonstrated that geochemistry of serpentinization can help to study concentration of platinum group elements in these rocks (Moreno et al. 1999), but this method can be

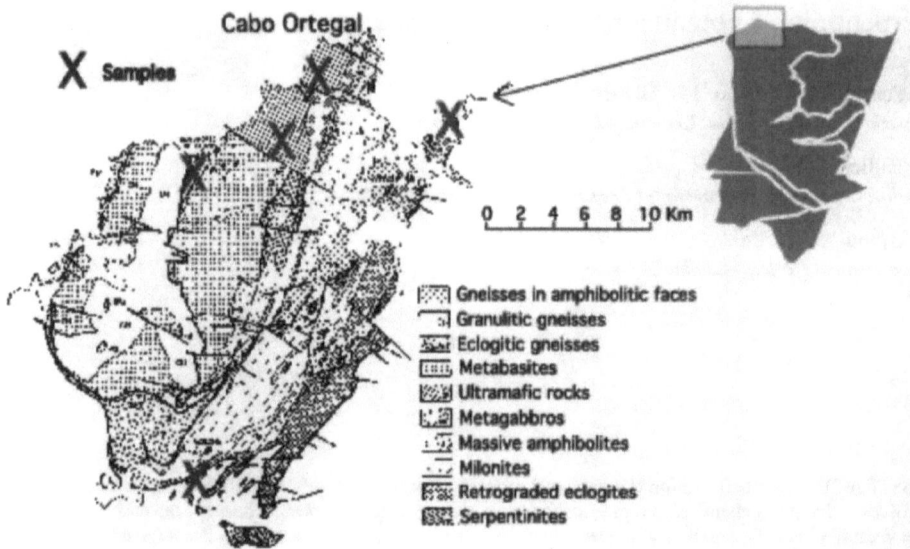

Figure 1.   Location of the studied area in Spain.

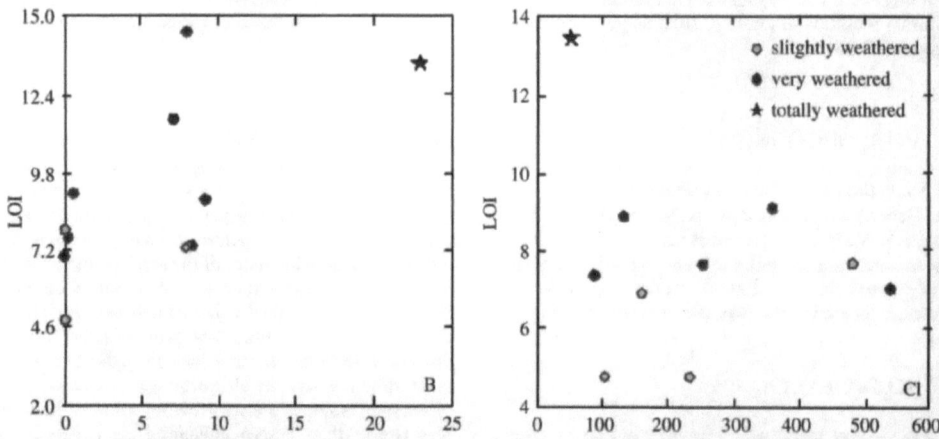

Figure 2.   Relationship between LOI and B, Cl content in serpentinite. Concentrations are in ppm.

used as well to characterize a rock as a potential dimension stone.

Previous studies on geochemistry of serpentinites from Ronda (SE Spain) demonstrated that there is a clear correlation between volatiles and other trace elements and the degree of serpentinization derived from shear structures (Pereira et al. 2003). The geochemistry of the serpentinite affected by shearing at Cabo Ortegal can lead to conclusions about their potential use as building stone.

The rocks we have studied are currently used as refractory material for industrial furnaces in power plants. However, some of them contain minerals that are not good for that use. There is a clear relationship between the degree of weathering and the content of metallic elements (i.e. Cr, Fig. 4) and this is a vital question for their use as refractory.

Other properties we consider for the study of serpentinites are their physical behaviour under compressive strength. Geomechanical studies on serpentinites

86

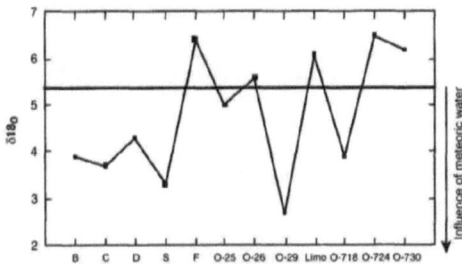

Figure 3. Influence of surface-derived water in the distribution of oxygen isotopes in serpentinite samples.

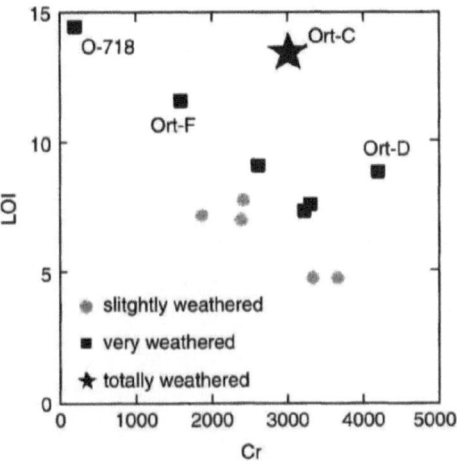

Figure 4. Relationship between LOI and Cr content in serpentinite. Concentrations are in ppm.

can help to distinguish between rocks with appropriate strength and those that do not agree with the physical requirements to be used as ornamental stone. We have selected samples to try compressive strength measurements, from the relatively fresh, to slightly weathered, to the completely serpentinized samples. We have compared the results obtained at our laboratories with the Standard Specification for Serpentinite Dimension Stone. Compressive strength has to be over 69 MPa (ASTM 2002). Our results for some of the samples are compatible with their use as dimension stone. Slightly weathered samples gave values of 67 to 84 MPa. Samples that were totally serpentinized (Ort-C and Ort-718) did not resist the minimum strength test.

4 CONCLUSIONS

Serpentinized peridotite is mined in Spain for use as refractory material. Metallic components as well as high water content (measured as lost on ignition) are bad attributes for rocks used for their refractory behaviour. Therefore, a quarry mined for fresh peridotite could be mined as well for serpentinized rock for other uses. Geochemistry of serpentinites shows that there is a clear correlation among some elements, and the study of a group of them could lead to the right characterization of these rocks. Relationship between mineralogy and geochemistry could help to define properly this kind of rock, and the geo-mechanical measurements will say whether it is feasible their use as ornamental stone or not.

The comparison among geo-mechanical, geochemical and petrographic studies could be used to select the outcrops that will not behave properly as refractory material but could be mined for dimension stone.

REFERENCES

ASTM International 2002. Standard Specification for Serpentine Dimension Stone. C 1526–02.
Acosta, A., Pereira, M.D. & Shaw, D.M. 2000. Influence of volatiles in the generation of crustal anatectic melts. *Journal of Geochemical Exploration* 69–70: 339–342.
Moreno, T., Prichard, H.M., Lunar, R., Monterrubio, S. & Fisher, P. 1999. Formation of a secondary platinum-group mineral assemblage in chromitites from the Herbeira ultramafic massif in Cabo Ortegal, NW Spain. *European Journal of Mineralogy* 11: 363–378.
Pereira, M.D., Shaw, D.M. & Acosta, A. 2003. Serpentinization of Ronda peridotite: A question of fluids. *The Canadian Mineralogist* 41: 617–625.

Dimension Stone 2004, Přikryl (ed)
© 2004 Taylor & Francis Group, London, ISBN 90 5809 675 0

Leithakalk: the ornamental and building stone of Central Europe, an overview

Á. Török & N. Rozgonyi
Department of Construction Materials and Engineering Geology, Budapest University of Technology and Economics, Budapest, Hungary

R. Přikryl
Charles University in Prague, Prague, Czech Republic

J. Přikrylová
Academy of Fine Arts in Prague, Prague, Czech Republic

ABSTRACT: Leithakalk is a porous, Miocene limestone that has been used as an ornamental and building stone in Central Europe. The stone was exploited from the Roman period onward and many important monuments and sites have been constructed in Austria, Czech Republic and Hungary from this limestone. The limestone is sensitive to weathering. In fresh state, it exhibits moderate strength and high water absorption capacity. The paper provides an overview of main geological features and physical properties of Leithakalk emphasizing the differences and similarities of various occurrences.

1 INTRODUCTION

Leithakalk is a Tertiary limestone that has been used as building and ornamental stone in Central Europe. The "Leithakalk" was named after Leithagebirge (Leitha Mountains), a small mountain range in East Austria. Besides its locus typicus, Leithakalk is also found in the Czech Republic and Hungary. The limestone is soft, porous and easy to work with, therefore it has been a very popular working stone from Roman period onward. Famous monuments of large cities such as Vienna, Sopron and historical sites like Lednice-Valtice (an UNESCO World Heritage Site) were constructed from Leithakalk. It has been also used in art: statues ornaments and engraved artefacts were made from various types of Leithakalk. The present paper provides an overview of Leithakalk occurrences in Central Europe and highlights the main textural and petrophysical properties of the Leithakalk limestones of Austria, Czech Republic and Hungary.

2 GEOLOGY OF LEITHAKALK

2.1 *Formation process of Leithakalk*

Various marine carbonates were formed on the shallow sub-tropical shelves of the Paratethys (a local sea that existed in the Tertiary period in Central Europe). Leithakalk belongs to these carbonates, being shallow marine sediment that was formed in the Miocene period. The depositional environment includes peritidal to subtidal marine systems. The most common sediment types and depositional forms were coastal carbonate sand dunes, shallow marine sand bars, oolitic shoals, small patch reefs. Sand and rarely gravel size non-carbonate particles are also present in Leithakalk, besides carbonate grains. As a consequence the average mineralogical composition of Leithakalk varies but quartz, feldspars and clay minerals can have a contribution up to 10%.

The Leithakalk sensu stricto is a limestone which is of Badenian age and found in Eastern Austria, southernmost part of the Czech Republic, and in Western-, Central- and Northern Hungary. Very similar limestones to Leithakalk have been found in other stratigraphic horizons but those are less commonly used as building stones. One exception is the Sarmatian oolitic limestone that was the main construction material of Budapest's monuments at the end of the 19th and at the beginning of the 20th century.

2.2 *Leithakalk of Austria*

Major Leithakalk limestones occurrences are bound to the Leitha Mts. (Leithegebirge, name derives from

Figure 1.   Leithakalk quarries in Hungary and east Austria.

Table 1.   Physical properties of Leithakalk from Czech Republic (data after Přikryl & Přikrylová 2004).

Density [kg/m³]	2643–2680
Bulk density [kg/m³]	2219–2405
Porosity [vol. %]	10–17
Compressive strength [MPa]	38–47
Tensile strength [MPa]	6–10
P-wave velocity [km/s]	4.3–4.8
S-wave velocity [km/s]	2.4–3.1

River Leitha) extending in East Austria – Niederösterreich (compare Fig. 1). The south-western core of Leitha Mts. is built of gneiss, the rest is composed of micaschists, surrounded by Leitha limestone. Leitha Mts. makes the south-eastern border of the Vienna Basin.

From numerous traditional quarries, only one remained active until present – Römmersteinbruch near St. Margarethen. In Leitha Mts. area, the older quarry sites include Deutsch-Altenberg, Hunsdheim, Mannersdorf, and Sommerein in Burck an der Leitha district, Zogelsdorf and Gross-Reipersdorf in Horn ditrict, Wöllersdorf in Wiener-Neustadt district, Breitenbrunn, Gross-Höflein, Loreto, Mühlendorf, Oszlop, and St. Margarethen in Eisenstadt district, Goysz, Kaisersteinbruch, and Winden in Neusiedel am See district, and Kroisbach in Ödenbrug district (Hanisch & Schmid 1901).

The second area of Leithakalk type limestone extend in SE Austria (Steiermark) in the so-called Graz basin. The quarries in Aflenz near Leibnitz (Hanisch & Schmid 1901) were operated from Roman times.

2.3   Leithakalk of the Czech Republic

The typical Leithakalk limestones belong to the Tertiary (Badenian) beds locally called hrušecké beds (Chlupáč et al. 2002). These beds form part of the NW part of the Vienna basin outcropping in the SE part of the Czech Republic (area extending from Brno to Breclav).

The following Moravian localities producing building Leithakalk-type limestones are mentioned in older literature – Holubice (Holubice limestone), Rousínov (Rousínov limestone), Řepka Lomnice, Sudice and Pamětice (Procházka 1911), Pratecký kopec (Procházka 1911), Hlohovec (Šob 1950) and Židlochovice – Výhon hill (Mrázek 1993). For all localities, the use as building material is documented; the Leithakalk from Židlochovice was probably used also for sculptural purposes (Mrázek 1993).

2.4   Leithakalk of Hungary

Leithakalk (sensu stricto) is represented by the Middle-Upper Badenian Rákos Limestone Formation. It is widespread in Hungary since outcrops are found from West Hungary (Sopron area) to Central (Transdanubian Bakony Mts) and North-Hungary. In some cases it is difficult to make any difference between the Badenian and the overlying Sarmatian limestone, since their geographic distribution overlaps and both have similar textural features. Two main occurrences have been exploited as building stones in Hungary. The western-most one in Fertőrákos (Sopron area) and the one that is found in the area of Budapest (Fig. 1.) There are other, but mostly younger porous "Leithakalk-type" limestones in Hungary (see Török 2004, for an overview).

3   FABRIC AND PHYSICAL PROPERTIES

3.1   Leithakalk of the Czech Republic

"Leithakalk" limestone from Moravian localities shows medium- to coarse-grained fabric in macroscopic view accompanied by variable pale colours from whitish through beige, rusty yellow, brownish to whitish grey.

Microscopically, the studied limestone is composed mostly of carbonate (calcite >95–97%) with an insoluble residue (minor quartz, gibbsite, goethite and clay minerals). It contains common subrounded fragments of dominant algae genus Lithothamnium, and minor fragments of corals, molluscs, and foraminifers. The calcitic cement varies from 10 to 20% including recrystallized micritic matrix cement (filling of inter-clast space). According to image analysis of thin sections, the size of bioclasts ranges from 0.1 to 1.5 mm with mean diameter of about 0.6 mm. The physical properties of Leithakalk from the Czech Republic are summarised in Table 1.

3.2   Leithakalk of Hungary

The limestone formation is 30 to 100 m thick and contains basal conglomerates, calcareous sandstones, calcarenites and molluscan limestones of reef origin

90

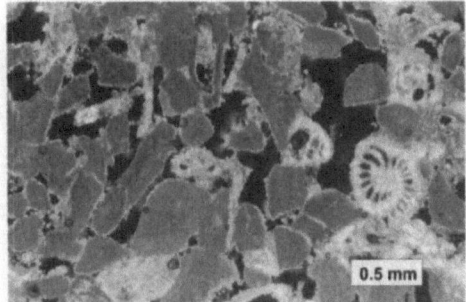

Figure 2. Thin-section photograph of Leithakalk, Fertőrákos (Hungary).

Table 2. Physical properties of Leithakalk of Hungary (data after Rozgonyi 2002).

Petrophysical state	Air-dry	Water saturated
Porosity [vol. %]	6–28	6–28
Bulk density [kg/m³]	1400–2500	1550–2550
Compr. strength [MPa]	8.9–41.6	7.3–26.7
Tensile strength [MPa]	1.8–7.0	1.5–3.2
Ultrasonic sound vel. [km/s]	1.7–4.5	2.0–4.9

Figure 3.   St. Stephan Church in Vienna (Austria).

(Hámor 2001). It is a whitish yellow medium to coarse-grained porous calcarenitic limestone. The main constituents of the Hungarian Leithakalk are Lithothamnium, molluscs, benthic foraminifers and corals. From textural point of view Leithakalk belongs to red algae, bioclastic grainstones (Fig. 2).

It is very prone to air pollution related weathering similarly to other Miocene limestones of Hungary (Török 2002).

The Leithakalk of Hungary has a high porosity (up to 28%). The stone quality in Fertőrákos varies between highly porous limestones with a coarse-grained fabric to more massive, compact, fine-grained lithotypes (Rozgonyi 2002). The bulk densities also reflect these differences in fabric ranging from 1400 to 2500 kg/m³. The stone is sensitive to water since its strength shows a drastic drop due to water saturation (Table 2).

4   LEITHAKALK IN BUILDINGS AND ARTEFACTS

4.1   Monuments in Austria

In the historic center of Vienna several Leithakalk monuments are found. Leithakalk slabs cover the facade of St. Stephan's Church (Kieslinger 1949) (Fig. 3), Opera House and Votivchurch. In the Castle of Schönbrunn staircases and ornaments were constructed from the stone that was quarried in Leitha Mountains (Kieslinger 1932). The most extensive overview of older building and monuments for which Liethakalk limestone from Austrian localities has been used is given by Hanisch & Schmid (1901).

4.2   Monuments in the Czech Republic

The most wide-spread use of the Leithakalk in the Czech Republic can be found in the Lednice-Valtice area. The generous construction activities of Liechtenstein family during 19th century (e.g. rebuilding of Lednice castle in 1840s, construction of the Colonnade near Valtice during 1811–1823, Rendez-Vous monument during 1810–1812 etc.) required application of noble dimension stone. The Leithakalk limestone was available from the Liechtenstein Moravina domain (three quarries) and was used for all monuments in the area. The application of domestic Leithakalk is documented form literature data (Anonymous 1959).

4.3   Monuments in Hungary

One of the most impressive monuments of Hungary, the Romanesque Church of Ják, was constructed from Leithakalk limestone (Szentesi & Ujvári 1999) between

91

Figure 4. Romanesque Church of Ják (Hungary).

1214 and 1256 (Fig. 4). It is the best-preserved Romanesque Church in Hungary. The provenance of the stone material has not been determined yet but the limestone blocks were most probably transported from the medieval quarries of Fertőrákos (Sopron area, Hungary) and/or from the quarries of Leitha Mts. in Austria (St. Margarethen).

In Sopron, the Roman city wall and other Roman historical buildings have been also constructed from this Miocene limestone (ICOMOS 1992).

Surrounding Sopron "Leithalkalk-type" limestone was used to build many castles e.g. Esterházy castle of Fertőd (early 18th century), Széchenyi Castle of Nagycenk (18th century) and churches e.g. St. Michael's parochial church, Benedictine church of St. Mary in Sopron and Romanesque Church of Sopronhorpács.

5 FUTURE RESEARCH

Common fabric features and variable physical properties of Leithakalk make finding of stone source localities difficult. This is important mainly during pre-restoration research of materials from monuments. In this context, the use of local vs. more distant resources makes still challenging research task.

The ongoing Czech-Hungarian intergovernmental research program is focused on the comparative study of Leithakalk taken from different geographical locations and stratigraphic levels. Amongst conventional petrographic, petrophysical, and geomechanical research, the advanced methods like fabric analyses using cathodoluminescence facies research, quantitative image analyses of bioclasts and research on variation of stable isotopes in carbonates have been started. The expected results should help discrimination of Leithakalk limestones coming from different localities thus improving localization of stone utilization in the past.

ACKNOWLEDGEMENTS

This work was partly financed by the Czech-Hungarian Intergovernmental Research Fund (grant no. CZ 5/2002). The financial support for some analyses from MSM 520000001 project is highly appreciated.

REFERENCES

Anonymous, 1959. *Vlastivědný věstník Moravský* XIV.
Chlupáč, I., Brzobohatý, R., Kovanda, J. & Stráník, Z. 2002. *Geological history of the Czech Republic.* Prague: Academia.
Hanisch, A. & Schmid, H. 1901. *Österreichs Steinbrüche.* Wien: Verlag von Carl Graeser & Co.
Hámor, G. 2001. Genesis and evolution of the Pannonian Basin, In J. Haas (ed.), *Geology of Hungary.* Budapest: Eötvös University Press: 193–242
ICOMOS Bands 1992. Fertőrákos. In *Hefte des Deutschen Nationalkomitees VII,* München.
Kieslienger, A. 1932. *Zerstörungen an Steinbauten.* Lepzig und Wien: Franz Deuticke.
Kieslinger, A. 1949. *Die Steine von St. Stephan.* Wien: Herold.
Mrázek, I. 1993. *Kamenná tvář Brna.* Brno: Moravské zemské muzeum.
Procházka, V.J. 1911. *Horniny průmyslové a užitečné Moravy.* Zprávy českých inženýrů v markrabství moravském.
Přikryl, R. & Přikrylová, J. 2004. "Leithakalk" limestones in the Lednice-Valtice area (southeast Moravia, Czech Republic): their occurrences and properties. In R. Přikryl, P. Siegl (eds), *Architectural and sculptural stone in man-cultivated landscape.* Prague: The Karolinum Press (in press).
Rozgonyi, N. 2002. *Verwitterungsbeständigkeit der groben Kalksteine an historischen Bauwerken unter dem Einfluß von Salzkristallisation und Schwarzkrustenbildung.* PhD Dissertation, Budapest.
Szentesi, E. & Ujvári, P. 1999. *Die Apostelfiguren von Ják.* Budapest: Balassi.
Šob, A. 1950. *Soupis lomů ČSR. Číslo 36. Okres Hodonín.* Praha: Československý svaz pro výzkum a zkoušení technicky důležitých látek a konstrukcí spolu se Státním geologickým ústavem ČSR, Vědecko-technické nakladatelství.

92

Török, Á. 2002. Oolitic limestone in polluted atmospheric environment in Budapest: weathering phenomena and alterations in physical properties. In: Siegesmund, S., Weiss, T., S., Vollbrecht, A (eds), *Natural Stones, Weathering Phenomena, Conservation Strategies and Case Studies*; *Geological Society London, Special Publications* 205: 363–379.

Török, Á. 2004. Leithakalk-type limestones in Hungary: an overview of lithologies and weathering features. In R. Přikryl, P. Siegl (eds), *Architectural and sculptural stone in man-cultivated landscape*. Prague: The Karolinum Press (in press).

93

Rock fabric and rock mechanics studies

Dimension Stone 2004, Přikryl (ed)
© *2004 Taylor & Francis Group, London, ISBN 90 5809 675 0*

The mechanical resistance properties of two limestones from France, Tuffeau and Sébastopol

K. Beck & M. Al-Mukhtar
CNRS-CRMD, Orléans, France
Ecole Polytechnique de l'Université d'Orléans cedex, France

ABSTRACT: The mechanical resistance properties of two porous limestones under saturated and dry conditions are studied in this paper. These properties are significantly different in spite of the similar values of total porosity for the two limestones. Analysis presented in this paper shows that this behavior can be attributed to the differences in mineralogical composition, distribution of the pores and the water retention capacity of the two limestones used in the study.

1 INTRODUCTION

Limestones are commonly used as a construction material in many engineering structures. The knowledge of the mechanical resistance properties such as compression and tension are important to characterize the suitability of the use of limestones for several engineering purposes. An understanding of the effects environment influences on the resistance properties can be useful for the mining industry in the operation of fracturing and extraction of limestones on a cost effective basis. Many properties such as the water content, the degree of saturation and macro and microstructure (porous spaces and textural organisation) influence the mechanical resistance properties of the limestones (Ando & Kosaka 1970, Dyke & Dobereiner 1991, Ojo and Brook 1990). The study presented in this paper forms part of a research program to better understand the geotechnical engineering behaviour of two lime stones, namely; Tuffeau and Sébastopol.

The first limestone, Tuffeau is used largely in constructions of the castles in the Loire Valley. Many elegant and aesthetic frontend sculptures have been constructed using these stones. One of the key advantages of the stone is its relative high ductility property. However, over a period of time these sculptures gradually deteriorate due to the effects of erosion. In some of these sculptures, gullies of a few centimetre thicknesses are typically formed due to problems associated with erosion. The second stone Sébastopol used in the study is selected to provide a comparison of mechanical resistance properties along with Tuffeau. The total porosity value of both the limestones used is similar. The mechanical resistance properties of the two limestones are tested in two extreme boundary conditions (i.e., completely dry and completely saturated studied) by controlling the moisture conditions. The water retention curves and pore size distributions were also determined for both the limestones. The mechanical resistance properties of both the limestones are discussed in this paper based on the results of these studies.

2 CHARACTERISATION OF STUDIED MATERIALS

The mineralogical composition of the two stones was obtained using various complementary tests in order to evaluate the qualitative and quantitative behaviour (see Table 1). These tests include: X-rays diffraction, thermogravimetric analyses and optical microscopy.

Table 1. Main characteristics of the two studied stones.

Parameter	Tuffeau	Sébastopol
Mineralogical composition	Calcite ≅ 50% Quartz ≅ 10% Opale CT ≅ 30% Clay and Mica ≅ 10%	Calcite ≅ 80% Quartz ≅ 20%
Skeletal density [g/cm³]	2.55	2.71
Bulk dry density [g/cm³]	1.31	1.58
Porosity	≅48%	≅42%

The Tuffeau used in the study was extracted from a quarry, which is located nearby St-Cyr-en-Bourg (Maine-and-Loire, France). This lime stone is originally white in colour but becomes grey-greenish when exposed to humid environment. The in-situ water content of the stone is about 21%. The colour of Sébastopol limestone quarried from an open mine St-Maximin (Oise, France) has a beige colour.

The two limestones Tuffeau and Sébastopol are sedimentary rocks. The principal constituents of these rocks are: calcite ($CaCO_3$) and silica dioxide (SiO_2). In addition, Tuffeau has quartz and opal cristobalite-tridymite in crystalline form. On the other hand, the same constituents in the Sébastopol stone are in the form of quartz. In Tuffeau, a small percentage of clays and micas are also available. The total porosity of two stones is rather close; however, there are significant differences in the mineralogical composition and measured density values.

3 EXPERIMENTAL PROCEDURE

3.1 Compressive strength tests

Compressive strength tests were carried out on a cylindrical test-tube (diameter 40 mm and height 80 mm) following the Standard AFNOR P94-420 procedures. An increasing compressive force is applied with an Instron 4485 press using a loading rate of 0.05 mm/min, along the axis of the test-tube, until the specimen ruptures. Tested samples were cut in the direction perpendicular to the bed of rock. The major compressive stress occurs along a plane in the direction perpendicular to the bed of rock. The tests were carried out in two extreme hydrous states: dry condition (after 48 hours of drying in an oven at 105°C) and saturated condition (after allowing the sample to imbibe water under vacuum for a period of 48 hours).

3.2 Tensile strength tests (indirect method – cylinder splitting test)

A series of splitting tensile strength tests called as Brazilian tests were conducted following Standard AFNOR P94-422 procedures on a cylindrical test-tube with circular cross section. An increasing compressive force is applied along two diametrically opposed directions. Tests were conducted with an Instron 4485 press using a loading rate of 0.05 mm/min, on a series of samples (diameter 40 mm and height 40 mm). Tested samples were cut in the direction parallel to the bed of rock and the compressive force is applied along the bed of rock. The tensile stress occurs along the direction perpendicular to the bed of rock. The tests were carried out under two extreme moisture state conditions: dry condition (after 48 hours drying in an oven

at 105°C) and saturated conditions (after allowing the sample to imbibe water under vacuum for a period of 48 hours). The indirect tensile strength σ_t is determined using Eq. 1:

$$\sigma_t = \frac{2F_{max}}{\pi dh} \qquad (1)$$

where: d is the cylindrical diameter of the sample, h its height and F_{max} the force at the rupture.

3.3 Mercury porosimetry test

The porosity and pore size distribution of porous materials such as the rocks are determined using this test procedure. The principle is to inject mercury under pressure (up to 210 MPa) in a degassed porous material. The volume of mercury injected corresponds to the cumulated volume of the pores accessible to mercury with a given pressure. Using Laplace equation, the radius r_c of access to the pores is determined:

$$P = \frac{2\sigma \cos \theta}{r_c} \qquad (2)$$

where: P is the applied pressure, σ is the surface tension of the fluid ($\sigma = 485 \cdot 10^{-3}$ N/m) and θ the capillary contact angle solid-fluid ($\theta = 130°$). Theoretically, the pores with a diameter between 350 μm and 4 nm can be highlighted with equipment used in the study, which is, Porosizer 9320 of Micrometics.

4 RESULTS AND DISCUSSION

The results obtained from the tests carried out are shown in Table 2 and Figures 1 through 4.

The compressive and tensile strength tests were carried out on at least three identical samples for each moisture condition. The precision of measurements obtained is ±7% in the case of saturated materials and ±3% in the case with materials in dry state. The results show elastic-plastic stress-strain behaviour. This type of deformation is characteristic for sedimentary rocks.

Moreover, results show that values of the compressive strength obtained for dry samples are about 12 MPa for Tuffeau and 10 MPa for Sébastopol stone (Table 2). Such compressive strength values are characteristic of weak rocks (Dobereiner & De Freitas 1986). The low mechanical strength is directly related to the macroporous structure in both these two stones.

Figure 5 summarizes the analysis of the mercury porosimetry results. The results suggest that Tuffeau is a multi-scales material as pore sizes extend over three orders of magnitude (7 nm to 20 μm). It also

98

Table 2. Compressive and Tensile strength for the tested samples.

Mechanical parameters	Stone			
	Tuffeau		Sébastopol	
	Moisture condition		Moisture condition	
	Dry	Saturated w = 37%	Dry	Saturated w = 27%
σ_c [MPa]	12.00	4.75	10.08	6.42
E [MPa]	2110	930	1982	1503
σ_t [MPa]	1.30	0.38	0.92	0.52
E [MPa]	162	58	126	55

σ_c: compressive strength
σ_t: Tensile strength
w: Water content
E: tangent Young modulus = $\Delta\sigma/\Delta\varepsilon$ (increase of axial stress/increase of axial strain)

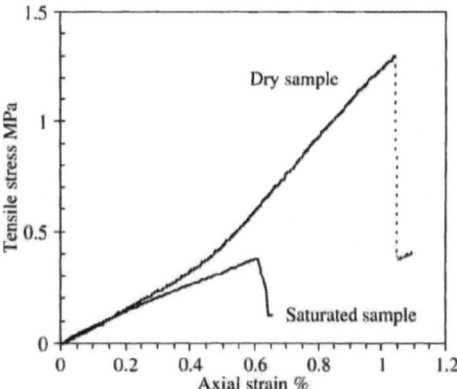

Figure 2. Tensile strength test (Tuffeau).

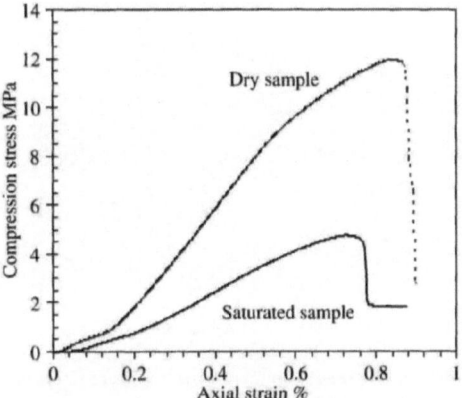

Figure 1. Compressive strength test (Tuffeau).

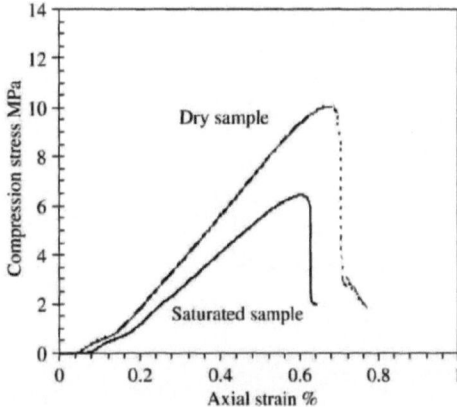

Figure 3. Compressive strength test (Sébastopol stone).

demonstrates that Tuffeau has a strong macroporosity in the range from 1 to 20 μμm representing 63% of pore space investigated. This porosity is dominated by the inter-particles pores of micrite and those generated by the arrangement of the spherical opal CT particles (Beck et al. 2003a). The microporosity represents a significant quantity of the porosity and fills more than one third of pore space. The pore sizes in Sébastopol stone (ranging from 1 μm to 40 μm) are mainly macroscopic with a main access diameter of about 20 μm. And indeed, in the case of Tuffeau, a part of microporosity is inaccessible to mercury intrusion. Due to this reason, contrary to Sébastopol stone, the injected mercury volume (N_{Hg} = 42.8%) is less than total porosity (48%). There is a significant difference in the particle

sizes, the mineralogical composition, the size of the pores and the arrangement of porous spaces in the two stones in spite of having similar total porosity value. The higher percentage of macro-pores in the Sébastopol stone explains the low mechanical resistance of this stone in compared with the Tuffeau.

The strength characteristics decrease in the both two stones with an increase in the water content. These results are in agreement with the studies published by West (1994). In Tuffeau, the reduction in compressive strength reaches 60% between the dry state and the saturated state, whereas in the Sébastopol stone, the loss is lower, which is about 36%. This behaviour can be attributed to the differences in the pore size distribution and consequently to the capacity of water retention

99

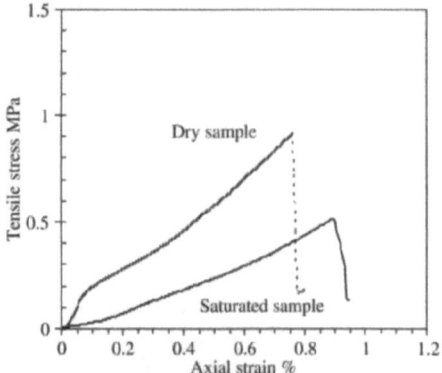

Figure 4. Tensile strength test (Sébastopol stone).

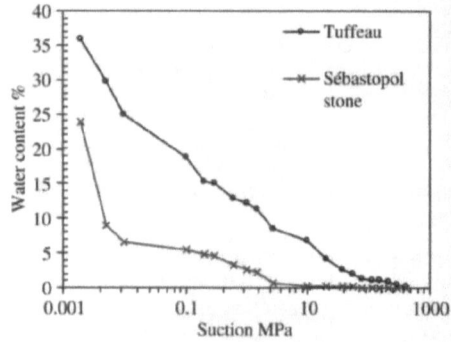

Figure 6. Water retention curves of the limestones.

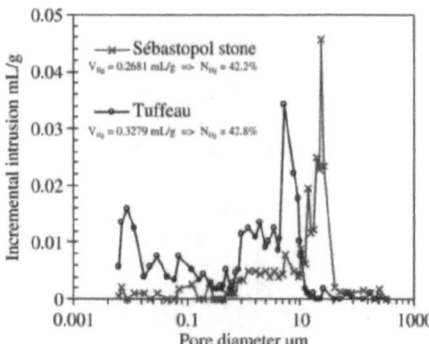

Figure 5. Pore size distributions of two limes stones from mercury porosimetry tests.

Figure 7. Backscattered SEM photograph of Tuffeau.

between two materials as shown in Figure 6 (Beck et al. 2003b). For all the values of applied suction, the water content is higher in the Tuffeau in comparison to the Sebastopol stone. This result shows the influence of the presence of water on the mechanical behaviour of the stone. The strength component due to cohesion in the stone decreases with an increase in the water content and due to this reason there will be a reduction in the mechanical resistance. It can also be explained using the lubricating effect of water involving a reduction in the energy of surface of contact between the constitutive particles and a modification of the inter-granular bonds by an increasing of pressure in the undrained pores. Moreover, the presence of clayey minerals in the Tuffeau, which are extremely sensitive to the presence of water, affects largely their mechanical resistance.

Tangent elastic modulus also depends on the moisture of the sample. The value of this modulus during compressive strength doubled when moisture reduces in the Tuffeau stone.

Tensile strength is very low in the two tested stones: it is about 8% for the saturated state and about 11% for the dry state of compressive strength. As for the compressive strength, in the dry state tensile strength is higher in the Tuffeau than in the Sébastopol stone, whereas, in the saturated state, tensile strength is higher in Sébastopol stone. The mineralogical character is of primary importance here. In the Tuffeau, the presence of clayey minerals induced cohesion (attractive forces between particles). These forces miss in the Sébastopol stone and only the forces of friction contribute in the mechanical resistance. It is well known that the cohesion value is significantly influenced even by small changes in water content; however negligible changes arise in forces of friction due to small water content changes. The backscattered SEM images of polished cuttings (Figs 7–8) provide additional information to those obtained by the mercury porosimetry technique. The particles of the

100

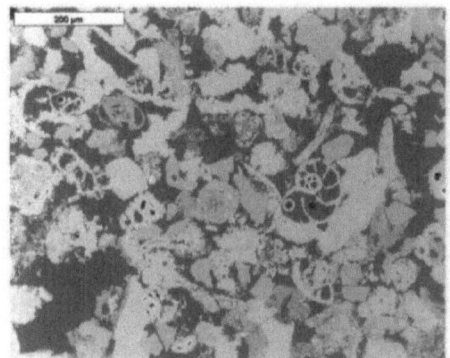

Figure 8. Backscattered SEM photograph of Sébastopol stone.

Sébastopol stone are rather homogeneous and large. They generate a texture influence by the structure contributed through macropores. In Tuffeau, the size and the shape of the particles are variable which induces a more uniform texture with both macropores and micropores. These images allow understanding the complexity of the porous network generated within thee particles of both the limestones studied.

5 CONCLUSION

This research paper summarizes in part, findings of an ongoing investigation study of the mechanical resistance properties of two porous limestones, Tuffeau and Sébastopol from France. The differences in the mechanical behaviour (in compression and tension) of the two examined stones with relatively the same total porosity are explained using their mineralogical composition, pore size distribution and water retention capacity behavior.

The Sébastopol stone mechanical resistance properties are significantly influenced by its macropore structure. The Tuffeau stone properties are influenced by a very wide range of the pore sizes, which mainly constitute amount of micropores. This difference in the pore structures of these stones, results in more conservation of water in Tuffeau in comparison to Sébastopol stone. The presence of clay and mica particles, which are very sensitive to water, induces cohesion in the limestone, Tuffeau. In other words, he texture and the pore structure of two limestones influence their geotechnical engineering characteristics. Some key observations of this study include:

– The tested stones can be characterized as weak rocks based on the compressive strength of dry samples

– The compressive and tensile strength decreases in both the stones with an increase in the water content.

However, the reduction in strength of Sébastopol is lower in comparison to Tuffeau. The reduction in tensile strength of the rock with increasing water content is a key property that can be used for furthering the efficiency of mining and tunneling machinery.

More data is required to establish a generalized approach to assess the geotechnical engineering behavior of sedimentary limestones. In-situ-conditions vary widely with the environmental conditions. A wide variation in the mechanical resistance properties can be expected even within a single day. The mechanical characteristics can be more rationally analyzed taking into account of suction and the quantity of water present in porous spaces of the limestones. More research is under progress towards the development of testing techniques to understand the influence of wet-dry and freeze-thaw cycles on the mechanical resistance properties of limestones. Such studies will be valuable to understand the long-term stability and durability properties of limestones.

REFERENCES

Ando, Y. & Kosaka, K. 1970. Effect of humidity on sound absorption of porous materials. *Applied Acoustic* 3: 210–206.
Beck, K., Al-Mukhtar, M., Rozenbaum, O. & Rautureau, M. 2003a. Characterisation, water transfer properties and deterioration in tuffeau: building material in the Loire valley-France. *Building and environment* 38(9): 1151–1162.
Beck, K., Al-Mukhtar, M. & Rozenbaum, O. 2003b. Pierres des monuments historiques: Caractérisations et mécanismes d'altération du tuffeau. *Colloque de l'Association française de Génie Civil (AFGC): Environnement, sécurité, Patrimoine: les nouvelles donnes*, 26–27 mai 2003, Actes 16 pages.
Dessandier, D, Antonelli, F. & Rasplus, L. 1997. Relationships between mineralogy and porous medium of the crai tuffeau (Paris Basin, France). *Bulletin de la Société Géologique de France* 186(6): 741–749.
Dobereiner, L. & De Freitas, M.H. 1986. Geotechnical properties of weak sandstones. *Géotechnique* 36(1): 79–94.
Dyke, C.G. & Dobereiner, L. 1991. Evaluating the strength and deformability of sanstones. *Quarterly Journal Engineering Geology* 24: 123–134.
Norme AFNOR: NF P94-420 2000. *Détermination de la résistance à la compression uniaxiale*.
Norme AFNOR: NF P94-422 2001. *Détermination de la résistance à la traction. Méthode indirecte – Essai brésilien*.
Ojo, O. & Brook, N. 1990. The effect of moisture on some mechanical properties of rock. *Mining science and technology* 10: 145–156.
West, G. 1994. Effect of suction on the strength of rock. *Quarterly Journal of Engineering Geology* 27: 51–56.

101

Dimension Stone 2004, Přikryl (ed)
© 2004 Taylor & Francis Group, London, ISBN 90 5809 675 0

Residual strain investigations using neutron-TOF-diffraction on marble building stone

C. Scheffzük

GeoForschungsZentrum Potsdam, Telegrafenberg, Potsdam, Germany
Frank Laboratory of Neutron Physics, Joint Institute for Nuclear Research Dubna, Dubna, Russia

S. Siegesmund & A. Koch

Geoscience Centre of the University of Göttingen, Department of Structural Geology and Geodynamics,
Goldschmidtstr. Göttingen, Germany

ABSTRACT: To describe and explain the effects of bowing on marble facade panels neutron time-of-flight diffraction was applied for residual macro-and microstrain determination on the mineral calcite. The results were supplemented by the determination of the crystallographic preferred orientation (texture) of calcite by neutron diffraction as well as studies on microfabric features of the specimens using optic microscopy.

1 INTRODUCTION

Durability is an important property to characterise natural rocks for exterior use. Marbles for instance frequently show a bowing of facade panels after a short time of exposure (Fig. 1). This bowing is generally accompanied with a reduction of strength properties (Koch & Siegesmund 2002). For a better understanding of the observed effect, three samples of calcite marble

Figure 1. Bowing of marble panels at the facade of the Oeconomicum building in Goetingen (Germany).

(CaCO$_3$) were investigated to examine the influence of locked-in stress on the bowing (Logan et al. 2004): a fresh broken marble (P1), a good conditioned facade panel (P2) and a strong deformed facade panel (P3).

2 PANEL BOWING AT THE BUILDING

Samples P2 and P3 are demounted from the Oeconomicum building in Goettingen (Germany). The facade panels of its marble cladding are characterised by a broad variation in bowing from convex (up to 11 mm/m) to concave (up to up to 23 mm/m) as reported in Koch & Siegesmund (2002). These differences are mainly caused by extrinsic effects as exposition (north, east, west, south, height above ground) and building physics.

Moreover, at the north facade of the building where the impact of extrinsic parameters is low another intrinsic parameter gets apparent. Different rock structures macroscopically visible on the panel surfaces are a result of marble slabs being cut in different orientations. The bowing is comparable for panels of the same block-cut direction, and is different when the cutting direction with respect to any metamorphic layering, foliation or macroscopic fold changes (Fig. 2). The degree of bowing of panels is associated with the orientation of these macroscopically visible fabric elements of the marble since all other influencing factors are relatively constant (dimension,

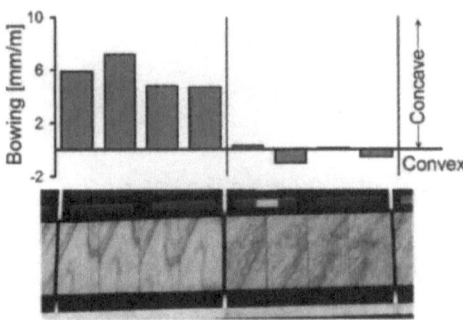

Figure 2. Example from a panel row at the north facade of the Oeconomicum building for the correlation between orientation of the macroscopically visible fabric elements like folds and foliation (bottom) and the degree of bowing (top). According to Koch & Siegesmund (2002).

exposition, microclimate, building physics). It varies between no bowing up to 11.5 mm/m within the same panel row.

Texture measurements reveal that the lattice preferred orientation is attributed to the foliation (see below), so that it is assumed that the amount of bowing and finally the degree of deterioration at the north facade of the Oeconomicum building is influenced by the preferred orientation of the calcite crystals within the facade panels.

3 MICROSTRUCTURE OF SAMPLES

The studied samples are characterised by a wide grain size distribution, from medium to coarse grained: the medium grain size is between 1 and 2 mm with a maximum of up to 6 mm. Domains with a coarser grain size exhibit a polygonal to interlobate shape, and straight to slightly curved grain boundaries (Fig. 3a). Evidence of crystal-plastic deformation is documented by deformation twins and undulatory extinction. Furthermore, the fabric is characterised by a preferred grain boundary orientation more or less parallel to the foliation (Fig. 3b).

In thin sections from strongly bowed panels, open grain boundaries, which are connected to intergranular microcracks, can be observed. The observed cracks are opened up to 0.5 mm with a length up to 5 mm (Fig. 3b). Intracrystalline cracks along twin planes are more rare. Apparently there is a correlation between the presence of microcracks and bowing. In contrast to the strong deformed sample, the fresh broken, undeformed sample does not show any evidence of open grain boundaries. For P2 a warping of 0.2 mm/m was measured, while for sample P3 17.1 mm/m was observed (Nordtest Method 2002).

Figure 3. Thin section images from demounted panels (Koch & Siegesmund 2002): (a) section vertical to the foliation (weak bowing of a good conditioned panel P2); (b) section vertical to the foliation (strong bowing of a strong deformed panel P3): the open grain boundaries are clearly visible (arrows).

4 TEXTURE

The fresh broken Peccia marble sample (P1) was measured using neutron time-of-flight diffraction at the texture diffractometer SKAT. The sample exhibits a distinct crystallographic preferred orientation. The (0006) pole figure of calcite shows an maximum normal to the macroscopic foliation with a weak tendency to form a girdle distribution around a maximum of the a-axis distribution in the foliation plane, while the $(11\,\bar{2}\,0)$ poles are arranged on a great circle around the (0006) pole maximum (Fig. 2). The crystallographic

104

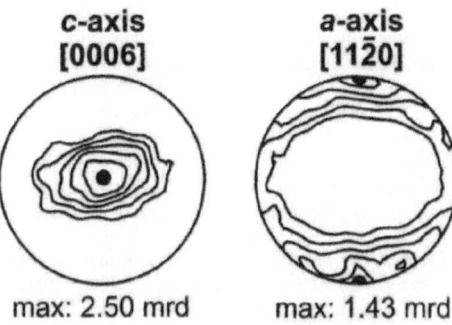

c-axis
[0006]

a-axis
[112̄0]

max: 2.50 mrd   max: 1.43 mrd

Figure 4. Crystallographic preferred orientation (texture) of the fresh broken Peccia marble (P1), measured by neutron-TOF-diffraction at SKAT, projection into the foliation plane.

a-axes corresponding to the (11-20) poles are oriented within the foliation plane.

The importance of calcite textures to the contribution of physical weathering has been widely discussed. A general observation is that the direction of maximum deterioration is closely linked to the orientation of the c-axis maximum. Only the texture of the fresh broken sample (P1) is shown, because the texture of the good conditioned plate (P2) and the strong deformed facade plate (sample P3) is similar.

5 RESIDUAL STRAIN

5.1 Experimental

The strain measurements were carried out at the diffractometer EPSILON-MDS at beam line 7A (Frischbutter et al. 2000, Walther et al. 2000). Six Bragg reflections of calcite (01-12), (10-14), (0006), (11-20), (11-23) and (01-18) were investigated. Macroscopic internal strains at all samples in relation to the stress free state were determined by analysing the position of the Bragg peaks. The stress free state as the reference value were determined by measuring rock powder, prepared by grinding up and annealing. Microscopic internal stresses, caused by dislocations and other microscopic defects, could be observed by peak broadening. Figure 5 shows the dependence of macroscopic and microscopic strain on the detected six Bragg reflections.

5.2 Results

Macro- and microstrain data for the acquired direction perpendicular to the foliation plane are shown in Figure 5. The (01-12) Bragg reflections for all samples are characterized by a positive strain. The microscopic

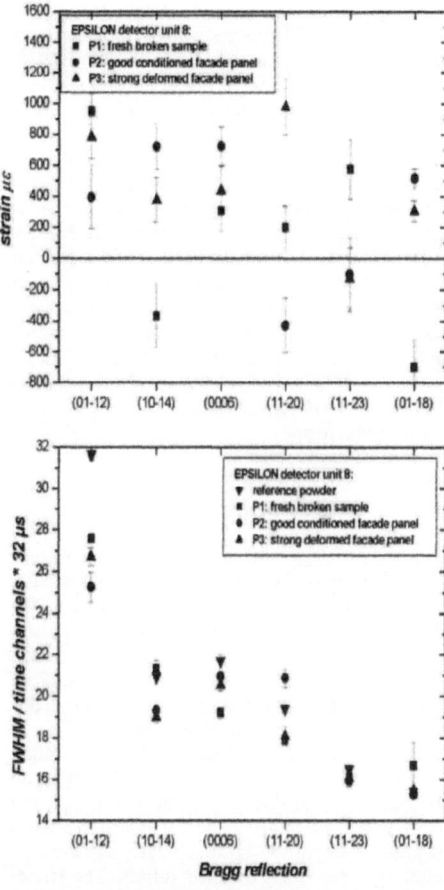

Figure 5. Macrostresses (top) and microstresses (bottom), measured by neutron time-of-flight diffraction at EPSILON-MDS, perpendicular to the foliation plane.

strain shows no significant differences between the three investigated samples. Only the good conditioned and the strong deformed facade panel show positive strain at the (10-14)-Bragg reflection, whereas the fresh broken facade panel shows a negative strain. The good conditioned sample shows a positive strain of $\gamma = +(720 \pm 150) \times 10^{-6}$, the strong deformed facade panel a lower tensional strain of $\gamma = +(380 \pm 140) \times 10^{-6}$. All three panels show comparable macroscopic positive strain values for the c-axis [0006]. The microscopic strain is characterised by a lower FWHM for the fresh broken sample in contrast to a little larger FWHM for good conditioned and the strong deformed sample. The a-[11-20]-axes are characterised by the highest positive strain value at

105

the strong deformed facade panel with $\gamma = +(980 \pm 170) \times 10^{-6}$, the fresh broken sample shows a tensional strain of $\gamma = (200 \pm 140) \times 10^{-6}$, whereas the good conditioned facade panel shows a negative compressive strain of $\gamma = -(420 \pm 180) \times 10^{-6}$.

In the acquired direction parallel to the foliation plane, also all three samples show comparable macroscopic strain values for the c-[0006]-axis, the a-[11-20]-axis, the lattice planes (11-13) and (01-28), but with a tendency to compression in the strong deformed facade panel. Microscopic strain by peak broadening were found at the strong deformed facade panel (P3) on the a-[11-20]-axis with a full width at half maximum (FWHM) of (21.8 ± 1.0) time channels * 32 µs in relation to a FWHM of (17.8 ± 0.3) time channels * 32 µs for the fresh broken sample and the good conditioned facade panel.

6 DISCUSSION AND CONCLUSION

The observed texture is mainly characterised by a preferred orientation of the basal (0006)-poles perpendicular to the foliation plane. The strong texture of the Peccia marble is a significant evidence for plastic deformation.

The measured Bragg peak positions are influenced by macro-and microstresses. The separate determination and interpretation of macrostresses is not possible, because the so-called microstresses should not be ignored, as diffraction methods like neutron stress analysis never give directly macrostresses, but only a superposition of macro- and microstresses (Pintschovius 1992). Microstresses can observed by broadening of the Bragg diffraction lines. The broadening may arise from a number of factors, including small crystal size, dislocations and stacking faults in the crystal structure and variations in lattice parameter on a microscopic scale due to internal strains or fluctuations in composition. Significant microstresses were found e.g. perpendicular to the foliation plane in the strong deformed facade plate by a considerable peak broadening.

Gross and Paterson (1965) reported a small broadening of X-ray powder diffraction lines of deformed marbles when compared with annealed ones which was attributed mainly to internal strains because during annealing at temperatures above 400 °C this broadening

is removed in two stages. In conclusion they favoured the hypothesis that such broadening could have survived for geological length of times without substantially complete removal by recovery and recrystallisation.

Since the investigation on the residual strain is very limited in the literature there is a lack on available data how common it is in marbles. Our first data give evidence that in all investigated samples which differ in the degree in bowing a residual strain could be detected.

Based on the observed strong texture, macro- and microstrain further investigations of microstructural features, weathering and thermal effects are necessary for a better understanding of the bowing of facade plates.

ACKNOWLEDGEMENTS

This work was supported by the BMBF grants 03-DUO3X4 and 03-DU03G1.

REFERENCES

Frischbutter, A., Neov, D., Scheffzük, Ch., Vrána, M. & Walther, K. 2000. Lattice strain measurements on sandstones under load using neutron diffraction. *Journal of Structural Geology* 22(11–12): 1587–1600.

Koch, A. & Siegesmund, S. 2002. Bowing of marble panels: On-site damage analysis from the Oeconomicum Building at Goettingen (Germany). In S. Siegesmund, T. Weiss & A. Vollbrecht (eds), *Natural Stone, Weathering Phenomena, Conservation Strategies and Case Studies*: 299–314. London: Geological Society.

Logan, J.M. 2004. Laboratory and case studies of thermal cycling and stored strain on the stability of selected marble. In S. Siegesmund, H. Viles & T. Weiss (eds), *Stone decay hazards*; Environmental Geology, in press.

Nordtest Method NT BUILD 500 2002. Cladding Panels: Field Method for Measurement of Bowing. *Nordtest project*: 1443–99/2.

Pintschovius, L. 1992. Macrostresses, Microstresses and Stress Tensors. In M.T. Hutchings & A.D. Krawitz (eds), *Measurement of Residual and Applied Stress using Neutron Diffraction*: 115–130. Kluwer Academic Publishers, NATO ASI Series E.

Walther, K., Scheffzük, C. & Frischbutter, A. 2000. Neutron time-of-flight diffractometer Epsilon for strain measurements: layout and first results. *Physica B, Condensed Matter*: 276–278: 130.

Dimension Stone 2004, Přikryl (ed)
© 2004 Taylor & Francis Group, London, ISBN 90 5809 675 0

Influence of the water saturation for the strengths of the Miocene limestone

B. Vásárhelyi & Á. Török

Department of Engineering Geology, Technical University of Budapest, Hungary

ABSTRACT: Several historical and public buildings were built from Miocene limestone in Budapest. This rock was quarried firstly by Romans (the majority of Aquincum – the capital of the Pannonia province, placed in North part of Budapest), later during the centuries nearly continuously used this rock. It was used for the House of Parliament, the Royal Palace, Citadella, the biggest churches and buildings, etc. The goal of this research was to investigate the influence of the water saturation for the strengths: uniaxial compressive strength, tensile strength and the deformation modulus. Due to the several big floods in Central Europe, this knowledge is very important for designing the prevention of the historical monuments.

1 INTRODUCTION

Several researches are investigated the rock strength (i.e. unconfirmed compressive strength, UCS) variation by a number of petrophysical factors e.g.: mineral composition (Price 1966), grain size and gain shape (Wong et al. 1996, Haney & Shakoor 1994, Přikryl 2001), density (Smordinov et al. 1970) and porosity (Dunn et al. 1973, Hoshino 1974; Onodera et al. 1974, Kelsall et al. 1986, Plachik 1999). The influence of moisture on the strength and deformability of sedimentary rock types was reported in detail by Colback & Wiid (1965) (shale lithologies and quartzitic sandstones), Van Eeckhout & Peng (1975) (shales) and Hawkins & McConnell (1992) (limestone).

According to the results of Přikryl (2001), in case of increasing porosity the UCS logarithmic decrease. However, investigating the relationship between the porosity and the strength of the rock Plachik (1999) found the following correlation:

$$UCS = m \ E/n \tag{1}$$

where m is a material constant, E is the elastic modulus and n is the porosity.

Firstly, Smordinov et al. (1970) applied exponential relationship of measured UCS in function of density in case of dry petrophysical condition for different dolomites and limestones. Non of these researches investigated the influence of the water.

Hawkins & McConnell (1992) determined a negative exponential function between the UCS and water

content (using the results of 35 different type of British sandstones):

$$UCS = a \ \exp(-bw) + c \tag{2}$$

where w is the water content, a, b and c are material constants. Their results were also analyzed by Vásárhelyi (2003). Recently, Vásárhelyi (2002) and Kleb & Vásárhelyi (2003) did statistical analysis of the influence of the water of different type of rocks. They showed that the relationship between the different petrophysical constants (i.e. Young's modulus-UCS, UCS-tensile strength) are independent on the water content.

This paper continuing the author's researching with the statistical analysis of the strengths of Miocene limestone (Vásárhelyi 2004).

2 LITHOLOGY OF THE MIOCENE LIMESTONE

The investigated Miocene oolitic limestone, "coarse limestone", is light yellow, yellowish white when it is unaltered. It consists of small well to moderately rounded calcitic ooids and micro-oncoids of 0.2–2.0 mm in diameter. Although calcite is the primary mineral a few percent of quartz and feldspars are also present mainly in the cores of ooids. Besides ooids red algae fragments, gastropods, bivalves and foraminfers are found in the rock. The ooids and other particles are surrounded by circumgranular acicular to bladed cement. Grain to grain contacts also occur often associated with thin cement rims. Most of the pores are connected intergranular ones ensuring a high effective porosity (up to 30%!). The pore size of the intergranular pores is in the order of 0.1 to 2 mm, while the intergranular pores in the foraminifers or within the ooids are generally smaller. The texture shows some varieties in the size of ooids and in the amount of other particles, but mainly belongs to ooid grainstone or bioclastic ooid grainstone. From textural point of view the ooidal limestone of Budapest is very similar to many British oolites such as Monks Park limestone or Bath stone.

It is also possible to differentiate the lithological varieties by visual inspection based on grain size and bioclast content. The following major types of coarse limestones were identified:

I.   very fine-grained (grain size is less than 0.5 mm);
II. fine-grained ooidal (average gain size is 1 mm);
III. medium-grained ooidal (grain size is 1–2 mm);
IV. coarser bioclastic ooidal (mostly with gastropods and larger pores).

In most buildings the second and third type of limestones were used while the first and last type are less common. The most common sedimentary feature of the limestone is the cross bedding, which is visible on some of the building blocks. It is also important to note that due to the depositional processes it is possible that the above listed rock varieties are found within one rock block (see more Török 2002).

3 LABORATORY INVESTIGATIONS

Specimen preparation and testing was performed in the Rock Mechanics Laboratory of the Technical University of Budapest. Right circular cylinders were prepared, according to the ISRM suggested methods (1978a), with a diameter of 54 mm and with height: diameter ratio 2:1. Standard values of the uniaxial compressive strength (σ_c) and of the tangent modulus of elasticity (E) were obtained in conjunction with the complete stress–strain curve.

Also, Brazilian tests were performed to determine the indirect tensile strength according to the ISRM (1978b) suggested method. The diameter of the samples was 54 mm, with a height : diameter ratio 1:1.

From a collection of the results of 45 blocks tested, the following data were analyzed for sample tested in both dry and saturated conditions: apparent density, uniaxial compressive strength, tensile strength and elastic modulus. The water saturation was carried out in a vessel which was full of water. Pressure was not applied for the saturation to model the real situation. It was assumed that after 48 hours the saturation was

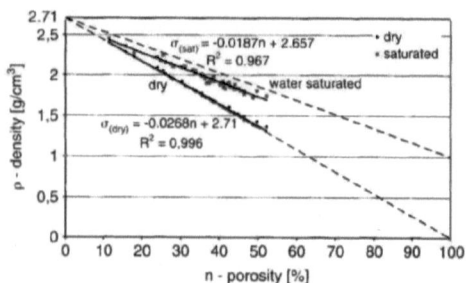

Figure 1.   The measured densities as a function of porosity (in volume %) with the theoretical lines (dashed-lines).

finished. The measured data can be read in the paper of Vásárhelyi (2004).

4 INFLUENCE OF THE WATER SATURATION FOR THE STRENGTH

Before investigated the influence of the water for the strengths of the limestone, the measured densities (dry and saturated) were plotted in function of porosity (Fig. 1). Considering, that the calcite is the main component of this type of limestone, the measured densities should be on a theoretical line:

$$\rho = -(\rho_{CaCO3} - w\,\rho_{water})\,n\,/100 + \rho_{CaCO3} \qquad (3)$$

where w is the saturation ratio ($w = 0$: compactly dry and $w = 1$: fully saturated) and n is the porosity (in volume %). The density of the calcite (ρ_{CaCO3}) and water (ρ_{water}) are 2.71 and 1.00 g/cm^3, respectively.

The experimental results in case of dry conditions were basically on the theoretical line, while the saturated densities were consistently below the theoretical line. This indicates the saturation was not complete.

Firstly, the influence of the water for the strengths (UCS and tensile) and the Young's modulus were investigated. Figure 2 plotted the measured results strength (uniaxial compressive and tensile) of saturated condition in function of dry condition.

According to the result the influence of the water can be written in a straight line and the slope of the line is the same in case of UCS and tensile tests, thus:

$$\sigma_{sat} = 0.659\,\sigma_{dry} \quad (R^2 = 0.933) \qquad (4)$$

The slope of line was similar in case of calculating the influence of the saturation for the Young's modulus in function of dry condition (it is $0.657 - R^2 = 0.866$).

The ratio of the UCS and the tensile strength was also examined in both conditions. Using linear correlation between the two petrophysical constants the

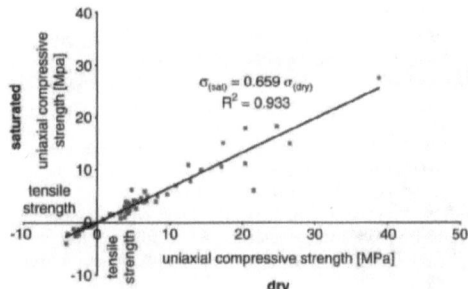

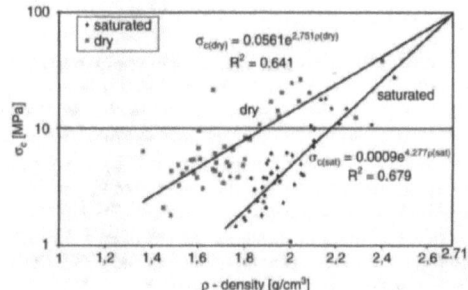

Figure 2. The measured saturated strength as a function of the dry strength.

Figure 3. The uniaxial compressive strength (σ_c) as function of density (ρ) in case of dry and saturated petrophysical states.

slope of the lines are basically the same in both cases, thus the water content is independent for this ratio and it is:

$$\sigma_t = 0.13\ \sigma_c \tag{5}$$

(the constant in case of dry and saturated conditions are $0.129 - R^2 = 0.742$ and $0.136 - R^2 = 0.886$, respectively).

The similar result was found for comparing the UCS and tensile strength vs. Young's modulus: the slope of the UCS-Young's modulus and the tensile strength-Young's modulus lines were basically independent to the water content. The following connections were found:

$$E = 380\ \sigma_c \tag{6a}$$

$$E = 2660\ \sigma_t \tag{6b}$$

The differences between the dry and saturated conditions were within the experimental errors.

5 THE PETROPHYSICAL CONSTANTS IN THE FUNCTION OF THE DENSITY

The uniaxial compressive strength is represented in function of the density in Figure 3. Similar relationships were found in case of tensile strength and Young's modulus vs. density:

$$\sigma = ae^{b\rho} \tag{7}$$

where a and b are material constants. Table 1 consists of these constants in case of dry and saturated conditions for UCS, tensile strength and Young's modulus.

Analyzing the results it can be mentioned that in case of dry condition b is nearly equal to the density of calcite, so it could be use as a density-dependent constants for other rock-types.

Table 1. The measured material constants for Equation (7) in case of dry and saturated conditions.

	a		b		R^2	
	Dry	Sat.	Dry	Sat.	Dry	Sat.
σ_c	0.0561	0.0009	2.751	4.277	0.641	0.679
σ_t	0.0039	0.00009	3.135	4.379	0.573	0.567
E	0.0088	0.0005	3.126	4.063	0.660	0.578

Theoretically, the two constants could depend on the water contents and the density of the minerals. Supposing, that the limestone consists on only pure calcite crystals the following equation can be suggested:

$$b = \frac{\rho^2_{CaCo_3}}{\rho_{CaCo_3} - w\rho_w} \tag{8}$$

where ρ_{CaCO_3} is the density of the calcite (2.71 g/cm³), ρ_w the density of the water (1 g/cm³) and w is the water content ($w = 0$: dry, $w = 1$: saturated). Thus Equation (8) in case of limestone:

$$b = \frac{7.3441}{2.71 - wl} \tag{9}$$

The intersection of the dry and saturated lines should be around the density of the calcite (i.e. 2.71 g/cm³): in case of compressive strength, tensile strength and Young's modulus are 2.71, 3.03 and 3.06 g/cm³, respectively, every cases within the experimental error. The theoretical pertophyisical constants of the calcite could be also determined with these equations. The average UCS, tensile strength and Young's modulus for the limestone which is not consisting of pores are 97.21 MPa, 15.97 MPa and 36.14 GPa, respectively.

109

6 DISCUSSIONS AND CONCLUSIONS

The goal of this research was to statistically analyzed the influence of the water for the Miocen limestones. 45 different blocks were tested, measuring the uniaxial compressive strength, tensile strength and the deformation modulus. Before the tests the densities were determined, as well. In this paper the author is continuing his researches of Miocene limestone (Vásárhelyi 2004).

Investigating the influence of the water for the strengths, in every case, the slopes of the trend lines were independent of the water content.

Also, an exponential relationship between the density and the measured strengths (uniaxial compressive and tensile) and elastic modulus was obtained for both dry and saturated conditions. The intersection of the two lines represents this relationship (dry and saturated) is near to the pore-free density of the rock, which is consistent considering that calcite is the main component of limestone (has a density of 2.71 g/cm^3).

It was supposed, that some material constants can be determined from the density of the mineral composition of the limestone. It is necessary more researches to show it. In the future, new tests in case of semi-saturated conditions should be carried out (similarly to Kleb & Vásárhelyi 2003).

ACKNOWLEDGEMENT

The author gratefully acknowledges the Hungarian National Research Foundation support (Grant No.: OTKA F043291).

REFERENCES

Colback, P.S.B. & Wikd, B.L. 1965. *The influence of moisture content on the compressive strength of rock.* Proc. 3rd Canadian Rock Mech. Symp. 65–83.

Dunn, D.E., LaFountain, L.J. & Jackson, R.E. 1973. Porosity dependence and mechanism of brittle fracture in sandstones. *J. Geophys. Res.* 78(14): 2403–2417.

Dyke, C.G. & Dobereiner, L. 1991. Evaluating the strength and deformability of sandstones. *Q. J. Engng. Geol.* 24: 123–134.

Haney, M.G. & Shakoor, A. 1994. *The relationship between tensile and comprehensive strengths for selected sandstones as influenced by index properties and petrographic characteristic.* Proc. 7th IAEG Cong. Lisbon, *Portugal,* II: 493–500.

Hawkins, A.B. & McConnell, B.J. 1992. Sensitivity of sandstone strength and deformability to changes in moisture content. *Q. J. Engng. Geol.* 25: 115–130.

Hoshino, K. 1974. *Effect of porosity on the strength of the clastic sedimentary rocks.* Proc. 3rd Int. Cong. ISRM, Denver, USA, II(A): 511–516.

ISRM, 1978a. Suggested methods for determining tensile strength of rock materials. *Int. J. Rock Mech. & Min. Sci. & Geomech. Abstr.* 15: 99–103.

ISRM, 1978b. Suggested methods for determining the uniaxial compressive strength and deformability of rock materials. *Int. J. Rock Mech. & Min. Sci. & Geomech. Abstr.* 16: 135–140.

Kelsall, P.C., Watters, R.J. & Franzone, J.G. 1986. *Engineering characterization of fissured, weathered dolerite and vesicular basalt.* Proc. 27th US Rock Mech. Symp. Tuscaloosa, USA, 77–84.

Kleb, B. & Vásárhelyi, B. 2003. Prediction of the rock strength for the wine-cellars in Eger. *Acta Geol. Hung.* 46: 301–312.

Odonera, T.F., Yoshinaka, R. & Oda, M. 1974. *Weathering and its relation to mechanical properties of granite.* Proc. 3rd ISRM Cong. Denver, USA, II(A): 71–78.

Palchik, V. 1999. Influence of porosity and elastic modulus on uniaxial compressive strength in soft brittle porous sandstones. *Rock Mech. Rock Engng.* 32: 303–309.

Přikryl, R. 2001. Some microstructural aspects of strength variation in rocks. *Int. J. Rock Mech. & Min. Sci.* 38: 671–682.

Prince, N.J. 1966. *Fault and joint development.* Pergamon Press, Oxford, p. 176.

Smorodinov, M.I., Motovilov, E.A. & Volkov, V.A. 1970. *Determinations of correlation relationships between strength and some physical characteristics of rocks.* Proc. 2nd ISRM Cong. Belgrade, Yugoslavia, II: Theme 3–6.

Török, Á. 2002. The influence of wall orientation and lithology on weathering of ooidal limestone building blocks. In R. Prikryl & H.A. Viles (eds), *Understanding and managing stone decay*: 197–208. Prague: The Karolinum Press.

Van Eeckhout, E.M. & Peng, S.S. 1975. The effect of humidity on the compliance characteristics of coal mine shales. *Int. J. Rock Mech. and Min. Sci.* 12: 61–67.

Vásárhelyi, B. 2002. *Influence of the water saturation on the strength of volcanic tuffs.* In C. Dinis da Gama & L. Ribeire e Sousa (eds), *Workshop on volcanic rocks*, Madeira. 89–96.

Vásárhelyi, B. 2003. Some observation regarding the strength and deformability of sandstones in case of dry and saturated conditions. *Bull. Engng. Geol. & Env.* 62: 245–249.

Vásárhelyi, B. 2004. Statistical analysis of the influence of water content on the strength of the Miocene limestone. *Rock Mech. Rock Engng.* (in press).

Wong, R.H.C. Chau, K.T. & Wanf, P. 1996. Microcracking and grain size effect in Yeun Long marbles. *Int. J. Rock Mech. & Min. Sci.* 33: 479–485.

Dimension Stone 2004, Přikryl (ed)
© 2004 Taylor & Francis Group, London, ISBN 90 5809 675 0

Microscopic and microstructural investigation and determination of anisotropy of building stones

P. Závada, K. Schulmann, F. Hrouda* & S.W. Faryad

Institute of Petrology and Structural Geology, Charles University, Prague, Czech Republic

*AGICO Ltd., Brno, Czech Republic

ABSTRACT: In our work we use microstructural analyses to characterize mechanical properties of building stones. We study textural relations, grain size and shape of individual minerals, their anisotropy, porosity and composition. By means of petrographic methods, EBSD, AMS and microprobe analyses we are trying to characterize physical properties of natural stones. These methods applied to volcanic rocks of trachyte and phonolite composition provide information on fabrics that resulted from magmatic flow and crystallization rate of rock-forming minerals. The crystal size is directly controlled by the velocity and thus on the viscosity of the ascending magma that penetrates through the host rock. Viscosities of studied rocks were calculated on the basis of chemical compositions and compared with the AMS. Planar pure-shear fabric and strongly linear fabric formed by sanidine crystals were recognized in the trachyte dome. The phonolite dome shows spectacular columnar jointing, which resulted from cooling of the body following its fast emplacement.

1   INTRODUCTION

Natural building materials are mixtures of a number of chemical compounds which either crystallize to form various definite shapes or solidify from the molten state as particles building micro-units. These are the physical and chemical building units, which constitute all building materials determining their physical and mechanical properties. In igneous rocks these properties mostly depend on viscosity, crystallization rate and PT condition during the magma emplacement.

In this paper we present the results of petrographic and microstructural investigation, mainly flow properties and emplacement mechanisms of silicate magmas of Tertiary volcanic rocks in the České Středohoří Mts. (NW part of the Czech Republic). Since the study of flow patterns in the past was restricted only to macroscopically measurable fluidity and time-consuming universal stage measurements (Jančušková et al. 1992), AMS analysis was used as a rapid and sensitive indicator of flow characteristics derived from orientation of magnetic minerals.

The AMS analysis of trachytic and phonolitic rocks was compared with calculated viscosities of silicate magmas based on their chemical composition. Detailed field study, supplemented by systematic AMS measurements and measurements of crystallographic preferred orientation using EBSD method of sanidine crystals was carried out on two particular intrusions of trachyte and phonolite to examine complex flow properties of magmas with contrasting viscosities.

2   ROCKS UNDER STUDY

Volcanic rocks of trachyte and phonolite composition occur in the SW–NE trending Ohře (Eger) Rift area situated in North Bohemia. Trachytes and phonolites of Miocene age penetrate Cretaceous sediments and Tertiary volcanosedimentary sequences. Two major groups of trachytic to phonolitic volcanic rocks exist in the studied area. The first group corresponds to trachyte with typical trachytic texture and contains sodalite (hauyne) or sodalite + alkali feldspar and accessory nepheline. The second group is represented by phonolite with abundant nepheline but only accessory sodalite.

The trachytic rocks form domes with well-developed cleavage (cleavage planes 2–10 cm apart), parallel with flow fabric at the apical part of the body. The flow fabric is defined by alignment of sanidine tabular crystals (200 mm) that form trachytic texture and

111

enclose phenocrysts of sanidine (up to 600 mm), aegirine augite (400 mm) or sodalite (up to 350 mm). Phenocrysts of sanidine and aegirine-augite are oriented parallel to sanidine crystals in the matrix.

The phonolites show smaller crystal size (10–30 mm) of matrix minerals (sanidine and nepheline) in comparison with trachytes. Glassy matrix shows trachytic texture. Rare phenocrysts of sanidine define flow fabric of the rock, while orientation of sanidine crystals in the matrix reflects the last increment of strain before the magma was cooled.

Cooling of the intrusive phonolite body Bořeň gave rise pronounced columnar jointing. Columns are straight and vertical in the apical part of the body and become continuously inclined on the slopes to horizontal position. The columns are approximately 80–150 cm in diameter forming polygons. The AMS and microstructural studies were performed on eleven volcanic bodies (Tab. 1). All the studied rocks are of alkaline, undersaturated felsic composition. They are basically composed of alkali feldspars and foid minerals (sodalite or hauyne, nepheline, rarely analcime); minor plagioclase is present only in several less developed trachytic rocks (e.g., at Kalich). Amount of mafic minerals is generally low and decreases rapidly from the plagioclase-bearing rocks to phonolites s.s.. The prevailing mafic mineral is clinopyroxene changing its composition from augite to aegirine-augite to aegirine reflecting continuing fractionation of the magma. Relics of hornblende are rare. Accessory minerals are sphene, magnetite, apatite, etc.

2.1 Analysis of the anisotropy of magnetic susceptibility

Around 70 oriented samples of trachytic and phonolitic rocks were cut to 8 cubes with edges 2 cm long. The AMS measurements were carried out with a KLY-3 Kappabridge (Jelínek 1981) at magnetic laboratory of

Table 1. Studied volcanic bodies.

Dome		Petrographical type
1 Kalich	ka	Sodalite-bearing trachyte
2 Majka	ma	Sodalite-bearing trachyte
3 Řetoun	re	Sodalite-bearing trachyte
4 Doubravská hora	do	Alkali trachyte (sodalite phonolite)
5 Věštanský vrch	ve	Trachyandesite
6 Želenický vrch	ze	Phonolite s.s.
7 Špičák	sp	Phonolite s.s.
8 Bořeň	bo	Phonolite s.s.
9 Zvon	zv	Sodalite phonolite
10 Ryzelský vrch	ry	Phonolite s. s.
11 Hradiště u Habří	hr	Sodalite-bearing trachyte to Sodalite phonolite

Agico, Ltd, Brno. Magnetic anisotropy data were evaluated using the ANISOFT and CUREVAL softwares. The mean susceptibility is $k_m = (k_1 + k_2 + k_3)/3$, where $k_1 > k_2 > k_3$ represents the shape of the ellipsoid of susceptibility. The corrected degree of anisotropy $Pj = exp2[(\ln k_1 - \ln k_m)^2 + (\ln k_2 - \ln k_m)^2 + (\ln k_3 - \ln k_m)^2]$ gives the intensity of preferred orientation of magnetic minerals in the given rock. The shape parameter $T = 2 \ln(k_2/k_3)/\ln(k_1/k_3) - 1$ (Jelínek 1981) indicates the symmetry of magnetic fabric, being linear when $-1 < T < 0$ and planar when $1 > T > 0$. Variations in magnetic susceptibility with temperature were determined using the KLY-CS3 instrument to identify magnetic minerals carrying the AMS in studied rocks.

2.2 Degree of anisotropy and its relationship to magma viscosity

The higher degree of AMS obtained for trachytic rocks (relative to basaltoids, Pj parameter ranging from 1.1 to 1.35) is related to high viscosity together with high crystallinity of trachytic magma (Fig. 1). Magnetic ellipsoids show mostly oblate shapes in volcanic domes with T values ranging from 0.6 to 1. Only at localities where magmatic feeder zones were studied, the T parameter ranges from −0.5 to +0.5 and Pj parameter decreases to 0.7–1.1.

In contrast, the phonolitic rocks in general exhibit lower degree of anisotropy and highly variable degree of anisotropy ranging from Pj = 0.3–1.2 and ellipsoid shapes showing T values from −0.9 to +0.9. Pattern of AMS fabric is very scattered indicating variation in flow directions on centimeters scale.

Although paramagnetic pyroxene is present in the majority of rocks with susceptibility higher than

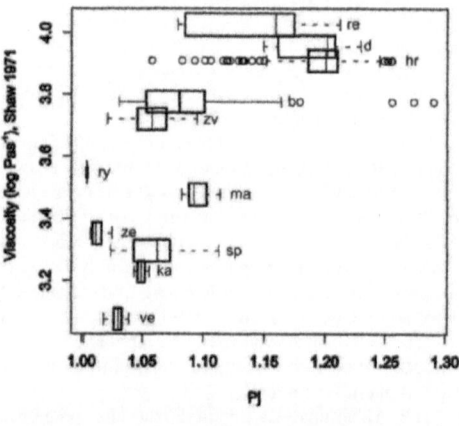

Figure 1. Diagram of calculated viscosities of magmatic silicate liquids for studied volcanics versus degree of anisotropy expressesed in Pj parameter of Jelínek (1981).

112

10 [SI] (pyroxene content is 5–23 vol. %), the only carrier of AMS was identified from thermomagnetic curves (Curie temperature from 440 up to 500°C) to be Ti-magnetite with approximately 10% of illmenite component (Nagata 1961). Titanomagnetite ranges between 3–7 vol. % in trachytes. The titanomagnetite is missing in phonolites.

The degree of anisotropy expressed by Pj parameter was correlated with viscosities of silicate magmas calculated on the basis of chemical compositions of studied volcanics using the KWARE Magma software of Ken & Wohletz. The viscosity is strongly dependent on the molar proportion of SiO_2 (Shaw 1972) so that alkali trachyte shows highest viscosity values of around 104 Pas^{-1}, while sodalite phonolites and sodalite trachytes exhibit significantly lower viscosities in a range of 103.6 to 3 Pas^{-1}. Figure 1 shows good correlation between the degree of anisotropy of magnetic susceptibility and calculated viscosity values so that highly viscous magmas reveal strong anisotropy of magnetic susceptibility. It is suggested that the high viscosity of magma controls the rate of extrusion (or emplacement), which changes magma rheology from Newtonian to dilatant/shear thickening. The increasing viscosity of magma is also related to higher crystallinity of akali trachytes with respect to sodalitic phonolites.

2.3 Field studies

The Hradiště sodalite-bearing trachyte, selected for this study, is located 14 km SE of the city of Teplice. It forms an elliptical dome with strong foliation formed by intense preferred orientation of sanidine tabular crystals. The trachyte penetrated Late Cretaceous sediments and was probably stopped by competent layer of volcanic rocks of the first volcanic phase (Ulrych et al. 2000). The foliation is very slightly inclined (weak) (10–20°) at the apical part but becomes progressively steeper (45–80°) towards the margin of the dome. The dip is sub-vertical or even reversed at the edges of the dome (Fig. 2). Distinct cleavage is parallel

to the magmatic fabric in the apical part in an abandoned quarry. This cleavage is less pronounced and oblique to the magmatic fabric on the slopes of the body. The fabric patterns are stable on the outcrop and the ellipsoid shape and degree of anisotropy variations depend on the site in the cupola.

Magnetic foliation (K3 axis) corresponds fairly with mesoscopic foliation, which is well-expressed by an angular difference lower than 10° between K3 direction and the pole. In SW part of the body the Pj parameter (=1.15) together with shape parameter T (=0.1) show lower values relative to those in the remaining part of the body (1.35 and 0.9, respectively). This is likely to indicate an occurrence of magmatic feeder zone in this area.

Measurements of crystallographic preferred orientation (CPO) of sanidine crystals from 7 samples revealed two major types of primary magmatic fabric. Samples from the apical part of the trachyte body show girdle-type distribution of poles to (001), (100) with sharp angle (70–80°) between two major (dominant) maxima (Fig. 3). This planar magmatic fabric originated during pure shear deformation. Samples from the margin of the body show strong linear fabric characterized by single maximum of all crystallographic directions. Abundant domains of sanidine crystals with CPO oblique to primary magmatic fabric formed during subsolidus deformation via fiber slip deformation mechanism that was generated by viscous drag of mobile magma, while CPO pattern of the domains is similar to that of primary magmatic fabric. The angular deviation between domains and fabric decreases towards the body margins as the viscous drag loses its strength.

The phonolite dome called Bořeň, located 2 km S of the Bilina town, shows lower ratio of diameter to thickness with respect to the Hradiště trachyte dome. It belongs to a group of intrusions (Zlatník, Želenický vrch, Bořeň) with similar geochemical character. These

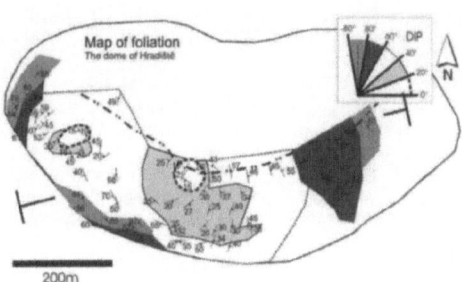

Figure 2. Map of macroscopically measured planar fabric in the Hradiště dome.

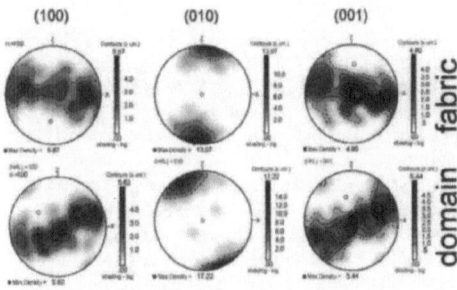

Figure 3. Lower hemisphere equal area stereographic projections of poles to crystallographic planes of sanidine crystals in samples from apical part of the Hradiště dome. Note similar pattern of CPO and oblique orientation of sanidine crystals from domains with respect to the primary magmatic fabric.

113

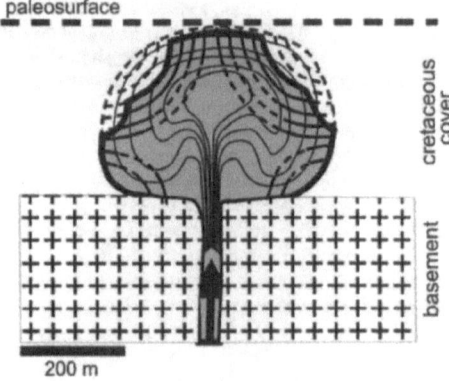

Figure 4. Schematic picture showing main structural features defining Bořeň phonolite intrusion. Thick lines define columnar jointing. Thin lines correspond to magmatic fabric. Thick contour surrounding the body reflects shape of the body after erosion.

intrusions are linearly arranged parallel to the major Bilina normal fault. Preliminary AMS study shows scattered K1, K2 and K3 directions without any systematic type of fabric patterns that vary at the outcrop scale indicating strongly inhomogeneous flow with fabric variations at centimeter scale. Microsanidine crystals (10–30 μm) within the glassy matrix of phonolite, which form the apical part of the body, show CPO pattern similar to that in trachyte rocks. This fabric formed at the very last increment in contemporary crystallization and deformation by upward compression of mobile magma.

Magmatic fabric in phonolite is defined by tabular cleavage that is not always parallel to alignment of sanidine phenocrysts. The whole dome is affected by pronounced columnar jointing (Fig. 4). Columns are vertical and perpendicular to subhorizontal fabric in the apical part of the body, and continuously become horizontal being perpendicular to the rock fabric at the body margins. The columnar jointing is consistent with interpretation by Long & Wood (1986) having resulted from lava contraction cooling. The direction of crack growth is perpendicular to the surface of the heat flow.

Cracks defining columnar jointing are terminated in peripheral parts of the body. Horizontal columns are cut by vertical magmatic fabric being here preferentially eroded. Magmatic fabric is less intense in comparison with the apical part of the body. The rock is free of cleavage and unaffected by alteration (zeolitization).

3 SUMMARY

The AMS and EBSD are very useful methods when investigating the flow patterns of volcanic bodies. The EBSD method can examine crystals of very small dimensions in matrix of volcanic rocks and establish precisely their orientation. These analyses done on trachytes indicate formation of distinct types of magmatic fabric, which control physical and mechanical properties of rocks. Magmatic fabric characterized by alignment of mineral phases, their shapes and crystal size distribution in the rock, resulted from complex flow of viscous magma during emplacement. The AMS fabric patterns of phonolite vary at the outcrop scale indicating strongly inhomogeneous flow with fabric variations in centimeters. The EBSD measurements of tiny sanidine crystals show trachitic texture. Thermomagnetic curves revealed titanomagnetite in trachytic rocks to be a major carrier of AMS. The increasing viscosity of magma is related to higher crystallinity of akali trachytes with respect to sodalite phonolites that were emplaced faster giving rise to columnar jointing perpendicular to the planes of heat flow. Massive rock without fractures is exposed in peripheral parts of the body where the columns were completely eroded. The cooling of trachyte dome resulted in more penetrative cleavage mostly subparallel to the magmatic fabric.

REFERENCES

Hejtman, B. 1957. Systematická petrografie vvyřelých hornin. *Nakl. ČSAV*, Praha.

Hibsch, J.E. 1899. Geologische Karte des böhmischen Mittelgebirges. Blatt II (*Ronstock-Bodenbach*). Wien.

Jančušková, Z., Schulmann, K. & Melka, R. 1992. Relation entre fabriques de la sanidine et la mise en place des magmas trachytiques. Example de massif de Hradiste. *Geodinamica Acta* 56: 235–244.

Jelinek, V. 1981. Characterization of the magnetic fabric of rocks. *Tectonophysics* 79: T63–T67.

Long, P.E. & Wood B.J. 1986, Structures, textures, and cooling histories of Columbia River basalt flows. *Geol. Soc. America Bull.* 97: 1144–1155.

Macháček, V. & Shrbený, O. 1973. Geochemistry of trachytic rocks of the České středohoří Mts. *Čas. Mineral. Geol.* 18(2): 131–161.

NAGATA, T. 1961. Rock magnetism. *Maruzen*, Tokyo

Shaw, H.R. 1972. Viscosisties of magmatic silicate liquids: an empirical method of prediction. *American Journal of Sciences* 272: 870–893.

Ulrych, J., Cajz, V., Pivec, E., Novak, J.T., Nekovarik, C. & Balogh, K. 2000. Cenozoic intraplate alkaline volcanism of Western Bohemia. *Studia Geophysica et Geodaetica* 44: 346–351.

114

*Computers, databases and GIS applied to
dimension stone studies*

Dimension Stone 2004, Přikryl (ed)
© 2004 Taylor & Francis Group, London, ISBN 90 5809 675 0

Marble classification using scale spaces

G. Dislaire & E. Pirard
Université de Liège, GeomaC, Géoressources Minérales, Liège, Belgium

M. Vanrell
Universitat Autònoma de Barcelona, Spain

ABSTRACT: Marble texture classification is an error prone undertaking when performed by humans. Therefore normalized methods are needed in order to obtain reproducible results. Technological advances in digital image acquisition and computing allows for the building of systems based on such methods. The classification will be represented here by dyadic scale-space models (powers of 2). We will take into account the functioning of the human visual system in reproducing its natural ability to extract the features of textures: opponent-colors space is used as well as dyadic approaches for both light-dark multi-scale feature detection and inter-pathway resolution ratios. The spatial organization will be captured with the use of features from the statistical sum and difference histograms, from a model-based blob-oriented morphological scale-space and from statistics of wavelet coefficients. Features will then be classified with a common method, a Learning Vector Quantization network.

1 INTRODUCTION

The requirement for aesthetic appearance constancy in marble products is essential to certify that slabs sold to the client are alike. In this way, a robust texture definition is important for the classification of slabs into homogeneous classes. Visual discrimination of the human expert can be translated into algorithms in order to reduce subjectivity of the human classification. Classification methods will be presented to illustrate the benefit of using scale-based models to improve the classification.

2 TEXTURE

It is always important to specify what we mean when discussing textures. Generally speaking, a texture is a repetition of pattern(s) with possible random variations in the primitive placement rules. To be more precise, we have to say that unlike structures or other organization types, the texture is strongly linked to the visual perception of this order. This explains the importance of the psychophysiology and of the translation into algorithms of the multi-channel frequency and orientation analysis performed by retinal and cortical neurons. In addition, a texture definition also depends on the observation scale. Therefore we will build a marble classification system based on a *visual perception* model of *spatial organization* of light intensities on a given *scale range*.

2.1 Texture definitions

In Tuceryan and Jain (1998) several texture definitions are proposed, definitions intrinsically linked to feature extraction methods chosen to identify the texture. They group methods into geometrical, structural, statistical, model-based and signal processing-based.

Structural methods assume that textures are composed of primitives – as textiles are composed of threads. Texture elements are first extracted and then the placement rule is analyzed. Elements can be blobs. Lindeberg's scale-space researches can be used to extract them at different scales (Lindeberg 1994). Morphology can be used to analyze them and placement rule can be defined as a tree grammar using symbols – the primitives – to form strings – the textures.

Texture is related to the spatial distribution of light intensities, so statistical methods such as the co-occurrence matrices are reasonable texture analysis tools. Autocorrelation captures repetitive placements and drops slowly or quickly depending on whether the texture is coarse or not. This last property can be linked to the power spectrum in signal processing models.

Model-based methods take out a set of parameters defining a model generally used as a constraint for synthesizing similar texture. Known models are based

117

on the Markov random fields or on the fractals. Portilla and Simoncelli introduce "A parametric texture model based on joint statistics of wavelet coefficients" (Simoncelli & Portilla 2000) that seems to capture the nature of the texture – its essential features. It binds model-based methods to signal processing ones.

2.2 Classification comparison

Randen and Husøy (1999) compare texture classification using statistical and signal processing approaches [2]. For multi-textured images the best classification performance is achieved with the highly complex Quadratic Mirror Filter f16b filter bank. Nevertheless the computationally more efficient DCT approaches or QMF critically sampled filter are of interest because, as the feature dimensionality decreases, the classifier complexity decreases too. They also conclude that the co-occurrence and the popular Gabor filter are not superior.

The processing time must not be forgotten in comparative methods. A classification system has to be viewed as a whole: the complexity of the feature model extracted will require an efficient classifier. Therefore some of the co-occurrence features such as mean, energy and contrast could already be enough in some cases and lighter for a classifier.

2.3 Visual perception – color and scale ranges

Psychophysiology has provided a very useful model for color reproduction. We are now aware of trichromacy, the ability to arrange a color match using three primary colors. Human oriented color spaces have also been constructed to reflect the opponent-colors pathways: the light-dark, the red-green and the blue-yellow channels. We will use the opponent-colors space presented in (Zhang & Wandell 1996) as a preliminary step to feature extraction:

$$X = 0.6067R + 0.1736G + 0.2001B$$
$$Y = 0.2988R + 0.5868G + 0.1143B$$
$$Z = 0.0000R + 0.0661G + 1.1149B$$

$$O_1 = 0.279X + 0.72Y - 0.107Z$$
$$O_2 = -0.449X + 0.29Y - 0.077Z$$
$$O_3 = 0.086X - 0.59Y + 0.501Z$$

Less known but also important is the difference of resolution within opponent-color spaces: the light-dark contrast achieves a maximum at 10 cycles per degree.

[1 degree corresponds to 0.89 centimeters on a screen viewed at 50 centimeters. In this configuration, 10 cycles per degree corresponds to $10/0.89 = 11.2$ cycles per centimeters.]

We will see that, more than the mean, especially when searching color-similar sub-class of a given

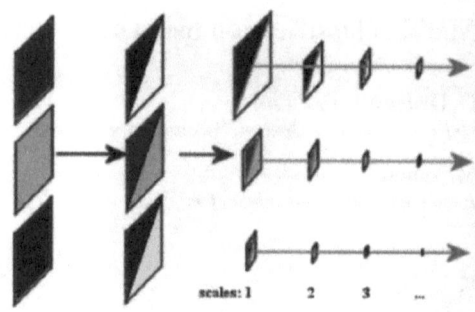

Figure 1. Using opponent color space model, we can link to a given scale different resolutions according to color pathways.

slab type, the contrast seems to be an effective discriminating factor in the human classification criteria. Actually this is due to an induction phenomenon intrinsically rooted in the filtering process performed by the retinal neurons: they compress the information by reacting especially to transitions.

Induction is the source of many illusions with two opposite forms: contrast for low frequency (0.7 cpd for blue-yellow), assimilation for high frequency (9 cpd for blue-yellow) (Vanrell & Baldrich 2003). Note that for 10 cpd, we have assimilation in blue-yellow but contrast in light-dark.

Light-dark contrast falls for resolutions above 30 cpd, red-green for 10–20 cpd when blue-yellow for 5–6 cpd (Wandell 1999). Undertaking a practical approach allows us to imagine dyadic scales (powers of 2) depending on the pathways: given a scale s_0 for the light-dark pathway we use a scale $s_1 = 2 \times s_0$ for the red-green and $s_2 = 4 \times s_0$ for the blue-yellow one.

Let's conclude the present contrast presentation with a numerical example: a 30 centimeter tile seen from 50 a centimeters distance will give a maximum light-dark contrast for variations of 0.89 millimeters – corresponding to $30 \times 11.2 = 337$ variations in the width of the field of view – and a resolution of 0.30 millimeters – corresponding to 1011 variations. A dyadic scale will give resolution of $0.30/2 = 0.15$ millimeters in the red-green pathway, of 0.075 millimeters for the blue-yellow one (see Fig. 1).

When we move away from a surface, we gain the larger scales – limited by the field of view – as we lose the smaller ones – limited by the retinal resolution.

3 EXPERIMENTAL MATERIAL

A set of Marfil slabs have been acquired using a tri-CCDs linear camera to obtain color images of the diffuse component of the light reflected by the slabs. Images have been pre-processed to remove shading

118

effects due to non-homogeneity of the lighting. This set of images has been used as the basis of the scale-based models discussed in the followings paragraphs.

4 METHODS

Dyadic scales have been used to classify the Marfil slabs using different image analysis methods. For a given acquisition we train color features at different scales by shifting dyadic pathway ratios from high to low resolution (see Fig. 2).

4.1 Spatial organization

In order to feed a vector of features to the classifier we have to capture the nature of the spatial organization in a digital form. In this way statistical models capture mean, energy, entropy, contrast and homogeneity. Other features, such as uniformity, density, coarseness, regularity, linearity, directionality... have various implementations. A "texture browsing descriptor" is considered by the MPEG-7 compression format.

But the most important is to keep relevant descriptors; thus depending on the texture, only certain ones are retained for defining a given model. We will compare the use of features from the statistical Sum and Difference Histograms (SDH), from a model-based blob-oriented morphological scale-space and from statistics of wavelet coefficients.

4.2 Marble classification

Ornamental stone textures are so varied that it is difficult to build a model classifying all the possible varieties found on the market. When granites seems to be easily classified due to a certain homogeneity of the repeated pattern at a given scale, marble often are characterized by the presence of veins that will produce a texture on a scale higher than the scale of the marble slab. Such slabs are evaluated by human experts with subjectivity and fatigue giving inconsistent results. Automatic classifications have been introduced by Martinez & Tomás (1999) to solve this problem using the SDH method which computes features on small neighborhoods.

Our experiments will focus on slabs of the type "Crema Marfil" coming from Murcia.

4.3 Scale-space variations

To improve results based on statistical methods, we will study the texture on different scales. Three methods are presented. For each one we work in the opponent-color space model as described in Figure 2. A Gaussian kernel is each time applied at a given scale and subsampled to produce the larger scale.

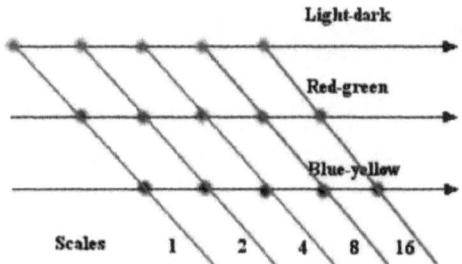

Figure 2. In a dyadic scale space decomposition of a color opponent model-based image, a given scale s_0 will be characterized by feature of the scale s for the light-dark channel, $2 \times s_0$ for the red-green channel, $4 \times s_0$ for the blue-yellow one.

4.4 Sum and Difference Histograms

The SDH algorithm is a powerful alternative to the usual spatial grey level dependence method or co-occurrence matrices: for a distance vector (d1, d2), the combination of two pixels zx, y and zx + d1, y + d2 forms the sum and difference vectors:

sx, y = zx, y + zx + d1, y + d2
dx, y = zx, y − zx + d1, y + d2

The normalized histograms are:

$$ps\,(i) = \frac{hs(i)}{N} = \frac{\#(sx,y = i)}{N}$$

$$pd\,(i) = \frac{hd(i)}{N} = \frac{\#(dx,y = i)}{N}$$

The statistical features used are:

$$\mu = \frac{1}{2}\sum_i i P_s(i) \qquad energy = \sum_i P_s(i)^2 \sum_j P_d(j)^2$$

$$entropy = -\sum_i P_s(i)\log(P_s(i)) - \sum_j P_d(j)\log(P_d(j))$$

$$contrast = \sum_j j^2 P_d(j) \qquad hom\,og = \sum_j \frac{1}{1+j^2} P_d(j)$$

The Figure 3 illustrates the mean and two contrasts at different scales. The use of a scale factor for this last feature will improve the classification from 75% (1 scale) to 88% (6 scales) – an improvement factor of 17%.

We have to notice the poor initial result for 1 scale related to the 90% with the same method used by Martinez & Tomás (1999). This is likely due to a different set of images and only the improvement factor should be retained.

4.5 Blob analysis in the Lindeberg's Scale-space

We explained that the contrast is a very discriminating factor in the human classification criteria. But

119

how to implement it to reflect this specificity to an image acquired by a static vision system? The answer is the Laplacian of Gaussian.

The Gaussian kernel and its derivates are one of the most precious tools in image analysis. For instance, filtering with a Gaussian kernel simulates the assimilation as a perceptual blurring; filtering with the second derivate is named the Laplacian of Gaussian and it simulates the contrast as a perceptual sharpening (Vanrell & Baldrich 2003).

The Laplacian filter gives a strong response in blob detection but is too sensitive to the noise, so a first Gaussian filtering has to be applied. Practically, it is the same to filter directly by the Laplacian of Gaussian:

$$I' = \nabla^2(I * G_\sigma) = I * (\nabla^2 G_\sigma)$$

The Lindeberg's scale-space theory introduces normalization to allow comparison of blobs responses between scales. It automatically selects the scale at which local image structures are better detected by differential operators (Salvatella 2002).

With a normalized Laplacian,

$$\nabla^2_{norm} = \sigma^2 \nabla^2$$

blobs responses are computed for all scales and the maximum over all scales gives all the image blobs no matter their size.

Basing ourselves on that fact, we will present a sharpening operator to not only find black and white blobs – by getting the minimas and the maximas over all scales but will produce an image with a flat background and contrasted response (see Figs 4 & 5).

We have classified the Marfil slabs using Blob analysis on the segmented blobs with extraction of features like the mean and maximum area, the mean and maximum ellipsoid major axes, the mean and maximum of the eccentricity weighted by the corresponding diameter. Results of classification of 82% are promising because the blobs features are still not fully exploited. More detailed feature distributions analysis would give better results. Luengo (2004) uses the size distribution to characterize structures; this distribution is estimated using a so called granulometry, which is the projection of a morphological scale-space on the scale axis.

The major advantage of this technique is its ability to extract veins. Indeed, statistical methods do not find such "non-textural" feature. Actually, veins are not

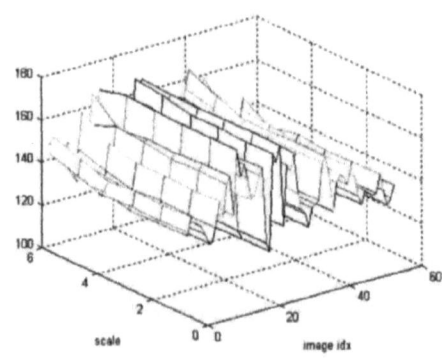

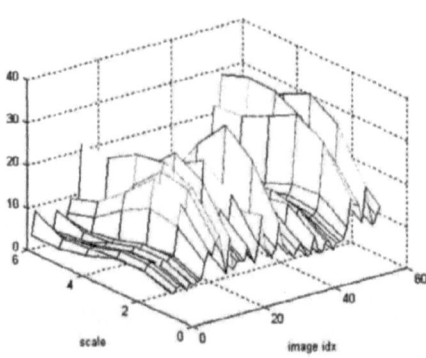

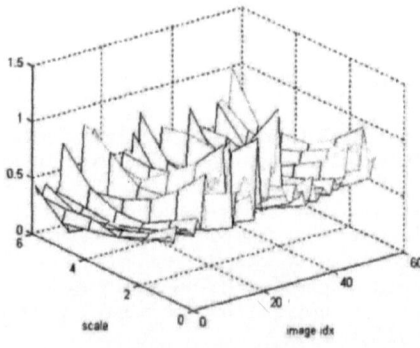

Figure 3. Light-dark mean, light-dark contrast and blue-yellow contrast for an image set and for 6 scales. The mean remains constant whatever the scale (what is expected) but the contrast present discriminent profiles.

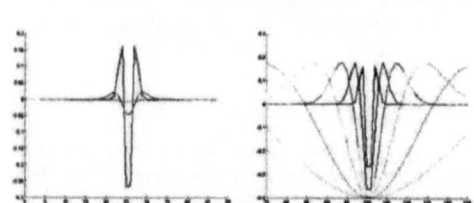

Figure 4. For $\sigma = 1, 2, 4, 8, 16$, profiles of Laplacian of Gaussian and normalized Laplacian of Gaussian.

120

repeated patterns producing a texture but produce a texture on a scale wider than the scale of the slab: that of the tiling.

4.6 Parametric texture model based on statistics of wavelets coefficients

Portilla and Simoncelli (2000) propose a universal model to capture important features of various texture images. It can serve as a high-level texture representation for characterization and segmentation applications.

It uses a pyramidal approach similar to the Laplacian pyramid but capturing orientations: a steerable pyramid. From this representation, key features are extracted to define four statistical constraints: capturing the pixel intensity distribution (marginal statistics), the periodic or globally oriented structure (raw coefficient correlation), the structural information such as edges, corners... (coefficient magnitude statistics) and illumination gradients due to 3D appearance (cross-scale phase statistics).

This complex representation summarizes the nature of the texture in 710 feature values that can serve for classification. An important property of this representation is the ability to synthesize texture from these features to verify their validity.

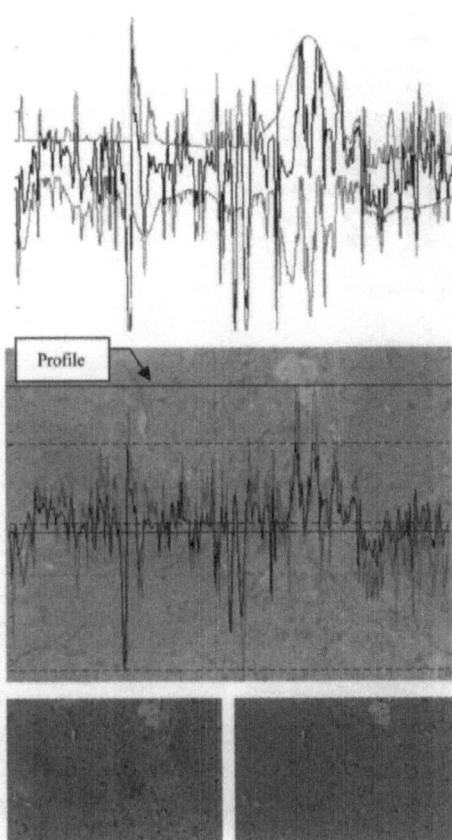

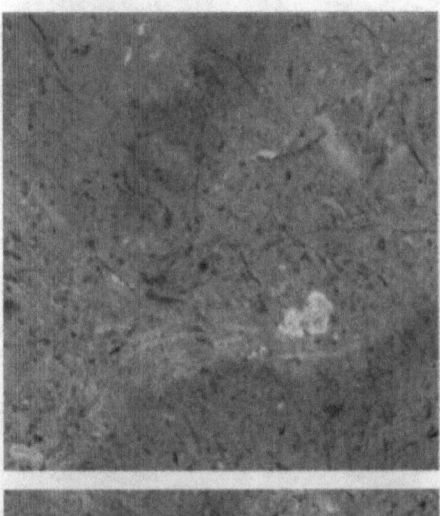

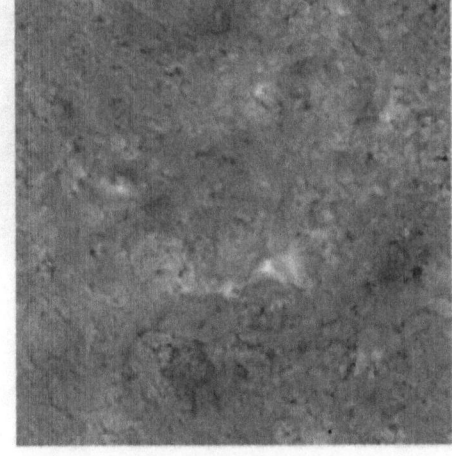

Figure 5. For given profile, the top picture illustrates the profiles of the maximums and minimums over all scales of LoG-normalized transformations giving top and bottom covers over these scales. Their difference gives the (LoG-normalized) sharpen profile. The middle picture shows, in overlay on a Marfil original image, the original profile (grey) and the sharpen one (dark). This sharpen operation flats the background and allows an easy segmentation by threshold. The bottom images illustrate a segmentation of the transformed original image.

Figure 6. A Marfil slab and a synthesized image from the set of statistics of wavelet coefficients.

121

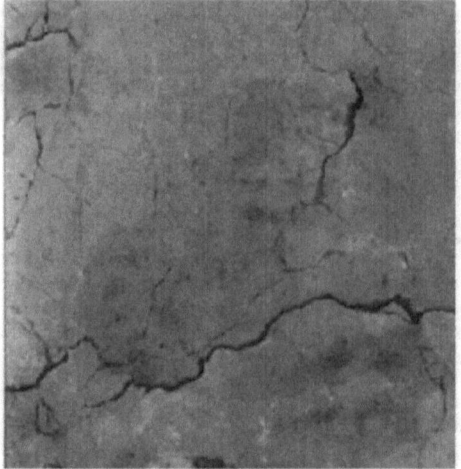

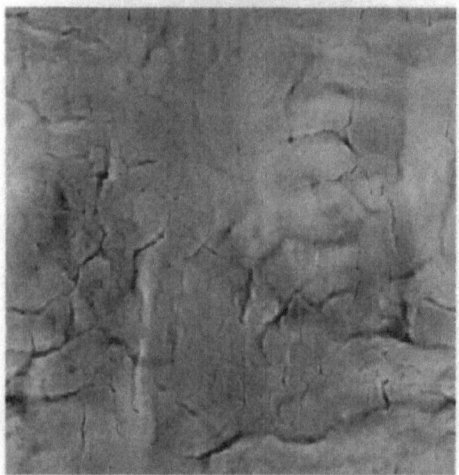

Figure 7. An other Marfil slab and its synthesized image illustrating a non-working case due to the veins extension not captured by the model.

5 CONCLUSION

Three scale-based models have been presented.

The SDH model gives interesting classification results on the Marfil slabs and we showed an improvement by introducing scale. This improvement seems to be essentially due to the contrast evolution over scales. This probably captures features of different sizes.

The Parametric Texture Model based on Statistics of Wavelets coefficients (PTMSWC) is very complete.

It seems to be the "holy grail" of the statistic-based models due to its fully automatic research of inter- and intra-scale features. Nevertheless the classification of the 710 parameters has to be improved by a reduction of these parameters to the relevant ones. In this way, principal component analysis or manual discriminating feature evaluation will be studied.

The blob-oriented classification is the only one that is able to identify veins. Indeed the statistical models fail to capture such features due to the "non-textural" nature of a vein. Therefore this method seems to be the more appropriated for veins analysis. Nevertheless, for background description the other approaches are more complete. Another problem of the blob-oriented model is its sensibility to the choice of the threshold used to extract blobs.

These three scale-based methods demonstrated the interest of scale in texture analysis. SDH over scales is an automatic and simple model that provides significant results. PTMSWC is more sophisticated and implies other studies to use the parameters as input to a classifier. The blob model catches the veins but is sensitive to the blob extraction threshold level.

In the three cases the scale is the only way to capture features of different sizes.

REFERENCES

Lindeberg, T. 1994. Scale-space theory: A basic tool for analyzing structures at different scales. *Journal of Applied Statistics* 21(2): 225–270.
Martínez, J. & Tomás, L.-M. 1999. Marble Slabs Quality Classification System using Texture Recognition and Neural Networks Methodology. *European Symposium an Artificial Neural Networks 1999 proceedings* 75–80.
Randen, T. & Husøy, J.H. 1999. Filtering for Texture Classification: A Comparative Study. *IEEE Trans. Pattern Analysis and Machine Intelligence* 21(4): 291–310.
Salvatella, A. 2002. *Texture Description based on Subtexture Components*.
Simoncelli, P. & Portilla, J. 2000. A Parametric Model Based on Joint Statistics of Complex Wavelet Coefficients. *International Journal of Computer Vision* 40(1): 49–71.
Tuceryan, M. & Jain, A.K. 1998. *Texture Analysis. Handbook of Pattern Recognition and Computer Vision*. 2nd ed.,: 207–248. World Scientific Publishing Co.
Vanrell, M., Baldrich, R. & Salvatella, A. 2002. *Induction operator for computational Color Texture Description*. Elsevier Science.
Wandell, B. 1999. Computational Neuroimaging: Color Representations and Processing. *Cognitive Neuroscience edited by Michael Gazzaniga*.
Zhang, X. & Wandell, B. 1996. A Spatial Extension of CIELab for Digital Color Image Reproduction. *Proceedings of the SID Symposiums, 1996*.

Dimension Stone 2004, Přikryl (ed)
© 2004 Taylor & Francis Group, London, ISBN 90 5809 675 0

Dimension stone collection at the Federal Institute for Geosciences and Natural Resources (BGR)

A. Ehling
Federal Institute for Geosciences and Natural Resources, Berlin, Germany

ABSTRACT: The Berlin-Spandau branch of BGR has one of the largest geoscientifical collections in Germany. It includes a Collection of Dimension Stones. It contains slabs, test cubes, thin-sections, pave stones and a number of rock samples of irregular shapes. The oldest specimen with proof of collection date is from 1892. More than half of the inventory was collected before 1945. The collection is quite valuable because of the numerous shutdowns of quarries, especially after 1945. Today most new entries to the collection are slabs of currently used dimension stones from all over the world. Although the facilities do not have the conveniences of a museum, the collection is open to the public during regular hour.

1 HISTORY OF THE GEOSCIENTIFICAL COLLECTION

The geoscientifical collection of the BGR is a valuable collection in terms of quality, quantity and history.

The first mineral collection in Berlin was started in 1770 within the Mining Academy. The earliest specimens in the Berlin-Spandau BGR collection date from the late the 19th century. The staff of the Royal Geological Survey of Prussia systematically expanded the collection, especially in connection with the geological mapping of Prussia. By the beginning of 20th century, material originated also from the former German colonies and from other countries. The specimens had been partly presented in a museum within the Geological Survey in the Invalidenstrasse 44 in the centre of Berlin.

Because of its location in the eastern part of Berlin, the collection was taken over by the State Geological Commission, which became the Central Geological Institute of the GDR in 1945. During this period, the collection expanded significantly with samples primarily from the GDR. The collection was not open to the public.

In 1990, the BGR became responsible for the collection and transferred it to its Berlin branch. Reorganization and computerized cataloguing began. Today the collection is in a specially remodeled building in Berlin-Spandau. Most specimens remain in shelves and boxes. Exhibits with easy access and explanations for

Figure 1. Example for an old showcase with limestone and marble plates (10 × 15 cm).

the general public are minimal, but anybody interested can visit the collection. There is even a site in the BGR Internet page to inform about the collection.

2 HISTORY OF DIMENSION STONE COLLECTION

2.1 *1873–1935*

In the first years after foundation of the Royal Geological Survey of Prussia (KPGLA), interest in

123

applied geology was minimal. Up to 1900, only some scattered documents and slabs have been properly cataloged. The earliest acquisition goes back to 1892 and is a weathered sandstone from the Magdeburg cathedral. Another early entry is a small collection of 25 French limestones, which came to Germany as part of reparation after the German-French War. By 1910, Leppla introduced a combination of petrographical investigation and technical examination of the dimension stones. This work resulted in the large number of test cubes in the collection.

During the 1920s, systematic rock collecting began in connection with the foundation of the Museum for Applied Geology within the Geological Survey. Burre and Dienemann studied dimension stones at buildings in Berlin and published a book about rocks of architectural interest in Germany. Some of the geologists did the petrographical characterization of the specimens, which were assayed at the Testing Institute for Materials (Burre & Dienemann 1968). All this work left behind a mark in the collection. Specimens came from all over former Germany and former German colonies.

2.2 *1936–1943*

Although there is not information for the date of admission to the collection for all specimens, clearly they were expanded in this period. It is connected with the development of a German Quarry Card File (Deutsche Steinbruchkartei). A large number of the quarries in the country were documented by acquiring information concerning the geology and exploitation; some specimens were used in petrographical and technical analysis. As a result, the collection was expanded with slabs, hand specimen, and thin sections. The work was still in progress when interrupted by World War II.

2.3 *1945–1990*

After the war and the division of Germany, the collection stayed in the GDR. The Museum remained closed and the collection was out of sight for the general public. Initially, the interest on dimension stones was minimal with very few exceptions concerning the most famous dimension stones at the GDR territory. Starting in 1975, a new estimation of the dimension stone deposits of the GDR took place. As a result, about 200 polished plates and several other specimens and thin sections expanded the inventory.

2.4 *Since 1990*

At the beginning of the 1990s, the BGR began computerized cataloguing of the collection. Since 1996, they can be visited in open and closed displays at special facilities. The stock of slabs at the BGR headquarters in Hannover had been integrated to the Berlin

Figure 2. Exhibition board with plates classified by petrographical criteria.

collection and every year the inventory expands with new specimens from all over the world. They are procured through dimension stone industry or in connection with our own excursions and projects. The current size of the slabs coming to the collection is 15 × 24 cm, standard that corresponds with the size of the famous dimension stone collection in Wunsiedel (Bavaria). It is interesting to note how this standard has change through time: 10 × 10 cm, 10 × 15 cm, 16 × 22 cm are some of the most frequent sizes for old slabs.

Today the collection includes 2660 slabs, 2000 thin-sections and about 2500 pave stones, test cubes and specimens of irregular shapes.

3 ACTUAL TASKS

3.1 *Computers*

Most specimens are well documented in an easily accessible data base. The input of thin sections into a data base is still in progress. There is a link to the collection at the BGR home page. The aim for 2004 is to have the entire collection online.

3.2 *Research*

Many of the older specimens in the collection are not adequately catalogued; some descriptions have poor to no information about the sampling locality. There is ample room for improvement. New findings are the subject of publications or new exhibits.

Another field is the petrographical characterization of the specimens. New specimens with only known trade name are analyzed concerning their mineralogical and geochemical properties.

124

Figure 3. Drawer with thin-sections dated from the beginning of 20th century.

No specimen that can be more properly described is left aside in ongoing efforts to obtain the most from the collection. Currently, we are analyzing building sandstones of quarries in Germany, which had been exploited in the past. The results will be used for a special publication.

3.3 Input

Currently we are increasing the inventory and try to be up-to-date concerning currently used dimension stones from all over the world in cooperation with the industry and other institutions.

Interesting specimens that are part of research projects commonly find their place in the collection once the projects are over. As a result, the number of thin sections of building sandstones has been significantly increased.

3.4 Public relations

The collection is accessible to all interested visitors. It is possible to study there, to borrow specimens, and in some cases to even take a sample home. Most of the visitors are geologists, conservators, or, sometimes, architects. Students from various fields are also a significant proportion of the visitors. Occasionally, we take part in special exhibitions, for example, an exhibition about building stones in Berlin in 2000.

4 CONCLUSIONS

The dimension stone collection at the BGR branch in Berlin-Spandau is a well documented and preserved collection with a large number of specimens which is of historical, actual and geological interest for all people working in this subject.

REFERENCES

Burre, O. & Dienemann, W. 1968. Die Arbeiten auf dem Gebiet der Steine, Erden und Mineralien. In H. Udluft, *Die Preußische Geologische Landesanstalt 1873–1939*: Beih. Geol. Jb., H 78; Hannover.

Dimension Stone 2004, Přikryl (ed)
© 2004 Taylor & Francis Group, London, ISBN 90 5809 675 0

GIS: a tool to re-planning extractive industry – The Estremoz Anticline case study

P. Falé, C. Vintém, P. Henriques, C. Midões & J. Carvalho
Instituto Geológico e Mineiro (IGM), Lisbon, Portugal

S. Sobreiro
Centro Tecnológico para o Aproveitamento e Valorização das Rochas Ornamentais e Industriais (CEVALOR), Borba, Portugal

ABSTRACT: Estremoz Anticline, located in southern Portugal, Alentejo region, is integrated in the so called Marble Zone, and it is the greatest Portuguese centre for marble extracting activity. Marble exploitation in Alentejo has some problems related with high geological complexity, conditioning, in some cases, exploiting activity and often leads to production abandoning. Extractive activity is considered to be of major importance for the Alentejo region economy. Regions where the extractive activity is concentrated have strong direct or indirect benefits in the population, with structural importance in the Ornamental Rock sub-sector, as an employment and richness generator. Nowadays, ornamental rocks exploitation in this region is faced with new challenges, related with legal environmental requirements and land use planning policies. Keeping these problems in mind, a Geographical Information System (GIS) model was created, comprising all geological and environmental information available, so that best zones for exploitation and expansion of extractive activity can be delimited, as well as zones that can be used for settlement of small facilities, able to support this industry. In this work, reference is only made to the studies conducted in the Vigária/Monte D'El Rei (Vila Viçosa) area.

1 INTRODUCTION

Ornamental Rocks exploitation in Estremoz Anticline (Alentejo region, Portugal) has been facing new challenges related with new legal environmental requirements and land use planning policies. Since extracting units are always located near the geological resources and it is not possible to eliminate biophysical degradation factors in the source, actions must be taken to minimize conflicts created by this activity.

Land Management is faced today from a prospecting point of view, where strategic planning assumes a fundamental role to the sustainable region development. Therefore, considering the need to combine mining activity with environmental preservation in land use planning policies, this study aims to delimit favourable areas to conduct and increase extracting activity, as well as favourable zones to create small facilities to support it, so that large scale re-planning of the considered area can be made.

Keeping in mind these problems, a GIS model was created with all the available geological and environmental information. This model was structured in four steps, culminating individually with a Geoeconomical Risk Map, Environmental Susceptibility Maps to each environmental descriptor, an Exclusion Zoning Map, and finally, a Re-Planning Map (Fig. 1).

2 MARBLE ZONE CHARACTERIZATION

The Estremoz Anticline is the most important Portuguese geological structure in marble outcrops. Several large areas located in Borba, Estremoz, Alandroal and Vila Viçosa regions constitute the so called Marble Zone.

Exploited marbles in Estremoz Anticline are part of the *Complexo Vulcano-Sedimentar Carbonatado de Estremoz (CVSCE)* (Oliveira et al. 1991), which have the following summarized lithological sequence: meta-volcanic rock from the top, marbles, marbles and metadolomite with intercalation of metavolcanic rocks, and metavolcanic rock from the base.

This Complex is part of the so called Estremoz-Barrancos sub-sector, of the Ossa-Morena Zone (tectono-stratigraphic unit) (Oliveira et al. 1991). It is

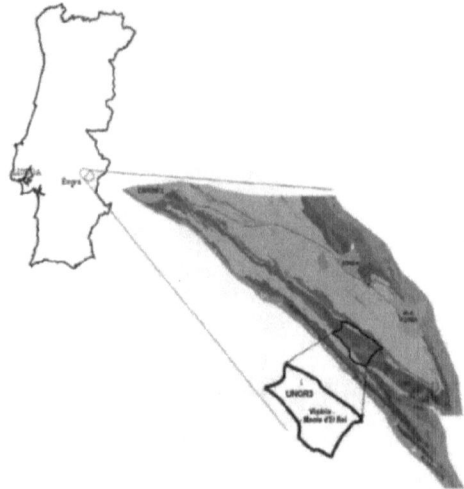

Figure 1. Land Use Planning Unit location (UNOR3).

underlined by a Cambrian Dolomite Formation and overlined by Silurian greenschist with liditic intercalation.

Geological studies in this marble rich structure have been carried by Instituto Geológico e Mineiro since the 1950s at various working scales. Nowadays a very detail geological cartography scaled 1:2000 exists as a direct response to the needs of the extractive industry, and based on a nature and colour marble classification:

- Pink Marble;
- Breccioted Pink Marble;
- White and Cream Marble;
- Breccioted Schistose White and Cream Marble;
- Dark Marble;
- Dolomitic Marble ("Olho de Mocho").

This new marble classification demanded an adequate knowledge about the complex structural disposition of the different lithologies. Together with information about rock fracturing degree, these studies were the data source for the extractive sector management and as a consequence to land use planning. The marble zone is associated with important ecological functions, regarding groundwater aquifers and high agricultural potential, where economical activities related to ornamental rocks are important.

Extractive activity represents an important socio-economical impact through the creation of several exploitation and transformation industries and related employment creation. These facts translate to an increase in economical vitality within the Alentejo region context.

3 A POSSIBLE STRATEGY TO PLAN AND RE-PLANNING EXTRACTIVE ACTIVITY

Land use planning can be seen as a process, scientifically and culturally based, involving a formal and functional composition, to organize uses and functions, in time and space, as a contribution to an integrated and sustained development of human communities. The main purpose of this process consists on achieving, with as little financial encumbrance and in a convenient timeline, the maximum productivity and well being to populations.

Information technologies, especially GIS, enhance natural resources planning and are of major importance to manage land use planning. Technology offers powerful tools to conduct inventories and high quality rigorous spatial analysis, supporting all the levels of administrative decisions. To achieve its objectives, a practical application of a GIS model was developed (in Software Geomedia Pro4 from Intergraph and ArcView 3.2 from Environmental System Research Institute (ESRII)).

Spatial information was collected based on ortophotomaps, field observations and all information needed to create a database, related with structured archive and descriptive attributes. The gathered information, geographical and alphanumerical types, was grouped in Aptitude, Planning and Field Data Maps. Later, most of the information – analogical format – was transformed into digital format, using Software Geomedia Pro4.

Information was geo-referenced using IGEOE coordinate system Hayford Gauss, Transverse Mercator projection system, Lisbon Datum.

This GIS model was structured in four stages. First stage created a Geoeconomical Risk Map, scaled 1:5000, based on the geological data, which defined best areas to extractive activity, as well as possible reclamation and slope stabilization areas. In the second stage, an Environmental Susceptibility Map was created for each analysed descriptor. In the third stage, the Exclusion Zoning Map was created, based on legal constraints.

The fourth stage was created by crossing information from the Geoeconomical Risk Map and the Environmental Susceptibility Map descriptor Hydrogeology, so that an intermediate map could be created, the Re-Planning Map. This is a relevant tool to take management decisions while planning extractive activity. This option was taken as a result of the environmental importance that hydrogeology assumes, in local and regional contexts; groundwater resources are not only the source for public water supply in the region, but are also used in agricultural and industrial sectors.

Finally we are able to cross reference the information from the previously created Re-Planning Map with the information from the others environmental

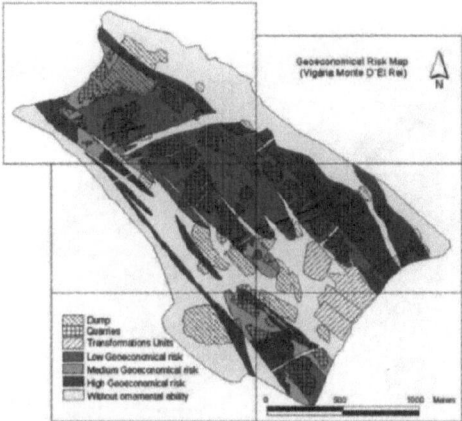

Figure 2. Geoeconomical Risk Map UNOR3 (resized). Adopted from Vintém et al. (2003).

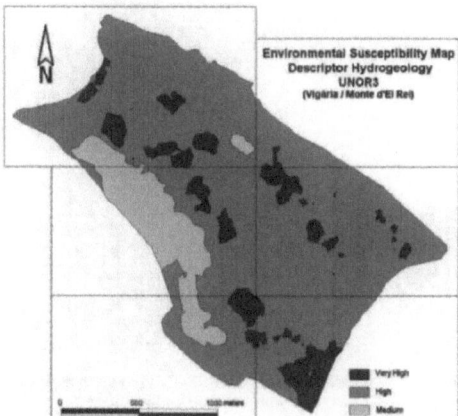

Figure 3. Environmental Susceptibility Map, descriptor Hydrogeology, UNOR3 (resized). Adopted from Vintém et al. (2003).

descriptors such as noise, dust, hydrography, slope, landscape, etc.

3.1 Some geological and environmental indicators: extracting industry application

Main purpose of Geoeconomical Risk analysis (1st stage) was to define areas with the best potential for ornamental rock exploitation, in function of lithology, geological structure and fracturing allowing extraction of blocks with commercial size. This analysis has also allowed for the definition of adequate areas to dump and transformation units, due to lack of resources or to their low potential.

To each mentioned parameter, classes were defined according to their good, medium, bad or without ornamental capability. These classes are a result of a critical analysis after several simulations, and the ones that give the best reality for the area were chosen.

After characterization and classification of the above mentioned parameters, and crossing all the information, zones with low, medium and high economical risk were defined (Fig. 2).

Environmental Susceptibility Analysis (2nd stage) corresponds to the current environmental characterization, through relevant biophysical indicators in areas under ornamental rock extracting activity influence, to understand eventual effects (adverse or good) that the activity may influence. Therefore, according to significance criteria related with this industry, the main environmental descriptor considered in this article was Hydrogeology (strongly dependent on several factors with regional expression, such as lithology, tectonics, climatology, etc.). Therefore, applied methodology

lead to the inclusion of involving areas of the studied planning units. A regional hydrogeological study was conducted in the SE part of Estremoz Anticline, followed by a more detailed hydrogeological study in the selected area.

This descriptor analysis was based on environmental susceptibility class definition: Very High and High Susceptibility, which allowed an interpretation based on cartographic representation, in Environmental Susceptibility Maps, 1:15000 scaled.

Every time the piezometric level is above topography, and where the underground water intercepted quarries the susceptibility class was considered to be very high. When the piezometric level is coincident with topography or is 60 m below the topographic surface, susceptibility class is high. If the piezometric level is more than 60 m below topographic surface, susceptibility class is considered to be medium (Fig. 3).

3.2 Land use planning

Re-Planning an industrial area requires a good definition of the legally protected areas, where extractive industry is totally forbidden. Keeping this in mind, applicable exclusion criteria can be considered as the "Defensive Zones" of the targets to protect, according with Portuguese law. Direct application of these exclusion criteria lead to the Exclusion Zoning Map (3rd stage), scaled 1:15000 to each production area, so that useful information can be used to open new quarries or enlarge existing exploitations. However, this map cannot be considered as concluded, since important information is missing.

129

Geoeconomical Risk		Descriptor Hydrogeology	
Classes	Values	Classes	Values
Low Risk	1	Very High	40
Medium Risk	2	High	30
High Risk	3	Medium	20
Without interest	4		
Urban zone	0		

		Descriptor Hydrogeology		
		40	30	20
	0	40	30	20
Geoeconomical	1	41	31	21
Risk	2	42	32	22
	3	43	33	23
	4	44	34	24

Figure 4. Resulting matrix from crossing information with Map Calculator. Adopted from Vintém et al. (2003).

In order to get a Re-Planning Map (4th stage) of the studied area, information from Geoeconomical Risk Map was crossed with Environmental Susceptibility Map descriptor Hydrogeology. Used methodology was based in a matrix, adding values classified in Geoeconomical Risk to the descriptor Hydrogeology. Values were given to the different classes in geoeconomical risk and environmental descriptor Hydrogeology as shown in Figure 4.

Re-Planning Map legend has resulted after this matrix analysis, through the definition of favourable or unfavourable to marble exploitation with hydrogeological constraints, as well as zones with no capability to marble exploitation as a resource, where abandoned or inactive quarries exist, in which reclamation is possible. Therefore, the key in Re-Planning Maps was structured as follows:

a) favourable exploitation zones with high hydrogeological constraints;
b) favourable exploitation zones with hydrogeological constraints;
c) reasonably favourable exploitation zones with high hydrogeological constraints;
d) reasonably favourable exploitation zones with hydrogeological constraints;
e) unfavourable exploitation zones with quarries for possible reclamation and slope stabilization;
f) zones without exploitation interest – other possible uses but with high hydrogeological constraints;
g) zones without exploitation interest – other possible uses but with medium hydrogeological constraints;
h) urban zone.

Zones classified as "without exploitation interest – other possible uses but with medium hydrogeological constraints" are considered to be the most favourable areas to transformation units settlement, crushed stone units and dump (Re-Planning Map scaled 1:5000, resized).

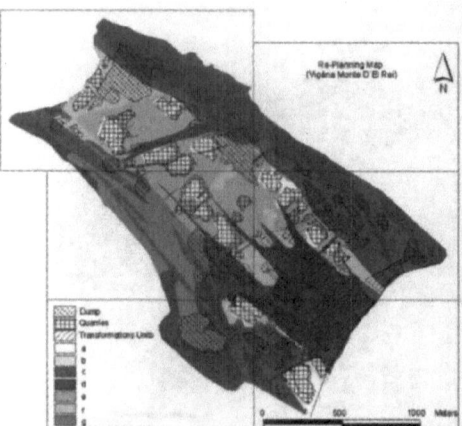

Figure 5. Re-Planning Map (resized). (a – favourable exploitation zones with high hydrogeological constraints; b – Favourable exploitation zones with hydrogeological constraints; c – Reasonably favourable exploitation zones with high hydrogeological constraints; d – Reasonably favourable exploitation zones with hydrogeological constraints; e – Unfavourable exploitation zones with quarries for possible reclamation and stabilization; f – Zones without exploitation interest – other possible uses but with high hydrogeological constraints; g – Zones without exploitation interest – other possible uses but with medium hydrogeological constraints; h – Urban zone). Adopted from Vintém et al. (2003).

4 REMARKS AND FUTURE WORKS

Extractive industry in this region is influenced by a set of factors depending on the existence and accessibility of marble resources. The importance to create innovating Geoeconomical Risk Maps is highlighted by the importance of the activity in this region.

The new concepts regarding sustainable development imply making mining activity compatible with environmental standards preservation and it must be a fundamental target to achieve, by using technical-administrative policies. Therefore, these studies are quite important to evaluate different areas, in terms of environmental vulnerability to the activity. Their final results, the environmental susceptibility maps, are crucial to a sustainable economical development in the region.

Crossing these information levels that have allowed creating Re-Planning Maps, in adequate scales, like those in this work, is an effective and fundamental tool to management decision, to conduct the most feasible land management and planning of the area.

A future work would be the crossing of information for all geological and environmental data using multivariate statistical methods such as principal component analysis or factor analysis. This study

130

allowing identifies the possible correlation between descriptors.

REFERENCES

Machado, R. 2000. *A emergência dos sistemas de informação geográfica na análise e organização do espaço. Fundação Calouste Gulbenkian*. Fundação para a Ciência e Tecnologia. Ministério da Ciência e da Tecnologia. 540 pp.

Martins, L. 2003. *Indicadores de desenvolvimento sustentável: sua aplicabilidade na indústria extractiva*. VI Congresso Nacional de Geologia. Universidade Nova de Lisboa.

Oliveira, J.T., Oliveira, V. & Piçarra, J.M. 1991. Traços gerais da evolução tectono-estratigráfica da Zona de Ossa Morena, em Portugal: síntese crítica do estado actual dos conhecimentos. *Comun. Serv. Geol. Portugal* 77: 3–26.

Vintém, C., Sobreiro, S., Henriques, P., Falé, P., Saúde, J., Luís, G., Midões, C., Antunes, C., Bonito, N., Dill, A.C. & Carvalho, J. 2003. Cartografia Temática do Anticlinal como Instrumento de Ordenamento do Território e Apoio à Indústria Extractiva, Instituto Geológico e Mineiro and Cevalor, internal report for AIZM – "Acção Integrada da Zona dos Mármores" (FEDER) do Eixo Prioritário 2 do PORA – "Programa Operacional Regional do Alentejo 2000–2006".

Dimension Stone 2004, Přikryl (ed)
© 2004 Taylor & Francis Group, London, ISBN 90 5809 675 0

The Virtual Tile Store and other digital applications for stone tile production

P. Laurenge
European Laboratory for Characterisation of Ornamental Stones, Bologna, Italy

S. Bonduà
Dipartimento di Ingegneria Chimica, Mineraria e delle Tecnologie Ambientali, University of Bologna, Bologna, Italy

ABSTRACT: The new harmonised standards referred to finished products like tiles or slabs impose a quality control of geometrical and visual properties. The control has to be done with reference to the specific European Commission Standards (CEN) that describes traditional techniques. However each standard is open to new techniques once they've shown their efficiency in an industrial application. The digital technology, already in use in the ceramic sector, allows an advanced control of such properties in continuous on the conveyor belt. Moreover, the digital images can be used for other applications applied to specific stone tile producers problem like the diversity of the aesthetical aspect of each tiles. This fact creates a lot of misunderstanding between the seller and the buyer that can be cancelled using the images of all the produced tiles in the aim to create the so called "Virtual Tile Store".

1 STONE TILE QUALITY CHARACTERISTICS

1.1 *European standards*

The European Standard Commission is introducing new standards that became obligatory for the natural stone final products and modular tiles (CEN Standard project prEN 12057: "*Natural stone – Finished products, modular tiles – Specification*"). These standards include a wide range of specification like the mechanical properties but also the visual appearance.

The producer has a lot of advantage to respect these standards but he can go further to obtain the maximum quality to its products.

1.2 *Quality control*

The stone tile or slab is a commercial product that carries an image of a high reliability material but it doesn't mean that its quality has to be neglected. What are the mean characteristics of a tile that has to be controlled? We attempt to give here a complete list of the main points that defined the quality of a product. These characteristics can be divided in two groups: the first one concerns the material in its entirety and the second one is relative to each single tile.

A Overall characteristics:
- chemical composition;
- physical properties: porosity, hardness, etc.;
- mechanical properties: flexural strength, abrasion resistance, thermal shock resistance, frost resistance, absorption by capillarity, etc.

B Single product characteristic:
- Dimension: width and height.
- Rectangularity: orthogonality of the tile (squareness).
- Planarity: same thickness in the entire tile (flatness).
- Surface: roughness and presence of bumps, holes and cracks.
- Edge: presence of chips on the edge and corners.
- Visual appearance: aesthetical valuation.

For the overall characteristics it is sufficient to do the measurement of these characteristics just on samples. These characteristics can be statistically considered general for the whole production doing measurements on samples. The methodology is already fixed by standards and accepts that the measurement can be done just on a few samples.

For the single product characteristics, some of them are defined by standards but the control is actually done just on few samples and not on the whole production. This is contradictory with the fact that these characteristics are linked to each tile. Common standards allow measurements on only a few samples because from the technical point of view the tile producers cannot control each single tile.

133

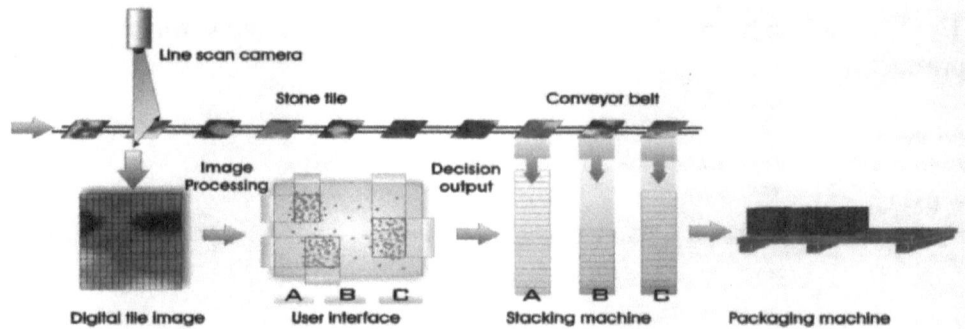

Figure 1. Schematic principle of an automatic tile selection system.

We will demonstrate in this article that this barrier can be break with the already existing technology that will be presented in the following paragraphs.

2 THE DIGITAL TECHNOLOGY APPLIED FOR THE STONE TILE PRODUCTION

2.1 *State of the art: what is already existing in the ceramic tile industry*

The ceramic industry produces also tile and the quality problems are equal to the stone tile production. So, having a look in this industry we can notice that the ceramic industry invests more in new technologies and especially in the online quality control. These controls are almost requested by the buyer who expects that the tiles are very homogeneous between themselves. For example, for a ceramic tiles pavement, the customer does not accept variability of tonality. So, the ceramic producers have been constrained to put on their production line a real time quality control. Even nowadays, before the tile packaging, an operator looks each tile and select the tiles in function of few aesthetical criteria. This is a supplementary cost for the company but the human factor cannot be regular in time. Fortunately, now it exists systems that allow replacement of the human operator.

The production speed could be very high in comparison of the stone tile production: 20 meters per minutes against 5 meters per minutes for the stone production. That's why in the ceramic tile factories, a specific system that scans each tiles and takes the decision instead of the human, has been conceived to speed up the production. This system is also linked to an automatic stacking machine that allows for the whole line being automatic. This kind of system gives good results in ceramic industry and there is no doubt that it could be working for stone tile production line.

We present in the following paragraph the principle of such equipment called "Automatic tile selection".

2.2 *Principle of the Automatic tile selection*

The tile on the conveyor belt enter inside the system where a high-resolution colour camera scans the tile (see Figure 1 for a schematic description). The digital image is then processed by an image processing software. A display shows the measurements set which characterises the tile and allows its classification in a given class. The system then outputs a signal to the sorting machine. The system is also able to detect a lot of physical defects like the cracks, glaze faults, holes, lumps, etc. In case of defects the tile is put in the garbage. A lot of parameters can be fixed through a user interface in the aim to adjust precisely the selection. At the end of a production phase it is possible to establish statistics to know exactly the quality of the different class selection and their intrinsic variability.

Having the image, in real time, it is possible to execute the measurement of the main characteristics mentioned at the beginning of this article like the rectangularity or squareness, the edge defects, the roughness quality or defects.

At the end with such equipment, instead having characteristics just on few samples, we can gather these information on the whole production. The digital image and the aesthetical characterisation.

3 VIRTUAL TILE STORAGE

3.1 *Why using a Virtual Tile Storage*

When a customer wants to buy an ornamental stone tiles pavement, the seller shows him only one sample of the material he wants to buy. Also in the classical catalogue, just one picture of the material

134

Figure 2. Examples of pavement projects. The seller is sometimes constrained by the customer to show what will be the final pavement because each tile is aesthetically different.

type is shown to the customer. Through the internet network, more and more ornamental stone producer sell on-line their material, but once more showing only just one picture of each material type of their catalogue.

This fact creates, after the sale, some problems between customer and seller. Very often the customer is deceived about the material he has bought: it doesn't correspond to what he expected. In extreme cases, the seller is constrained to spread all the tiles on the ground to show what will be the final result of the pavement (Fig. 2). Obviously this fact is due to the natural variability of aesthetical properties, today not fully controlled. Nobody can blame the seller to simplify its production using just one sample per material because technically and practically any cheap solutions have been proposed until now.

We propose here a new system that characterises the general product as individual product elements. The digital technology that exists nowadays fully and cheaply allows this possibility. As we presented in the previous part, the automatic selection system takes an image of each tiles in the aim to decide to which category it belongs. By taking digital photographic images, the whole production can be memorised, so that the seller can exploit these data to improve its commercial strategy. Obviously, if the customer has the possibility to see all the tiles he wants to buy, he will be much more confident. The digital images avoid the seller unpacking the boxes if the customer wants to verify the quality of the box's tiles. Moreover this allows the buyer to choose what he wants among the boxes.

The company Elcos S.r.l. is working on this project in collaboration with Surface Inspection Ltd. The project is called: "Virtual Tile Storage". The project can be divided in two parts: the hardware and the software that we present briefly now.

3.2 The hardware

First of all, we need a system which is able to take the images at a sufficient speed. A simplify version of the system is able to take images to feed a "Virtual Tile Storage". This is done thanks to a line scan colour camera. The system includes also a lightning part which allows an homogeneous light in the aim to have a good response from the digital sensor.

The camera is linked to a computer that saves the result of the scanning inside a digital image. This last one can be considered like a simple file. From the point of view of the electronic components, this kind of system is already working since ten years on industrial production lines and has already given good results. The hardware has already been tested in extreme conditions: dusty ambient, hot temperature.

Given the low speed of the conveyor belt and the relative high cost of a line scan camera, the acquisition camera can be replaced with a classic matrix CCD camera, much more economic.

The whole system has the important advantage that can be embedded on a production line without modifying it.

3.3 The software

On the base of the tile images availability, we can browse backward the entire tile production. Therefore we have conceived user interface software which allows for visualisation of the tile images. The complete information concerning each tile is stored in a database. The information is typically:

- Name of the tile: the name is given automatically by the acquisition software that gives an ID number to the tile.
- Path to the stored image: the file image is not stored in the database but just the path to reach this important file.
- Physical information: size, type, surface finishing, etc.
- Acquisition information: date, production line, etc.
- Aesthetical quality information: we can retrieve all the aesthetical measurements from each image that characterise the tile quality; this could be the tone, the pattern characteristic, etc., there the same algorithm is used for the automatic selection (see part 1 of this article).
- Tile location information: we can store also the information where the tile has been put: which box and which pallet. For example we can store the number ID of the tile inside a box. If the system is linked to a stamping machine each tile can be marked with their ID number and storing also its box number, it is the afterwards possible to retrieve a given tile (Fig. 3).

135

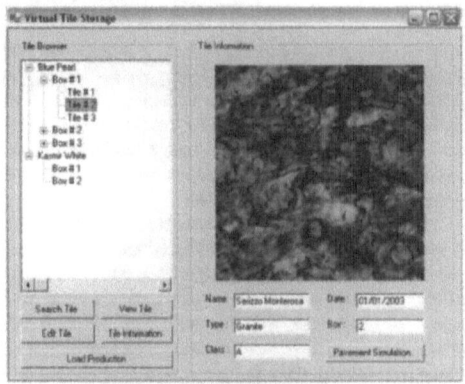

Figure 3. Example of a user window interface used in the aim to browse the tile production. Here the visualisation of a single tile is shown.

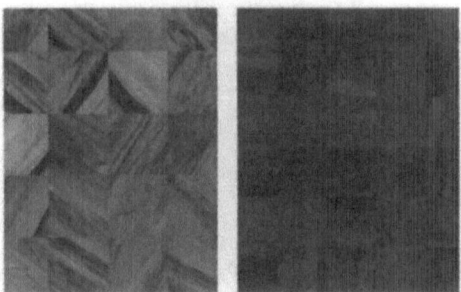

Figure 4. Examples of pavement simulations. The virtual tile store is able to show to a customer what kind of pavement he will really have and this without unpacking any boxes.

All this information is saved in a database. We decided to store this information in a simple text file which is structured in a tagged XML file. This choice avoids the use of classical database software that requires always a database connection.

We present in the Figure 3 one example of user interface window that allows for a user to unpack the tile storage in a virtual way. For example, the user can make by choosing one box, simulate a pavement (see Figure 4). Another important application of the pavement simulation is the capacity of the software to provide for Computer Aided Design of surface coverings by tiles and slabs. The fully identification of each tile allows also for the paving project.

3.4 Applications: on-line tile shop

We show now that the virtual tile storage can have an immediate application in the case of an e-commerce.

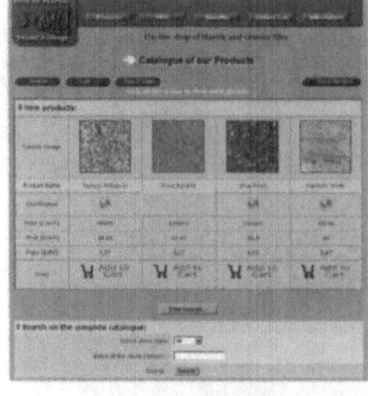

(a) The surfer can choose the material type. A catalogue search is also implemented.

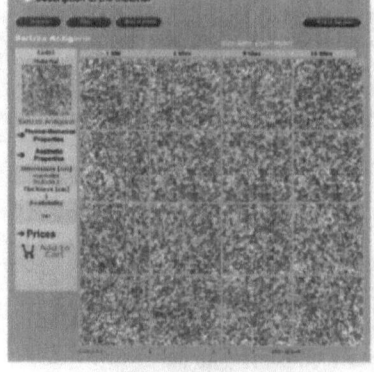

(b) The web site offers the possibility to simulate the pavement using the images of the different tiles.

Figure 5. The web site www.bestofmarble.com shows a direct application of the "Virtual Tile Storage".

This has been already done with the web site called "www.bestofmarble.com".

This site conceived by Elcos S.r.l. uses "Active Server Pages" (ASP). This consists to elaborate dynamic web pages through a programming language that creates web pages according to the surfer's requests. We show in the Figure 5b that the customer, once he has chosen a material, he can elaborate a pavement simulation and verify his choice. The simulation method used there is simpler than the previous presented software: each tile of the pavement is taken randomly in the image database. Obviously, the previous Windows application can be also implemented like a web application.

136

4 CONCLUSION

This article has shown briefly that the digital technologies are able to supply stone producers with tools that can help them giving an added value to their products. The technology is already working and the investment will be amortised very quickly.

Concerning the automatic tile selection is already working in industrial environments. The system is not so diffuse inside this kind of industry but will be surely, in the next years, be a necessary element of each stone process line. This will be also a necessary tool to increase the quality and face the competition.

Concerning the "Virtual Tile Storage", each plant has its own structure and organisation, so this project has to be adapted to the existing process. Therefore each industrial implementation is a small autonomous project which investment will be supported for 2004 by Elcos and Surface Inspection.

REFERENCES

European Laboratory for Characterisation of Ornamental Stone Web site. *http://www.elcosnet.com.*

Best of Marble Web site. *http://www.bestofmarble.com.*

Surface Inspection Ltd. *http://www.surface-inspection.com.*

Bruno, R., Persi Paoli, S., Laurenge, P., Corsi, C., Coluccino, M., Muge, F., Ramos, V., Pina, P., Sottomayor, L., Chica Olmo, M., Serrano, E. & Querela, J.M. 1999. Characterization and classification of ornamental stone by image classification: EUROTHEN '99: *Proceedings of the Second Annual Workshop p. 335–351, Laboratory of Metallurgy, NTUA.* Athens, Greece.

Bruno, R., Persi Paoli, S., Laurenge, P., Corsi, C., Coluccino, M., Muge, F., Ramos, V., Pina, P., Mengucci, M., Chica Olmo, M. & Serrano, E. 1999. *Image Analysis of Ornamental Stone Standard's Characterization.* Geovision '99, Liège, 6–7 May 1999.

Dimension Stone 2004, Přikryl (ed)
© 2004 Taylor & Francis Group, London, ISBN 90 5809 675 0

Monitoring color alteration of ornamental flagstones using digital image analysis

V. Lebrun, C. Toussaint & E. Pirard
Université de Liège, GeomaC, Géoressources Minérales, Liège, Belgium

ABSTRACT: This paper presents a methodology for quantifying color alteration of dimensions stones as caused by weathering using digital image analysis. Specific imaging methods are suggested for color calibration and geometric repositioning and practical solutions are provided in order to account for spatio-temporal drift and optical aberration. The procedure consists in grabbing and comparing images of polished granite flagstones before and after accelerated ageing tests. This method is non-destructive, allowing pre-weathering and post-weathering image acquisition on the same sample after repositioning with an accuracy of 0.4 millimeters. Color alteration is computed using a Euclidean distance in a (pseudo)-L*a*b* color space. The results of a practical study on three selected granites are presented and discussed.

1 INTRODUCTION

The aim of this paper is to propose technical solutions and analytic methods for monitoring the color decay caused by weathering using digital image analysis. The mineralogy of three selected granites is first briefly described, followed by an overview of the complete testing protocol. The two main sections develop the calibration procedure for the imaging system and the image analysis methodology. Finally, the most significant results are presented and commented. A brief discussion about potential extensions of the proposed techniques and its limitations conclude the subject.

The quantitative description of the behavior of ornamental stones submitted to natural or artificial weathering is a crucial challenge not only scientific but also economic. This is indeed one of the most important characteristics of these materials, determining their field of use in the building market. Mineralogical or geochemical mechanisms of rock alteration have been widely studied (Delvigne 1998). Some works have been performed trying to quantify the degradation of mechanical and physical properties (De Cleene 1995), but very little information is available about the aesthetic alteration of stones after weathering. However, from the consumer's point of view, this aesthetic point of view is probably the most important one. For example, any deviation, even slight, in the predominant color of adjacent tiles in a paving badly affects the aesthetic of the work as a whole by emphasizing the color discontinuity induced by the cement joints.

Of course, standard colorimeters or spectrophotometers are available for quantitative measurements of color. But both devices integrate very limited fields of investigation (typically = 1 cm²). They are thus largely irrelevant for monitoring color variations in textured materials like granites.

The idea of using digital image analysis to characterize or control the quality of ornamental stones (Muge et al. 1997) or ceramics (Baldrich et al. 1999) is not a revolutionary idea. For the last five years various consortiums have been managing ambitious research programs at European level with the aim of integrating automatic inspection systems for ceramic tiles (Maccari 1998, Donia 2000). This recent interest for quantitative color quality control by digital video camera confirms the potentiality of the technique and reflects the need for an accurate and objective tool to control the visual aspect of tiles in the stone industry.

The present work tries to go one step further by integrating color image analysis into a wider time-scale, which leads to the notions of color alteration and color durability. The testing methods presented here are first available in laboratory but could be extended, with some limitations, to "in situ" paving on facades and to natural outcrops of stones.

2 MATERIALS AND METHODS

2.1 *Selection of granites*

For this study, three granites have been selected especially according to their high propensity to chromatic

139

Table 1. Name, origin and mineralogy of the selected granites. Albite (Ab), Microcline (Mc), Quartz (Qz), Biotite (Bt), Hornblende (Hb), Kaolinite (Kl), Chlorite (Cl).

Name	Nature	Origin	Mineralogy		
			Princ.	*Second.*	*Access.*
Tarn	Granite	France	Ab		Kl
			Mc		Cl
			Qz		
			Bt		
Azul Paveno	Gneiss	Brazil	Ab	Bt	Kl
			Mc		
			Qz		
Baltic Brown	Wiborgite (granite)	Finland	Ab	Bt	
			Mc	Hb	
			Qz		

alteration (Elsen 1996). Table 1 presents the names, nature, origin and mineralogy of the granites.

Three polished flags (300 × 300 mm) of each granite have been sampled and submitted to the analysis procedure. The tiles have been randomly selected among a daily production of the quarries.

2.2 *General testing protocol*

For future normative tasks one has defined a rigorous protocol of test having a care for generality and reliability. Each fresh sample is first imaged and its initial color statistics are computed and stored. The tiles are then submitted to defined accelerated ageing tests. After precise repositioning under the imaging system, a second color image is acquired from which final color statistics are extracted. The color alteration is computed from a pixel to pixel comparison of the pre-weathering and post-weathering images.

2.3 *Accelerated ageing tests*

The ageing chamber used is a KSE-300 designed for the realization of corrosion tests following the DIN50017 and DIN50018 norms (DIN 1982, 1997) and for alternating condensation controls.

The following normalized tests have been performed:

1. Acid test (two samples of each granite): 21 days at 20°C in a SO_2 saturated atmosphere.
2. Water and heat test (one sample of each granite): 21 heating cycles during 20 hours in a ventilated drying-room at 105°C followed by 4 hours in a water bath at 20°C.

These tests have been chosen for two main reasons: First, Elsen (1996) observed some visually significant color changes in the same granites submitted to the acid test. Second, both tests are using oxidant atmospheres suitable for modeling the urban environment to which external wall covering tiles could be submitted during their life.

2.4 *Imaging devices*

A specific acquisition system has been designed taking into account the optical properties of the observed samples. The three main characteristics of polished granite tiles are:

- macroscopic dimensions (300 mm by 300 mm);
- high reflectivity;
- textured color.

These properties induce technical constraints on the imaging system (Lebrun et al. 1999) which needs to capture diffuse reflectance from a daylight source while maintaining a stable response in time. In practice, the final solution adopted for this study is an imaging chamber equipped with high frequency "daylight" fluorescent tubes, diffusing walls and a high resolution black and white CCD sensor fitted with a filter wheel bearing red, green and blue gelatin filters.

3 IMAGE CALIBRATION AND ANALYSIS

Like for any other scientific instrument, each step of the image acquisition with a CCD-camera has to be controlled and calibrated. The lighting system, the optics and the electronic devices all induce noise and color deviations that must be checked for stability and uniformity of response through time and space.

3.1 *Corrections for temporal noise and drift*

A low frequency drift of the video signal with time is due to a progressive heating up of the CCD sensor. In order to minimize this, it is recommended to switch on the camera for about one to two hours before operating (Pirard et al. 1999).

A high frequency component in the signal variation with time is due to both electronic and thermal noise (Holst 1998). Such noise can reasonably be filtered by time-averaging a sequence of sixteen images.

3.2 *Corrections for spatial noise and drift*

Even with an almost ideal acquisition system, non-uniform illumination and optical aberrations will induce a low frequency drift across the field of view.

This can be corrected for by taking two reference images: the first one is a "black noise" image obtained by grabbing a picture from the camera when preventing any light to hit the sensor; the second one is a "white reference" image obtained by picturing a

140

surface sprayed with barium sulfate or alternatively a plate of glossy white high density polyethylene (PEHD) under exactly the same imaging conditions as those used for the whole study.

From both the black and white reference images it is possible to achieve a numerical gain and offset adjustment of each individual pixel so as to obtain a perfectly homogeneous response from a mineral no matter where it lies in the scene (Pirard et al. 1999).

3.3 Color calibration

Each optical component of the acquisition device induces deviation of the output red green and blue channel compared to the intrinsic color of the observed material. These errors can be numerically corrected by calibrating the system with standardized color charts. Various solutions are available to perform this calibration (Connolly & Leung 1995, Chang & Reid 1996, Marszalec & Pietikäinen 1997). The common goal is to compute the transfer-matrix of the imaging tool, which converts the device dependant color channels (RGB) into a calibrated color system like CIE-RGB or CIE-L*a*b*.

In this study, the main objective is to quantify relative shade differences rather than absolute color values. It is thus possible to work directly with the output R, G and B value given by the camera. Nevertheless, an original procedure is used to optimize the color rendering of the system and to ensure its reliability. Each image is acquired with a standard grey scale (Kodak Q-14) present in the scene. The integration time, the gain and the offset of each channel are tuned individually in order to obtain identical mean intensities in the grey patches of the scale. Although the obtained color co-ordinates remain device dependant, this calibration ensures the reliability of the relative color deviation measure, as the same device is used for both the pre-weathering and post-weathering acquisition steps.

3.4 Positioning calibration

In order to achieve precise repositioning of the tile under the image acquisition system after having suffered the ageing test, it is necessary to stick three targets on each tile. These spatio-referenced targets allow combining images of fresh and altered tiles with a spatial accuracy of a single pixel (0.4 mm). The image grabbing geometry is kept identical during both acquisition phases (the system being locked during the ageing tests). In this manner, the images are affected in exactly the same way by eventual lens distortion.

3.5 HSL global statistics

Global color deviations are measured on entire tiles in the hue-saturation-intensity space, HSI (Gonzalez &

Woods, 1992). This is a bi-conical color space, directly deduced from RGB by linear transforms. Of course, it is non-linear and device dependant, but it presents the great advantage to be probably the most intuitive manner of specifying color. The relative pre-(IN) to post-(OUT) differences between average values of each channel are computed as follows:

$$dX = 2 \cdot \frac{\left(\overline{X}_{Out} - \overline{X}_{In} \right)}{\left(\overline{X}_{Out} + \overline{X}_{In} \right)} \qquad (1)$$

where $\overline{X}_{In}$ denotes the global mean estimator of the desired channel (H, S or I) in the input image.

These values are convenient for explaining global color deviations, but they must not be considered as quantitative measures of these changes. Even a Euclidean distance in HSI space would not give such an objective quantification as it is not a linear space regarding to visual perception (Mac Adam 1942). Nevertheless, for the industrial transfer of the methodology, the utilization of such a self-understanding color decomposition might appear necessary.

3.6 L*a*b* local distances

In order to map the quantitative color variation undergone by each pixel during the accelerated ageing operation, one needs to use a perceptually linear space like CIE-L*a*b*. It is a uniform spherical color space especially created by the CIE for color difference computations. The transformation from R, G, B channels to L*a*b* is non-linear. For the complete description of the transfer equations, the reader is referred to the CIE recommendations (CIE 1986).

These relationships are only relevant for calibrated CIE-RGB values. Applying the same formulas to the device dependant red, green and blue channels could make poor sense. Nevertheless, the pseudo L*a*b* has been preferred here because it will allow using the same method for CIE-L*a*b* computation in future works with a calibrated CIE-RGB input.

Once the L*a*b* co-ordinates are computed, the difference between two measured colors is given by the following relationship:

$$\Delta E_{Lab} = \sqrt{\left(\Delta L^* \right)^2 + \left(\Delta a^* \right)^2 + \left(\Delta b^* \right)^2} \qquad (2)$$

In the present application, ΔL^*, Δa^* and Δb^* are the pixel-to-pixel differences within each of the L*, a* and b* channels before and after accelerated ageing tests. This resulting distance function is then mapped into a new image in which the intensity of each pixel is proportional to the computed L*a*b* color distance.

The mean and maximum values of the distance map give a more quantitative, but less intuitive, idea of the global color deviation than the HSL statistics.

141

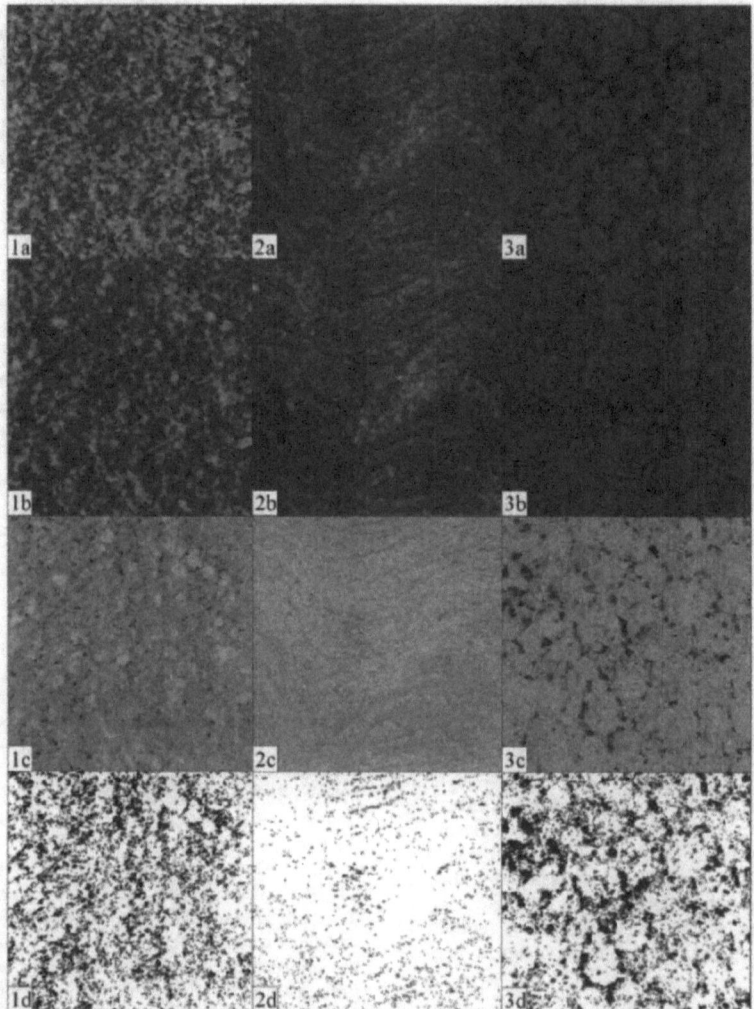

Figure 1. Selected granite tiles from Tarn (1), Azul Paveno (2) and Baltic Brown (3) imaged before (a) and after the ageing test (b). The color distance image (c) is thresholded to identify regions affected by significant changes (d).

By choosing an appropriate threshold in the distance map, one can extract the most altered regions in the tiles and compute some useful statistics on them:

- number of regions;
- relative cumulated area of all regions;
- average size of individual regions;
- mean color distance within a region.

Figure 1 shows images from the sound and altered selected granites, the corresponding color distance images and the binary images resulting from the thresholding operation.

4 RESULTS

4.1 Macroscopic observations and mineralogy

The "water and heat" test did not cause any perceptive chromatic alteration. No significant color change could be visually detected. However, the parameters of this test (heat, wet, O_2 and CO_2) are expected to favor the oxidation of pyrite into limonite or goethite (Costagliola et al. 1997). The absence of this phenomenon can be due to a very low pyrite content or to a very slow kinetic of oxidation.

142

Table 2. HSI relative differences between altered and fresh samples.

	ΔH (%)	ΔS (%)	ΔI (%)
Tarn	−33	+74	−28
Azul Paveno	−12	−27	+2
Baltic Brown	−26	+37	−10

Table 3. Global and local analysis of the CIE-L*a*b* color distance images with, A_A: cumulated surface fraction; μ(Size): average area of individual regions; $\mu(\Delta E)$: mean color distance within individual regions.

	Global statistics		Statistics on regions		
	Mean	Max	A_A (%)	μ(Size)	$\mu(\Delta E)$
Tarn	10.4	66.8	21.4	8.3	15.3
Azul Paveno	7.0	35.2	4.1	1.9	15.1
Baltic Brown	10.5	74.4	23.4	8.9	15.6

On the contrary, some of the samples exposed to the acid test locally underwent drastic color changes, especially those of Tarn and Baltic Brown.

Yellowish white efflorescence appeared on the three granite types. These crusts were identified by optical microscopy as composed of gypsum ($CaSO_4 \cdot 2H_2O$) and are more abundant in Baltic Brown. Elsen (1998) explains this relative abundance by the presence, in the latter, of calcic hornblende, which probably acts as a catalyzer of the crystallization. He argues that this mineral contains Ca^{++} ions needed for gypsum formation and does not take a good polish, thus offering a larger specific surface for alteration.

Rust-colored patches (limonite and goethite) appear on Tarn and, to a lesser extent, on Baltic brown tiles. In both cases, a careful macroscopic observation reveals that muscovite is the preferential alteration site. Table 2 contains the relative HSI differences between fresh and altered samples computed following equation (1).

Global and local statistical parameters computed from the distance images are presented in Table 3. The first two columns contain the mean and maximum distances as obtained for the entire surface of the tiles. The following columns concern local information about the most altered regions obtained from thresholding the distance image at a level of 13.

5 DISCUSSION

The HSI differences presented in Table 2 are not really representative of color changes, mainly because they indicate global information when color alteration is most often localized. Nevertheless, they allow expressing visual observations into numerical terms and checking that the digital imaging system is able to model human eye perception, at least up to a certain level.

Table 3 confirms that the most important chromatic variations are observed in the Tarn and Baltic Brown granites. The intensity channel is affected by two opposite contributions: the surface state alteration tends to decrease the overall brightness, whereas the apparition of white efflorescence pushes the intensity upwards. By comparing the three granites based on ΔI, one can deduce that the tarnishing effect is predominant in Tarn and, to a lesser extent, in Baltic Brown. In Azul Paveno, both surface state and efflorescence effect compensate mutually. The sharp increase in saturation observed for Tarn and Baltic Brown is probably due to the rust colored patches having high saturation levels compared with the duller shades of natural granites.

The L*a*b* distances maps and their related parameters give a much more precise description of the intensity of color deviations. One can summarize the information from Table 3 as follows:

- The mean color distances are in complete accordance with the HSI differences concerning the global alteration of each granite.
- The lower global color deviation observed in Azul Paveno is not due to a lower mean color distance in the altered sites. It results from a lower amount of altered regions combined with a smaller mean size, thus a lower surface proportion, of these regions.
- If Baltic Brown and Tarn present similar global color alterations, the spatial distribution of this alteration is somewhat different. The Tarn granite contains fewer patches, with lower mean size and lower mean distance than Baltic Brown.

6 CONCLUSIONS

From the results presented above, it can be said that digital cameras and image analysis are able to perform quantitative color measurements of large textured samples. Combined with accelerated ageing test this proves to be a powerful tool to evaluate the color deviations caused by weathering on ornamental stones.

More generally, calibrated digital image analysis appears to be a suitable tool for geologist and material science engineers in order to tackle problems of quality control of building materials, durability of stones, aesthetic alteration, etc.

6.1 Advantages and possible extensions of the method

Digital imaging is a non-destructive analytical tool. Compared to other tools used for durability assessment

143

such as mechanical or chemical testing, measures can be achieved exactly on the same sample before and after weathering.

Questioning the adequacy and reality of the weathering tests is out of the scope of this paper. It suffices to assume that useful information about the durability and resistance of a material against weathering can be derived from such tests.

In the present case, one can claim that calibrated digital image analysis is the only one method to obtain a meaningful, quantitative and reliable evaluation of the color modification.

Moreover, color is not the only physical property than can be monitored by digital imaging. Under low-angled light, pictures can be taken in which the intensity will be inversely proportional to the roughness of the surface. Using the adequate modus operandi it will be possible to quantify the loss of luster induced by weathering on polished stones.

Additionally, the exact position of altered regions can be stored in a GIS-like geographical information system together with mineralogical, chemical, micro-hardness and other data. Such a multi-data system could open a wide range of possible scientific studies on stone alteration.

Finally, the use of digital imaging devices is not restricted to laboratory experiments. Calibrated video imaging can be used to observe "in situ" the degradation of wall covering materials. Real scale studies can be set up by grabbing pictures of facades at the reception of works and comparing them with pictures taken one or two years later under the same conditions.

One step further concerns the application of image analysis to field imaging of natural outcrops (Lebrun et al. 1999). In this case, digital imaging proves to be the only way to store and retrieve quantitative information related to geometry and color of geological bodies.

6.2 Limitations

The main limitations of the imaging techniques are linked with the reliability of the image acquisition itself. If it is not possible to take pictures in exactly the same conditions (geometry, field of view, lighting), calibration will be a very delicate problem. In any case, spending time to optimize the image acquisition is never losing time. But, the information lost by using an inadequate imaging system will be lost forever.

Techniques suitable for observing flat surfaces are not readily transferable to surfaces with severe topographical gradients. Numerical shading corrections do exist, which can attenuate the problems but never eliminate them.

In a future work, it would be interesting to improve the method by computing the transfer-matrix of the system using a large reference color set (Marszalec & Pietikäinen 1996). The color set should be chosen in order to cover the entire shade gamut of the studied granites. This would allow standardizing the measure of color alteration in the CIE-L*a*b* system.

REFERENCES

Baldrich, R., Vanrell, M. & Villanueva, J.J. 1999. Texture-color features for tile classification. EUROPTO/SPIE Conference on Color and Polarisation Techniques in Industrial Inspection, Munich (Germany).

Chang, Y.-C. & Reid, J. 1996. RGB calibration for color image analysis in machine vision. *IEEE Trans. Image Proc.* 5(10): 1414–1422.

CIE 1986. *Colorimetry-Official recommendations of the International Commission on Illumination*. CIE Publication No. 15.2, Vienna

Connoly, C. & Leung, T. 1995. Industrial color inspection by video camera", IEE Int. Conf. On Image Processing and its Applications, Conference publication No. 410, IEE, pp. 672–676.

Costagliola, P., Cipriani, C. & Manganelli del Fa, C. 1997. Pyrite oxidation: protection using synthetic resins. *European Journal of Mineralogy* 9: 167–174.

De Cleene, M. 1995. *Interactive physical weathering and bioreceptivity study on building stones, monitored by computerized x-ray tomography (CT) as a potential non-destructive research tool*. Commission of the European communities, directorate-general for science, research and development, Bruxelles, 286 p.

Delvigne, J. 1998. *Atlas of micromorphology of mineral alteration and weathering*. Mineralogical association of Canada, Ottawa, 494 p.

DIN 50017-Ausgabe:1982-10 1982. Klimate und ihre technische Anwendung; Kondenswasser-Prüfklimate. Deutsches Institut für Normung e.V. Berlin.

DIN 50018-Ausgabe:1997-06 1997. Prüfung im Kondenswasser-Wechselklima mit schwefeldioxid-haltiger Atmosphäre. Deutsches Institut für Normung e.V. Berlin.

Donia, G. 2000. Inspector 2000 project. *Ceramic World Review* 37(10): 216–219.

Elsen, J. 1996. *Diagnostic des pierres naturelles et optimisation de leurs techniques de mise en œuvre, T.2 : Le comportement des granits sous l'influence de l'exposition en atmosphère acide*. Centre scientifique et technique de la construction (ed.), Bruxelles.

Elsen, J. 1998. Durability of Granites for Construction. *Aardk. Mededel.* 9: 35–40.

Gonzalez, C. & Woods, R. 1992. *Digital image processing*. Addision Wesley (ed.).

Holst, G.C. 1998. *CCD Arrays, Cameras and Displays*. SPIE Optical Engineering Press, Washington, p. 127–131.

Lebrun, V., Bonino, E., Nivart, J.-F. & Pirard, E. 1999. Development of specific acquisition techniques for field imaging – Applications to outcrops and marbles. Geovision, Int. Symp. on imaging applications in geology, pp. 165–168.

MacAdam, D.L. 1942. Visual sensitivities to color differences in daylight. *J. Opt. Soc. Am.* 32: 247–273.

144

Maccari, A. 1998. ASPECT: an intelligent sorting system. *Ceramic World Review* 28(8): 138–141.

Marszalec, E. & Pietikäinen, M. 1996. Some aspect of RGB vision and its applications in industry. *International Journal of Pattern Recognition and Artificial Intelligence* 1(10): 55–72.

Marszalec, E. & Pietikäinen, M. 1997. Color measurements based on a color camera. In Rolp-Jürgen Ahlers, Philippe Réfrégier (eds) *New image processing techniques and applications: algorithms, methods and components* II., pp. 170–181, Proc. SPIE Vol. 3101, June 1997.

Muge, F. et al. 1997. Caracterization of ornamental stones standards by image analysis of slab surface (COSS)

Eurominerals '97, II Int. Cong. of Natural and Industrial stones, Lisboa.

Munsell, A.H. 1975. *A Color Notation: an illustrated system defining all colors and their relations by measured scales of hue, value and chroma.* Munsell Baltimore: Color Cy.

Pirard, E., Lebrun, V. & Nivart, J.-F. 1999. Optimal acquisition of video images in reflected light microscopy. *Microscopy and Analysis* 60: 9–11.

Russ, J.C. 1999. *The Image Processing Handbook.* 3rd ed., Boca Raton: CRC press.

Soille, P. 2000. *Morphological Image Analysis.* Berlin: Springer Verlag.

Dimension Stone 2004, Přikryl (ed)
© 2004 Taylor & Francis Group, London, ISBN 90 5809 675 0

World's quarries of commercial granites – localization and geology

D. Pivko

Comenius University, Faculty of Natural Sciences, Bratislava, Slovakia

ABSTRACT: The article presents localization and geology (tectonic unit, age, petrographic name) of the most used commercial granites on market. Origin of the most frequent stones in internet was searched out according commercial names in web pages, catalogues, geological maps and articles. The commercial granites are exported mainly from India, Brazil, USA, Finland, South Africa, Norway, Italy, Canada, China, Spain, Saudi Arabia, Zimbabwe, Ukraine, and Sweden. The most used stones are granites and migmatites, less gabbros, syenites, gneisses and granulites. Granites were formed especially about 2.5 Ga, 1.6 Ga, 1.1 Ga, 0.6 Ga and 0.35 Ga ago, migmatites in Archean and Proterozoic.

1 INTRODUCTION

Granites in the commercial sense are hard natural stones that are polishable and need to be worked on by harder tools than for marble for cutting, shaping and polishing. They are usually suitable for interior and exterior use. Petrographically, they are either magmatic or metamorphic rocks.

Commercial granites are mixtures of minerals and are composed of visible multicoloured mineral grains. Grains of one colour are typically encircled by grains of other colours, e.g. grey quartz is close to pink orthoclase, white plagioclase and dark mica.

Natural stones for decorative purposes show today great development. The stones from the whole world are imported and many buildings in city centers are covered by them. They even penetrate to houses and flats. Big areas of polished stones offer excellent opportunity for studying them, e.g. for walking tours.

Sellers and stone masons do not know petrography and origin of the natural stones, only commercial names. Some catalogues present also petrographic name and age of stone (e.g. Müller 2001).

There are many hundreds of commercial natural stone names. Firstly the names had to studied to create system of natural stone deposits (quarries). Choice of the most used commercial granites were statistically calculated from frequency in web pages in Google search. Many catalogues, scientific articles and maps were searched out to obtain sites of quarries, geological units, age and petrography of stones. The most frequent commercial names in internet were arranged to tables in alphabetical order (Pivko, 2002, 2003b,

2004), arranged according to geological age (Pivko 2003a) and petrography (Pivko 2004).

In the article the commercial granite quarries are ordered according the continents where the quarries are sited, then according tectonic unit of quarry areas. Description of tectonic unit, list of commercial names and petrographic types are made for each area. More references can be found in Pivko (2003a).

2 WORLD

The most offered and so the most sold commercial granites were analysed from web pages and then added up. Among continents commercial granites from Asia present 38% of world, further South America 22%, Europe 19%, North America 14% and Africa 7%. First place among countries belongs to India 32%, Brazil follows with 22%, then USA 10%, Finland 5%, South Africa 5%, Norway 4.5%, Italy 4%, Canada 4%, China 3.5%, Spain 3%, Saudi Arabia 2%, Zimbabwe 1.5%, Ukraine 1% and Sweden 1%.

Sequence of petrographic types are as follows: granites (including alkali granites) 32%, migmatites 17%, gneisses 13%, gabbros 7%, syenites (including alkali syenites) 7%, dolerites 7%, granulites 6%, charnockites 6% and 5% of anorthosites, granodiorites, quartzites and metaconglomerates.

The most productive times of Earth history are Proterozoic about 60%, Archean 20% and Paleozoic 15%. Magmatites used as natural stones comes from Neoarchean shields (2.7–2.6 Ga) 18%, Hercynian orogeny (Devonian and Carboniferous) 17%, Pan-African

147

orogeny (0.7–0.5 Ga) 16% and Grenvillian orogeny (1.2–1.1 Ga) maybe more than 10%. Further periods are 1.7–1.6 Ga with about 10%, 0.3 Ga with 7% and 2 Ga with 6%. Metamorphites used as commercial granites were formed almost exclusively in Proterozoic (74%) and Archean (24%).

3   ASIA

India has dominant position among Asian granite producers, about 85%. The second position with 9% belongs to China, but its production rapidly increases. Third level is reserved to Saudi Arabia with about 6%. Sri Lanka, Iran and Turkey are smaller exporter of commercial granites. The most of Indian granites comes from south India: Dharwar craton with Closepet granite and its vicinity, south and less central and north parts of East Ghat Belt in Tamil Nadu, Andhra Pradesh and Karnataka states.

3.1   Granites and migmatites from Dharwar craton of south India

Dharwar craton in south India is made up of classical triplet of Archean terrains: Peninsular gneisses, volcano-sedimentary greenstone belts and younger Late Archean (2.5–2.6 Ga) granitic intrusions. The most impressive magmatic body is huge Closepet granite (batholith) of linear shape with approximately N-S-trending (Jayananda et al. 2000). The quarries of New Imperial Red, Ruby Red and Sira Grey granites in Karnataka state are connected with the Closepet granite. The Closepet granite and similar granites caused partial melting of gneisses and so migmatites were formed. There are Multicolor Red, Paradiso Classico, Paradiso Bash, Juparana India, Himalayan Blue, Lillet, Imperial White, Viscon White, Kuppam Green, Golden Fantasy and Green Rose from southern Karnataka, northern Tamil Nadu and southern Andhra Pradesh states.

From northern Andhra Pradesh state there are Sapphire Brown, Sapphire Blue and Tan Brown granites to orthogneisses. Dharwar craton is locally cut by large dykes from which dolerites are quarried with commercial names as Absolute Black, Jet Black, India Black, Premium Black (especially in Andhra Pradesh, less in Karnataka and Tamil Nadu states) and Hassan Green from Karnataka.

3.2   Gneissic rocks and gabbros from Eastern Ghats Belt of south India

In Eastern Ghats Belt, Madurai block and Kerala Khondalite Belt there are original rocks from the late Archean and early Proterozoic, which were several times metamorphosed during the Proterozoic orogenies.

The Grenvillian orogeny was very important in southern India and formed Eastern Ghats Belt. Some rocks were also reworked by Pan-African orogeny (Bartlett et al. 1998, Mezger & Costa 1999, Rickers et al. 2001). From the Eastern Ghats Belt many granulites come as Kashmir White, Kashmir Gold and Ghibli from Tamil Nadu, Tropical Green from Kerala state, and White Galaxy from eastern Andhra Pradesh, gneisses and migmatites as Raw Silk, Madura Gold, Vyara, Tiger Skin, Shivakashi, Ivory Brown, Shiva Gold, Juparana Colombo and Golden Oak from Tamil Nadu, probably Colonial Dream from Sri Lanka, and bluish migmatites with charnockite composition from eastern Andhra Pradesh and Orissa states as Vizag Blue, Lavender Blue, Orissa Blue, Bahama Blue and maybe Seaweed Green.

In Eastern Ghats Belt (Andhra Pradesh, Ongole) there are basic magmatic bodies with gabbro composition – Black Galaxy (gabbro with bronzite), Silver Pearl and Black Pearl.

3.3   Neo-Proterozoic granites from western India, Rajasthan

Intracratonic rift was formed in Marwar craton about 750 Ma ago that was accompanied with mainly felsic volcanism and plutonism – Malani Igneous Province in south-western Rajasthan (Sharma 2003). Jalore, Barmer and Siwana granites of the province are quarried in many quarries and sold as Rosy Pink, Mokalsar Green, Raniwara, Merry Gold, Chima Pink and Copper Silk variegated commercial granites.

3.4   Cretaceous granites from south-eastern China

In Fujian state there are Upper Cretaceous granites to granodiorites of Zhangzhou and Fuzhou igneous complexes (Chen et al. 1995), from which commercial granites with names e.g. Almond Mauve, Misty Mauve, Padang are quarries.

China is source of many commercial granites, but their geological position has been determined not yet. Black dolerites from Shanxi are very popular on market (Shanxi Black, Absolute Black).

3.5   Granites from Arabian Shield, southern Saudi Arabia

Late- to post-tectonic granites (650–530 Ma) of Pan-African orogeny (Nehlig & Salpeteur 1999) are exploited as dimension stones with commercial names: Violetta, Tropic Brown, Silver Sea Green and Golden Leaf.

4   SOUTH AMERICA

Brazil produces almost all South American granites from Ribeira Belt in Espirito Santo, Rio de Janeiro

148

and Bahia states, and Sao Francisco craton in Minas Gerais and Bahia states. The small amount of commercial granites comes from Argentina and Uruguay.

4.1 Metamorphites and magmatites from Sao Francisco craton, eastern Brasil

Sao Francisco craton is one of the Archean-Proterozoic parts of South America consolidated about 2 Ga ago (Engler et al. 2002). Many very used commercial granites come from the craton. There are *Kinawa, Jacaranda* and *Giallo California* migmatites, and *Cafe Imperial* syenite from Minas Gerais state, *Verde Bahia* charnockite, *Azul Bahia* sodalite syenite, *Crema Bahia* alkali granite, *Azul Macaubas* and *Azul Imperial* dumortierite quartzites, and *Verde Marinace* metaconglomerate from Bahia state. Originally Archean migmatites from Campo Belo Metamorphic Complex were epimetamorphosed about 1.9 Ga (Carneiro et al. 2000). These mainly green migmatites are sold with commercial names: *Verde Maritaca, Verde San Francisco, Verde Lavras, Verde Tropical, Verde Candeias, Lilla Gerais* and *Gran Violet.*

4.2 Gneisses, granulites, charnockites and granites from Ribeira Belt, eastern Brasil

Sao Francisco craton is edged by Ribeira belt (Aracuai belt) from SE which was formed during Pan-African orogeny (about 650–500 Ma) (e.g. Pedrosa-Soares et al. 1999). There are typical yellow garnet gneisses from Espirito Santo state: *Giallo Veneziano, Santa Cecilia, New Venezian Gold, Giallo Topazio, Amarelo Real* and *Giallo Ornamental*, migmatites from Rio de Janeiro vicinity: *Juparana Classico, Juparana Fantastico* and Espirito Santo: *Golden Beach*, granulites from Espirito Santo: *Verde Eucalipto, Bianco Romano*, and southern Bahia state: *Samba White* and *Bianco Regina*, granites from Sao Paulo state (*Capao Bonito*), from Rio de Janeiro vicinity (*Carioca Gold*), from Espirito Santo and Minas Gerais states (*Giallo Napoleone, Giallo Antico* and *Giallo Fiorito*), syenites from Espirito Santo (*New Caledonia*) and from Parana state (*Tunas Green*), and charnockites from Espirito Santo (*Verde Ubatuba, Verde Labrador and Verde Butterfly*).

5 EUROPE

Half of commercial granites come from Finland (27%) and Norway (23%). They are succeded by Italy (22%), Spain (19%), Ukraine (5%) and Sweden (5%). Among further countries there are France, Switzerland, Portugal, Germany, Czech Republic, Poland and Bulgaria. Proterozoic rapakivi granites (Finland, Ukraina and Sweden) prevail over Paleozoic Hercynian granites (Italy, Spain) and Paleozoic alkali syenites (Norway).

5.1 Rapakivi granites from southern Finland

Great rapakivi granite (anorthosite) massifs were formed in Baltic shields after the finishing of the Svecofennian orogeny 1.7–1.5 Ga ago (e.g. Sharkov 1999). From Wiborg batholith *Baltic Brown, Baltic Green, Carmen Red, Eagle Red* granites to alkali granites, *Spectrolite* anorthosite and from Vehmaa batholith *Balmoral Red* granite are quarried.

Kuru Grey granite and *Amadeus* garnet-cordierite gneiss from Finland, *Vanga Red* orthogneiss and *Swedish Mahogany* granite from southern Sweden are results of Svecofennian orogeny.

5.2 Rapakivi granites and anorthosites from Ukraine

Great anorthosite-rapakivi granite massifs were intruded earlier than in Finland, 1.8–1.7 Ga ago (e.g. Yashchenko & Shekhotikin 2000). From Ukraine *Rosso Santiago* granite and *Volga Blue* anorthosite are well-known.

5.3 Anorthosite of Rogaland igneous complex from southern Norway

Layered intrusion of Rogaland igneous complex was intruded about 930 Ma ago (Schärer et al. 1996) which is source of *Labrador antique* anorthosite.

5.4 Granites of Hercynian mountains from Spain and Italy

During Hercynian orogeny in Carboniferous when Europe collided with Gondwana great granite massifs were intruded as Gallura granites in todays Sardinia, Italy (Boni et al. 2001) (*Bianco Sardo, Rosa Beta, Grigio Sardo, Rosa Limbara, Rosa ghiandone*), in Iberian massif of Spain (Castro et al. 2000): Galicia (*Rosa Porrino, Mondariz*), Madrid (*Blanco Cristal, Blanco Castilla*) and Extremadura (*Azul Platino*).

Sidobre granite in southern France is also of Hercynian age, where *Tarn* granite is extracted. *Azul Aran* is probably Hercynian pegmatite incorporated to Pyrenees. Hercynian granites are quarried also in Portugal, Germany, Czech Republic, and Poland.

5.5 Alkali syenites of Oslo rift from southern Norway

At the Carboniferous – Permian boundary, continental rift was formed in southern Scandinavia, where Larvik Plutonic Complex (299 Ma) was intruded (Dahlgren et al. 1996), from which well-known larvikites – alkali

149

syenites are quarried with *Blue Pearl* and *Emerald Pearl* commercial names. In southern Sweden extension dykes as a response to Oslo rift were created filled by dolerites – *Swedish Black*.

6 NORTH AMERICA

USA sells about 72% and Canada 28% of North American commercial granites. Granites come from Archean (Ontario, South Dakota), Proterozoic (Quebec, Ontario, Minnesota, Texas) and Paleozoic (Eastern coast of USA).

6.1 Granites from Superior province of Canada and USA

Superior Province is very old core of North America continent. Migmatites from Minnesota commercially named as *Rainbow* are 3.6 Ga old. Superior province contains also Neoarchean granites of 2.7–2.6 Ga old (e.g. Dalrymple 1991). They are quarried in Ontario (*Vermilion Pink, Crystal Gold, Red Deer Brown, Pine Green*) and Manitoba (*Lac du Bonnet*) of Canada and in South Dakota (*Dakota Mahogany, Carnelian*) and Minnesota (*Agate*) of USA.

6.2 Granites of Penokean orogeny from Minnesota, USA

On the margin of Superior province Penokean orogeny was formed 1.9–1.8 Ga ago (e.g. Dalrymple 1991), when large bodies of granites to granodiorites were created. They are exploited in Minnesota as *Rockwille White* or *Charcoal Black*.

6.3 Magmatites of Grenville province from eastern Canada

Grenville province was built during large Grenville orogeny in Proterozoic 1.4–1 Ga ago (e.g. Levin 2003). Nain Plutonic Suite as a part of the province was formed 1.3 Ga ago (Ryan 2001) from which *Blue Eyes* anorthosite is quarried (Newfounland state). Charnockites (*Caledonia, Polychrome, Autumn Brown* and *Atlantic Green*) and gabbro (*Cambrian Black*) of 1.1–1 Ga come from Quebec.

Mountain Green gneiss from New York state (USA) is also of Grenvillian age.

6.4 Granites of Llano uplift from Texas, USA

Part of Grenvillian orogeny was manifested also in Llano uplift with Town Mountain Granite 1.1 Ga ago (Reed 2002). The most used are *Sunset Red* and *Texas Red*.

6.5 Granites and migmatites of the Appalachians from eastern USA

The Acadian orogeny in the Appalachians was connected with creation of granite massifs. There are *Solar White, Salisbury Pink* and *Mount Airy White* granites from North Carolina, *Bethel White* granite with zoisite and *Barre Gray* granodiorite from Vermont, *Deer Isle* granite from Maine and *Georgia Gray* granite from Georgia (USGS 2004).

Terranes of Pan-African age (boundary between Neoproterozoic and Paleozoic) are included to or into the Appalachians. *Silver Cloud* paragneiss is from Georgia, *Milford Pink* granite from Massachusetts. *Stony Creek* migmatite from Connecticut is connected with Permian granites intruded to Avalonia terrane of Pan-African age (Sullivan & McHone 2002).

The Appalachians were locally cut by dolerite dykes as a evidence of the Atlantic opening at the boundary of the Triassic and the Jurassic *Jet Mist* from Virginia and *American Black* from Pennsylvania (USGS 2004).

6.6 Magmatites of Sierra Nevada batholith from California, USA

Sierra Nevada batholith is huge magmatic body intruded to Northern Sierra Terrane in Jurassic and Cretaceous (e.g. Irwin & Wooden 2001). *Academy Black* gabbro and *Sierra White* granite are quarried from the batholith.

Younger granites of Terciery are quarried in British Columbia, western Canada.

7 AFRICA

The most of commercial granites comes from South Africa (73%), less from Zimbabwe (22%). The rest is from Angola, Zambia, Egypt, Namibia and Nigeria.

7.1 Gabbros and granites of Bushveld complex from South Africa

Bushveld complex – a huge magmatic body was created by processes in thinned continental crust about 2 Ga ago (e.g. Knoper et al. 1999). In layered magmatic intrusion from NE South Africa, heavier mafic rocks are in the bottom (*Impala Black* gabbro-norite) and felsic rocks are on the top (*African Red* granite). Intrusion was accompanied by dolerite dykes as a *Belfast Black* dolerite.

7.2 Charnockite and migmatite of Namaqua – Natal belt from western South Africa

Namaqua – Natal metamorphic belt is intruded by charnockites of Grenvillian age, 1.2–1 Ga ago (Robb

150

et al. 1999) from where *Verde Fontaine* charnockite comes from. *Olive green* migmatite is from the same area.

7.3 Dolerites of Mashonaland igneous event from Zimbabwe

Zimbabwe craton was cut by basic dykes and sills 1.9–1.8 Ga ago from which *Zimbabwe Black* dolerite is quarried (Evans et al. 2001).

8 CONCLUSION

Commercial granites as a hard natural (dimension) stones are today very used in public buildings, but they even penetrate to private houses. They are petrographically either magmatic (homogenous appearance) or metamorphic rocks (wavy or banded appearance).

Commercial granite names in hundreds web pages, many catalogues, scientific articles and maps were analysed to obtain sites of quarries, geological units, age and their petrography. The most used commercial granites were chosen from hundreds commercial names on basis of frequency in Google search.

Among continents commercial granites from Asia present two fifths of world with India domination. Second place belongs to South America with Brazil prevailing, followed by Europe, North America and Africa. Stones from Australia are disregardable.

One third of commercial granites come from India, one fifth from Brasil, followed by USA, Finland, South Africa, Norway, Italy, Canada, China, Spain, Saudi Arabia, Zimbabwe, Ukraine and Sweden.

Granite is the most used petrographic type (about one third of commercial granites). Further sequence of petrographic types are as follows: migmatites, gneisses, gabbros, syenites, dolerites, granulites, and charnockites. Magmatic to metamorphic rocks ratio is 2:1.

The most productive times of Earth history are Proterozoic about 60%, Archean 20% and Paleozoic 15%. Magmatites used as natural stones come mainly from Neoarchean shields (2.7–2.6 Ga), Hercynian orogeny (0.4–0.3 Ga), Pan-African orogeny (0.7–0.5 Ga) and Grenvillian orogeny (1.2–1.1 Ga), less from 1.7–1.6 Ga, 0.3 Ga and 2 Ga ago. Metamorphites used as commercial granites were formed almost exclusively in Proterozoic and less in Archean.

ACKNOWLEDGEMENTS

Many thanks KEGA 3/0023/02 Grant for support, Prof. M. Mišík from Slovakia, V. Somani from India, M. Bertoli from USA, my wife and God most of all.

REFERENCES

Bartlett, J., Dougherty-Page, J.S., Harris, N.B.W., Hawkesworth, C.J. & Santosh, M. 1998. The application of single zircon evaporation and model Nd ages to the interpretation of polymetamorphic terrains: an example from the Proterozoic mobile belt of south India. *Contributions to Mineralogy and Petrology* 131: 181–195.

Boni, M., Iannace, A., Villa, I.M., Fedele, G. & Bodnar, R. 2001. Multiple fluid-flow events and mineralization in SW Sardinia: from Variscan onwards. In R. Cidu (ed.), *Water-Rock interaction*. Lisse: Balkema.

Carneiro, M.A., Teixeira, W., Oliveira, A.H. & Fernandes, R.A. 2000. Recent Advances Concerning the Tectonic Evolution of the Southern Portion of the Saõ Francisco Craton, Brazil, 31st International Geological Congress, Rio de Janeiro, Abstracts Volume.

Castro, A., Corretgé, L.G., El Biad, M., El Hmidi, H., Fernández, C. & Patiño Douce, A.E. 2000. Experimental Constraints on Hercynian anatexis in the Iberian Massif, Spain. *Jour. Petrol.* 41: 1471–1488.

Chen, C.H., Tien, J.L., Lo, C.H. & Kelong, W. 1995. Fission Track and ~L40~KAr/~L39KAr Dating of the Late Cretaceous Zhangzhou Igneous Complex, SE China. Poster. International Union of Geodesy and Geophysics, XXI General Assembly, International Association of Volcanology and Chemistry of the Earth's Interior, earth.agu.org/iugg/vb_ prog.html, 1995.

Dahlgren, S., Corfu, F. & Heaman, L.M. 1996. U-Pb isotopic time constraints, and Hf and Pb source characteristics of the Larvik plutonic complex, Oslo paleorift. Geodynamic and geochemical implications for the rift evolution. *Journal of Conference Abstracts* 1: 120.

Dalrymple, G.B. 1991. *The Age of the Earth*. Stanford University Press.

Engler, A., Koller, F., Meisel, T. & Quemeneurd, J. 2002. Evolution of the Archean/Proterozoic crust in the southern Saõ Francisco craton near Perdões, Minas Gerais, Brazil: Petrological and geochemical constraints. *Jour. South Amer. Earth Sc.* 15: 709-723, www.unileoben.ac.at/~chemie/Engler.pdf.

Evans, D.A.D., Gutzmer, J., Beukes, N.J. & Kirschvink, J.L. 2001. Paleomagnetic constraints on ages of mineralization in the Kalahari Manganese Field, South Africa. *Econ. Geol.* 96: 621–631, www.gps.caltech.edu/users/jkirschvink/pdfs/EconGeol.pdf.

Irwin, W.P. & Wooden, J.L. (eds.) 2001. Plutons and accreted Terranes of the Sierra Nevada, California, with a Tabulation of U/Pb isotopic Ages. USGS, geopubs.wr.usgs. gov/open-file/of01-229/of01-229.pdf.

Jayananda, M., Moyen, J.F., Martin, H., Peucat, J.J., Auvray, B. & Mahabaleswar, B. 2000. Late Archaean (2550–2520 Ma) juvenile magmatism in the Eastern Dharwar craton, South India: constraints from geochronology, Nd-Sr isotopes and whole rock geochemistry. *Precamb. Res.* 99: 225–254.

Knoper, M., Jordan, T.H. & Ashwal, L.D. 1999. Thinspot origin for the Bushveld Complex? Rockscapes, www. geocities.com/Yosemite/Trails/1453/faf4bic2.htm.

Levin, H.L. 2003. *The Earth Through time*. 7th ed., New York: John Wiley & Sons.

Mezger, K. & Costa, M.A. 1999. The thermal history of the Eastern Ghats (India) as revealed by U-Pb and $^{40}Ar – ^{39}Ar$

dating of metamorphic and magmatic minerals. Implications for the SWEAT correlation. *Precambrian Research* 94: 251–271.

Müller, F. 2001. *INSK kompakt. Die internationale Natursteinkartei für der aktuallen Markt.* Ebner Verlag.

Nehlig, P. & Salpeteur, I. 1999. The Mineral Potential of the Arabian Shield: A reassessment. IUGS/UNESCO Meeting on the "Base and Precious Metal Deposits in the Arabian Shield", Jeddah, gisarabia.brgm.fr/Publications/RESUMNEW.doc

Pedrosa-Soares, A.C., Wiedemann, C.M., Fernandes, M.L.S., Figueiredo de Faria, L. & Ferreira, J.C.H. 1999. Geotectonic Significance of the Neoproterozoic Granitic Magmatism in the Aracuai Belt, Eastern Brazil: A Model and pertinent Questions. *Revista Brasileira de Geociências* 29: 59–66.

Pivko, D. 2002. The World's most popular Granites. FindStone.com – Marketplace for Building stones, India, www.findstone.com/daniel1.htm.

Pivko, D. 2003a. Natural Stones in Earth's History. *Acta Geol. Univ. Comen.* 58: 73–86, www.fns.uniba.sk/prifuk/casopisy/geol/200358/pivko.rtf.

Pivko, D. 2003b. The World's Most Popular Granites. Workshop – India Stone Mart 2003. Centre for Development of Stones, Rajasthan: 8–28.

Pivko, D. 2004a (in press). The Most Popular Granites and Marbles. The National Training Center for Stone & Masonry Trades, North Carolina.

Pivko, D. 2004b (in press). The World's Most Used Natural Stones – Overview of Geology. *Acta Geol. Univ. Comen.*, 59.

Reed, R.M. 2002. Town Mountain Granites. University of Texas, uts.cc.utexas.edu/~rmr/tmg.html.

Rickers, K., Mezger, K. & Raith, M.M. 2001. Evolution of the Continental Crust in the Proterozoic Eastern Ghats Belt, India and new constraints for Rodinia reconstruction: implications from Sm-Nd, Rb-Sr and Pb-Pb isotopes. *Precambrian Research* 112(3–4): 183–210.

Robb, L.J., Armstrong, R.A. & Waters D.J. 1999. The History of Granulite-Facies Metamorphism and Crustal Growth from Single Zircon U-Pb Geochronology: Namaqualand, South Africa. *Jour. Petrol.* 40(12): 1747–1770, library.iem.ac.ru/j-petr/12–4099.

Ryan, B. 2001. A Provisional Subdivision of the Nain Plutonic Suite in its type-area, Nain, Labrador (NTS Map area 14C/12). Current Research, Newfoundland, Report 2001 – 1: 127–157, www.geosurv.gov.nf.ca/curres/CR2001/Ryan.PDF.

Schärer, U., Wilmart, E. & Duchesne, J.-C. 1996. The short duration and anorogenic character of anorthosite magmatism: U-Pb dating of the Rogaland complex, Norway. *Earth Planet. Sci. Lett.* 139: 335–350.

Sharkov, E.V. 1999. Vnutriplitnye magmaticheskie sistemy serediny proterozoya na primere anortozit-rapakivigranitnych komplexov Baltiyskogo i Ukrainskogo shchitov. *Rossiyskiy zhurnal nauk o Zemle* 4: eos.wdcb.rssi.ru/rjes/rje98013/rje98013.htm.

Sharma, K.K. 2003. Overview on Malani Magmatism of India.www.mantleplumes.org.

USGS, 2004. National Geologic Map Database. US Geo- logical Survey, GEOLEX – Search for lithologic and geochronologic units, ngmsvr.wr.usgs.gov/Geolex/geolex.html, 2002.

Sullivan, K. & McHone, N.W. 2002. Connecticut Geology Illustrated Stories Carved in Stone Land Links to the Sound The Coastal Drainage Basins. A Workshop with Exercises. Geological and Natural History Survey of Connecticut. www.wesleyan.edu/ctgeology/LISproject/connecticut_geology_illustrated.htm.

Yashchenko, N.Y. & Shekhotikin, V.V. 2000. New data on the tectonomagmatic history of the Ukrainian Shield (Ingul-Ingulets region). *Litasfera* 12: 76–84, www.ac.by/publications/litho/litho12.html.

Dimension Stone 2004, Přikryl (ed)
© *2004 Taylor & Francis Group, London, ISBN 90 5809 675 0*

Internet database of decorative and building stone

B. Schulmannová & H. Skarková
Czech Geological Survey, Klárov, Praha, Czech Republic

ABSTRACT: The Czech Republic has a rich tradition in the extraction and processing of natural stone. The increasing demand for good-quality material that is suitable for decorative and building purposes makes a considerable contribution to the development of stone-working branches of industry. A project was established at the Czech Geological Survey last year, targeted towards the creation of an accessible internet database of decorative and building stone in the Czech Republic. The database contains basic types of rocks extracted in the Czech Republic for these purposes in the past and present, together with their properties, sources and uses. Future connection of the Czech database with similar systems in Europe will lead to the creation of a complex international system that can be employed in a number of branches of science and industry.

1 INTRODUCTION

In recent years, there has been an increase in interest on the part of the lay and professional public in architectural memorials, buildings and other historical objects from the viewpoint of the use of stone. This is also related to the increasing demand for materials employed for building and decorative purposes and thus development of the stoneworking branches of industry.

The extraction, working and use of decorative and building stone has a long tradition in the Czech Republic and thus a number of geologists and geological workplaces have been concerned with this subject. On the basis of information on similar activities at geological workplaces in Europe, a project was created at Czech Geological Survey in 2003, entitled "Internet database of decorative and building stone" in an attempt to present information collected to date on geological sources suitable for these purposes on the web site of the Czech Geological Survey, www.geology.cz.

2 METHODS EMPLOYED AND TECHNICAL BACKGROUND

In the first stage of creation of the database, it was necessary to base out work on the results of work to date, which were gradually supplemented with updated information and photographic documents. A working version of the database was created at the Informatics Department of the Czech Geological Survey in Prague, in the Oracle system, which is designed so that it could potentially be connected in the future with similar database systems in Europe.

3 BASIC INFORMATION

3.1 *Scheme of the database*

The data part of the database of decorative stone is located in the data stores of Czech Geological Survey. The decor-kamen table, into which are inserted petrographic data on the rocks (decorative stone) is connected through a primary key to tables on mineralogical composition, physical-technical parameters, quarries and use in buildings. The user must be registered in the Oracle Portal system during entering data and editing. Thus the systemic name of the user is recorded and is employed for internal control of the progress of work on the database. At the present time, only the employees of Czech Geological Survey have access to the registration forms; it is expected in the future that the records part will be made available to authorized professionals.

3.2 *Structure of the database*

Basic information on rocks in the territory of the Czech Republic, employed in the past and present as decorative stone, are included in the database. These data include petrographic, mineralogical and physical-mechanical properties of rocks, assigned to regional geological units, together with their structure, grain-size and colour. The "trade name" employed by Czech stoneworkers is used for the rocks. The database also includes a list of stone objects (e.g. gravestones, pedestals, window ledges, etc.) and specific examples of sites where these objects have been used. An important part of the newly created database consists in a survey and the locations of quarries with all the available information on them, related to the current condition of the quarry, the purpose of extraction and information on its properties. A brief history and contemporary photographs are included.

The newly created database will also include a list of the properties of the rocks that can be considered to be "stone diseases", their geochemical characteristics and also photographic documents of samples, thin sections and selected localities.

In relation to the potential for choice, the database is prepared in a version intended for professionals from amongst stoneworkers, architects and specialists in the sphere of the stoneworking industry (selections are concerned with the technical parameters of decorative stone, the location of extraction or the site and manner of use) and also in a version intended for geologists (in this case, selections are more concerned with the petrographic properties of the rocks).

4 CONTENT OF THE DATABASE

The following groups of rocks, as differentiated by Rybařík (1994), are currently part of the database (Fig. 1). These are light and dark intrusive rocks, trachytes, sandstones, marlstones (the Czech term is "opuka"), marbles, travertines, serpentinites and shales. From a petrographic standpoint, each of these groups encompasses one kind of rock or related kinds, which are used in a similar manner in practice.

4.1 *Intrusive rocks*

The first of these groups includes granites and related light and dark (called "syenites" by Czech stoneworkers) eruptive rocks. Granites form the major part of the geological basement of the Bohemian Massif, from extensive complexes (the Central Bohemian pluton, Moldanubian pluton) to smaller bodies. Because of their strength and resistance, they have been favoured as building material and are used only rarely for decorative purposes because of the difficulty entailed in working this material. It was not until the 2nd half of the 19th century that new granite quarries were established and granite began to be used for more demanding stonework. This was manifested mainly in larger cities; this material is employed in Prague for facing and paving material for number of important buildings, for pedestals of monuments, and

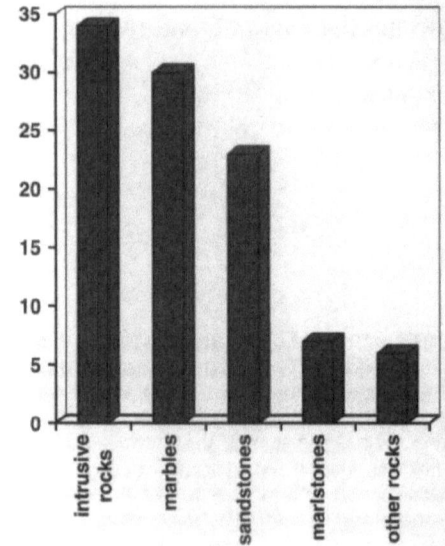

Figure 1. Proportional diagram of decorative stone included in the database.

for gravestones and was very popular as facing and paving material in the construction and reconstruction of the stations on all three lines of the Prague Metro (Březinová et al. 1996).

4.2 *Sandstones, marlstones, marbles*

From a regional geological point of view, sandstone and related rocks (arcoses, arcose sandstones, etc.) occur more frequently in the territory of Czech Republic. Because of the ease of working them, Cretaceous and Permocarboniferous sandstones were used in architecture in stone sculptures mainly from Gothic times to 19th century.

Marlstone and marble were used to a greater degree for decorative purposes in cities and rural areas in this country. Marlstones constitute the oldest building stone in Czech Republic and can be found in all preserved Romanesque structures. Marbles became an important decorative stone, mainly in the architecture of large cities, as early as in Gothic times; during the Baroque and Rococo, they were popular because of their variegated colour and technical properties.

4.3 *Other rocks*

Rocks in other groups named above were employed either only locally or to a limited degree; nonetheless, they occupy an important place amongst kinds of decorative stone. (Hanzl et al. 2003).

154

Figure 2. The Barrandian diabas fountain created by J. Plecnik (Prague castle, Prague).

In addition, this classification also encompasses rocks of different character, which were used for stoneworking purposes, but cannot be classified amongst the above groups. These consist in the Barrandian diabas (Fig. 2), Teplice ryolite, phonolite ("basalt") from Kunětická hora near Pardubice, basalt from Panská skála near Kamenický Šenov, crinoid limestone from Stránská skála at the eastern edge of Brno, greisen from Prameny (Sangersberg) near Marienbad and "krupník" from the Šumperk area (Rybařík 1994).

5 CONCLUSIONS

The varied geological structure of Czech Republic provides a wide range of decorative stone, which has become part of the cultural heritage of this country in the form of historical and modern structures and architectural objects. Although imported materials were increasingly used in the 20th century, it should be recalled that there are still a number of localities in the Czech Republic from which rocks could be used for decorative purposes because of their interesting appearance and suitable properties. The newly created database of decorative and building stone will thus become a complex information system that is readily available on the Internet and that can be employed by stoneworkers, architects, construction contractors and restoration workers, as well as professionals in the area of the geology.

REFERENCES

Březinová, D., Bukovanská, M., Dudková, I. & Rybařík, V. 1996. *Praha kamenná. Přírodní kameny v pražských stavbách a uměleckých dílech.* Praha: Národní museum.
Hanzl, Z., Gába, Z., Procházka, L., Sedlická, K., Slouka, J. & Traxler, J. 2003. *Kámen v rukodělné výrobě českého venkova.* Praha: Nakladatelství Lidové noviny.
Rybařík, V. 1994. *Ušlechtilé stavební a sochařské kameny české republiky.* Hořice v Podkrkonoší: Nadace Střední průmyslové školy kamenické a sochařské.

Dimension Stone 2004, Přikryl (ed)
© *2004 Taylor & Francis Group, London, ISBN 90 5809 675 0*

Introducing an international Register of Natural Stone

J. Stein
Ornamental Stone Investigations, Waldsee, Germany

ABSTRACT: A new method for an international register of natural and dimension stone will be presented. By the development in consumption, exploitation, industrial producing, international trading of dimension stone since of last 60 years and the development of technical progress in building construction it is necessary for an neutral and ageless system for communication. This system gives a guarantee at communication for all parts in the different levels of dimension stone industry.

1 REQUIREMENTS OF AN INTERNATIONAL REGISTER OF NATURAL AND DIMENSION STONE

1.1 *Definition of dimension stone*

The term dimension stone designates all products made of natural stone (i.e. general term for naturally existing rock in differentiation from artificial rock) that are put into a defined form and dimension in an industrial, manufacturing or artificial work process. This prohibits significant alterations in/of:

- substance or texture of matter (minerals);
- size, shape, configuration of minerals (structure, texture, colour);
- formation and cohesion of minerals (physical and technical features).

Basis of the manufacture of dimension stone products are raw blocks extracted from a congregation of rocks of a natural stone deposit. Further processing effects a broken and dimensioned end product or a panelled semi-finished good (raw panel, tranche). In the further manufacture, an end product will be machined or crafted consciously dimensioned, shaped and finished (surface treatment).

1.2 *Background*

Since of the 1950s, mining, processing and using natural and dimension stone has been characterised by 3 decisive processes. These effected a radical change of the stone sector in a historically short time:

(1) The traditional craftsmanship stonemason changed into an industrial process spanning mining up to manufacture. Constructional workmanship still remains in terms of craftsmanship, yet integrated into an industrial process. As a result, prices for the end product declined and the product DIMENSION STONE became a commodity. Since the end of the 20th century, industrialisation has reached the formerly craftsmanship sector of ornamental architecture.

(2) General globalisation has led to an intensive interdependence of primary producers (mining for raw blocks and intermediary products), places of manufacture (sawmills, manufactures), and consumers. Almost all the countries in the world are supplying raw blocks on the international stone market; raw blocks and intermediary products are turned into convenience products bearing no reference to their place of origin and place of utilisation. Consumers and architects assess their requests exclusively to pattern and price regardless of national traditions.

(3) A trade independent from the actual product and manufacture is emerging. Having only one general performance feature – natural stone is hard and durable – the orientation for the actual utilisation is made on the price.

This trend has already affected extensive anonymity of the mining origin, the manufacture, and the constructional processing of the building material dimension stone. On one hand, this anonymity is conditioned objectively by industrial process. On the other, it is partly done for subjective purposes of trade policy. As the anonymity of a given dimension stone shows 2 aspects, the traditional craftsmanship experiences of a limited number of types of natural stone from pre-industrial times are no longer valid:

(1) The place of origin of a given rock can no longer be distinguished making evidence of genuineness

157

of the supplied product impossible. At the same time, technical (i.e., geo-scientific and structural engineering) enquiries are prevented.

(2) The features – technical and physical (petrographical) figures, mineralogical and chemical composition, constructional performance – of a rock/ trade quality are unknown.

This situation causes insecurities on all levels of the industrial sector in question that can lead to faults in buildings. Most apparent expression of this anonymity and a thread to all participants is the trade name of a natural stone. The effects of this situation may be summed up in levels of action:

- Legal level. The origin of a rock cannot be retraced with its intermediary or end product by name. Generally acknowledged technical quality features are not transferable (in general). The deduction of a name-bound authorisation is doubtful.
- Planning level. An unambiguous fixation of a requested product is impeded on all levels of planning and execution.
- Level of production line, processing and trade. A communication of all partners does not allow for an unambiguous fixation on the requested trade quality. A self-evident correlation between a trade name and the characteristics of a rock (as in frequent use in the past) is prevented.

2 IDENTIFICATION AND REGISTRATION EFFORTS

The increasing demand for and the rise of industrial work techniques in the mining industry effected a descriptive classification and categorisation of dimension stone in Italy, Switzerland and Germany. Up to the beginning of the 20th century, this division was probably sufficient for the trade, manufacturing and processing industries connected to the chisellers guild.

From the 1950s on, rapid internationalisation of the natural stone industry corresponded to an increasing demand for the advanced description of rocks for all experts on all levels of activity. Therefore, surveys on acknowledged and nationally mined types of rock were published by national institutions on different levels (administrations of government, trade associations, private organisations.)

In every directory, identification took place using a trade name. Very few surveys contained information on the geographical conditions of the mining company or the geological features of the deposit.

Mostly, a description often complemented with a depiction of the rock was given. Currently there is a tendency to also include geological parameters and directions on the constructional performance. On this basis, Italian promotion companies listed internationally traded natural stone, partly with random numbering unfortunately giving no reference to place of origin or rock genesis.

The first international and systematised register of natural and dimension stone was introduced by Müller (1973) in co-operation with the INSK – Internationale Naturstein Kartei (International Natural Stone Card Index) This systematisation according to genetic-petrographical view points and colour supplies every rock with 3 numbers:

- genetic-petrographical basic feature;
- colour characteristics;
- consecutive numbering.

Search key and basis for registration remains a trade name. This file records and systemises rocks as natural stone independent of their current mining status. It contains rough notes on the mining area, a description of the rock, seldom chemical, mineralogical, and petrographical specifications. A number of authors are presently imitating this way of registering dimension and natural stone.

As a result of the growing pressure on the international market caused by a continuing introduction of new types of stone, 1996 saw a singularly attempt to systematising the world of rocks according to a biological principle. Yet, this systematisation was not designed for resolving the problem of identifying a trade quality, and was not meant to remain within the realm of petrology. This was taken over as the Linnean classification of the world of rocks in the EN 12670 (2002) "Terminologie Natursteine".

In terms of trade and in a technical sense, this classification may not be very handy as it assumes certain petrological expertise. In addition to that, it does not contain any reference to the place of origin.

Since the end of the 1990s, the first attempt for a trade quality specific classification has been running by state authorities in the People's Republic of China. This classification does not rely on a trade name and consists of one Latin letter with 4 Arabic numbers. It is illustrated in the Chinese norm GB/T 17670–1999, and comprises the following features:

- rough genetic classification: all rocks are subdivided into marble (M), granite (G), and slate (S);
- the first 2 numbers represent a specific Chinese province;
- the last 2 numbers are designated to the respective quarry in a consecutive way.

This classification facilitates a one-to-one identification of any Chinese natural stone independent from its trade name. The only disadvantage here is the rather rough genetic classification containing too little information even for non-geo-scientists, not to mention geological and petrographical experts. The 2-digit

158

number of the numeric numbering will soon come to its limits. This way, an internationalisation of the classification cannot be realised.

As a European classification, the EU published the DIN EN 12440 (2001) "Naturstein-Kriterium für die Bezeichnung". This EN lists all trade qualities reported by national associations of the natural stone industry or by owners of quarries up to a certain deadline. The registration contains:

- trade name (reported name);
- petrographical family according to EN 12670;
- typical colour;
- place of origin.

Under 3.1., the term "name of the natural stone (trade name)" is defined as: "...corresponds to a specific rock from a specific place of origin. Geographical names that do not refer to the actual place of origin of a rock/stone shall be avoided."

Naming the place of origin is subject to the reporting authority. In some cases, simply provinces or regions are denoted. Sample surveys revealed that reported trade names do not correspond to the actually used names throughout the country, esp. in terms of the mining company. The assignment to a certain petrographical family (petrographical assignment) is not necessarily correct. Thus, the trade name remains a decisive criterion for registration. The current and historical dilemma of a one-to-one identification of a type of natural stone is caused by clinging to a proper name of all systems of registration.

This naming does not imply:

- neither country nor place of origin of a natural stone;
- the type of rock (apart from the historically coined rough classification structures in Italy and France);
- the actual identity of the rock as a trade quality;
- the opportunity for a targeted enquiry.

3 REQUIREMENTS IF AN INTERNATIONAL REGISTER

An international classification of natural stone should only work as a world-wide means of registration and one-to-one designation of types of natural and dimension stone. It should not aim at becoming a new petrographical or genetic classification but fall back on already established regulations in a simplified way. A type of rock should be registered in terms of type/trade quality. The allocation will be decisive for recognising a type of rock or natural stone under all possible circumstances (historical and geographical). This allocation has to be independent from subjective influences of co-operating partners in the natural stone industries, esp. in the field of trade.

In an actual geographical and mining situation, **one** type of rock corresponds to **exactly one** trade quality. A single geographical location may produce a number of trade qualities in different mining situations (e.g., mining of layers of the same type of rock with different surface patterns or layers with different types of rock).

No matter how reasonable, any appellation of a trade quality referring to a type of rock in a given geographical and mining situation can be subject to alterations caused by ignorance or alleged competitive advantages – at any point of its industrial processing.

Administrative documentations are assumed to have a long-term continuance. The location of a mine has always been an administrative matter documented in various ways. In modern times, these administrative actions are responsibility of the national mining authorities. These authorities should conduct registrations appropriately, and should serve as a contact person in case of enquiries. The specification should immediately inform about the country any deposits are in. The actual specification of the place of origin of a trade quality should remain subject of the mining authorities and their data file system. The latter should be open to enquiries in a yet to be defined mode of response. Using co-ordinates in the identification number would be too voluminous. The respective countries should be in charge of providing consecutive licence numbers should be provided by the respective countries in the easiest way possible.

The provision of an identification/licence number only would then be connected to an administrative action. In addition to that, an additional number denoting minor differentiation within a trade quality is needed. These differentiations are mostly considered in mining or later in production and trade.

A brief technical and genetic classification of the type of rock giving the expert (not the geo-scientist) a clue for further working would appear sensible. This rock classification should comply with traditional experience and, by showing essential classes of rock, should imply microstructure and technical features.

The identification/licence number can however not contain a comprehensive genetic and petrographical analysis as a number of natural stone that are being offered will be provided with a geo-scientific analysis only in the long run.

Marking the colour without nuances (regarded from 1 m distance under shadowy daylight) should serve as an aesthetic agent. Allowing EDP-supported analysis and internationalisation of data, all classifications shall be made with Arabic numbers.

All necessary petrographical and mineralogical findings and evaluations should be restricted to the file sheet of the national authority. An identification/licence number thus corresponds to the file sheet of a specific trade quality. This trade quality is then characterised by national authorities according to its

159

place of origin/mining instead of its trade name. This implies that any trade quality has its own country identifier and a nationally provided identification/licence number corresponding to the place of origin.

4 MODEL OF AN INTERNATIONAL REGISTER

The register captures 17 digits (Fig. 1). The first 2 digits marked with letters are restricted to the country. The digits 3 to 8 show the identification/licence number as provided by the state/country.

The state number and the digits 6 to 9 suffice for an unambiguous identification. Any entry in the register must contain these numbers. All other digits can be marked with =0=.

4.1 Numbers 1/2

Common tangeability and on-the-spot availability are the advantages of this identification in form of internationally acknowledged state numbers. In case of changes to state territories, future use of the numbers 1 to 9 is uncertain. A new registration number could be published under the condition of keeping the former number in addition.

In principle, every country has a **national mining authority**. Which institution corresponds to this authority within federal structures as in Germany would be subject of further investigation. Parallel information structures at the central national geological services could be used here.

4.2 Numbers 3/4 and 5/6/7/8

The design of these numbers should be the responsibility of the respective authorities. Beside inner-state classification, internationally used consecutive numbering would be sensible. Examinations would have to be made on whether the number of digits will suffice for 9,999 trade qualities, or whether one more digit would have to be inserted. A way of recording the place of origin of a single trade quality in case of different quarries or deposits would also have to be examined. Providing each quarry with its unique registration number, or differentiation via numbers 10/11 could be possible solutions. This would necessitate an agreement of all owners of the deposits/quarries in question. Particularly uniform layers of sedimentary rock (e.g., the Franconian Jurassic limestone "Jurassic marble") or mining in large plutons are typical of the above-mentioned situation.

4.3 Numbers 9/10

These should serve as a means of distinguishing identical trade qualities from different places of origin/quarries/mining areas. This figure facilitates a continuing, quarry-specific registration of developments in the deposit while mining, without executing an administrative act. This registration shall be reported to the administering authority. In case of significant buildings with a specific sorting within trade qualities, this figure also facilitates identification.

4.4 Number 11

The technical main type of rock – hard or soft rock – facilitates an internationally acknowledged traditional identification of any type of rock in question.

4.5 Number 12

This figure allows for a general evaluation of rocks concerning their mineability and fissility.

4.6 Number 13/14

The entire world of rocks should be arranged in accordance with the classification of the simplified register in the EN 12670 (2002). This allows for a genetic-petrographical classification of a given type of rock. This classification then serves as a basis from which to draw genetic, constructional, and aesthetical information. The "Linnean classification" however shall not be considered here. In principle, schematised identification will cause objections by geo-scientists. This means of identification is not meant as a petro-genetic basis for geo-scientific discussion. It shall mainly serve as a basis for recording essential information concerning a group of rocks for the experts of the natural stone industry involved. Adding *charnockites* (plutonites), to the EN 12670 (2002) due to their specific content of mafites is a subject that also needs to be examined. In case of metamorphic rocks, stating specific types of rock (serpentinites, Cipollino marble, Dolomite marble, and calcite marble) as they are being used extensively in the natural stone industry shall be introduced. *Pyroclastites* and *migmatites* are however not considered here. The missing specifica of dykes may also be of disadvantage.

4.7 Number 16

Particularly in query functions, rough identification of colour facilitates a selection. Yet, identification by colour also implies subjective sources of fault. Using the MUNSELL scale of colour would result in a number of figures too large to handle.

4.8 Number 17

This figure is meant to list crucial features of texture: dense/hyaline/micritic, fine grained, medium grained, coarse grained, porphyrotopic, breccia, conglomerate.

160

Digit	Type of identi-fication	Meaning of the digit	Sub-division	Provided by
1 2	Letter	International state number	Common combination of letters	State
3 4	Figure	Internal country identification	Provinces, Federal states, other administrative units 00-99	State
5 6 7 8	Figure	Internal country identification	Numbering responsibility of state: 0001 – 9,999 types of rock	State
9 10	Figure	Internal type differentiation	Differences within a type of rock (variety, mining etc.)	Quarry
11	Figure	Main type of stone	0 = no record 1 = hard rock 2 = soft rock	Quarry
12	Figure	Mining	0 = no record 1 = en block 2 = fissile product	Quarry
13 14	Figure	Type of rock = petrological and genetic classification (interchangeable)	0 0 = no record 1 X = magmatite 2 X = magmatite 3 X = magmatite, mafite > 90% 4 X = magmatite, mafite > 90% 5 X = vulkanite 6 X = metamorphite 7 X = metamorphite 8 X = sediments 9 X = sediments	Quarry
15 16	Figure	Colour identification	Rough colour identification 0 = white 1 = light grey 2 = grey 3 = black 4 = yellow 5 = pink 6 = red 7 = brown 8 = blue 9 = green	Quarry
17	Figure	Texture	Grain size	Quarry

Figure 1. International register of natural stone.

5 IMPLEMENTING THE REGISTER

Implementing such an internationally acknowledged and identical register may not be effected via national, multinational and international boards. For every state, this register means additional administrative effort. In parallel, a mutual agreement without state-specific special requests is inevitable.

Whether the national associations of the natural stone industry will overcome national particularities (influenced by the trade guilds) on an international level may be doubted.

It also remains unclear, whether they can achieve competitive edge with tradition-based type specific identification. Actual trade organisations are not likely to accept such a register being a label of origin and trade quality, as this would imply losing competitive advantages.

The implementation of a generally accepted international register may only be accomplished by factual means. In principle, EU standardisation provides proper preconditions for that. This standardisation, translated into national laws, commands the national record of trade qualities and their further classification. A starting point would be reached as soon as singular states could agree upon a true registration on a neutral, numeric basis.

In a number of states, current jurisdiction on planning laws and consumer protection command neutral reference classification to producer and to product. In the case of dimension stone, the architect could serve as the consumer's mediator in form of the claiming party.

Introducing an internationally valid numeric register would not be difficult with regard to the People's Republic of China. With the national norm, contextual and administrative preconditions have been accomplished here. Applying the international register would be sensible in case of states with minor or developing deposits/mining of natural stone. The publication of such a means of designation would be a competitive advantage on the international market as a specific national production would be recognisable independent from a trade name. Moreover, a record of resources that are examined or noted but not currently in mining could be made for the national or the international market.

6 COMMENTS ON THE APPLICATION

The proposed register is not meant as a tool for employees of a company in every day work. As a matter of fact, established in-house numbering and denotations remain in use. No master would want his apprentice to fetch him GER 123456!

Yet, a type specific identification/licence number should become commonplace from quarry, to processing, to all levels of trade and up to the manufacturer as an additional trade quality identification. Especially engineers and architects are thereby provided with a safe tool for qualifying capabilities lists and requirements.

This identification/licence number also puts primary producers (quarries) before the task to executing the descriptions of certain trade qualities more carefully. The case of the Portuguese sandy to red marble with differentiated grain may serve as a reminder here.

The goodwill of national mining authorities would initiate immediate recording of inner-state identification/licence numbers. Adding further tags from figure 9 onwards can always be done at any future time.

REFERENCES

Müller, F. 1973. INSK – Internationale Naturstein Kartei (International Natural Stone Card Index), Ulm: Ebner Verlag.
EN 12670 2002. Terminology of natural stone, Berlin: Beuth Verlag.
EN 12440 2001. Naturstein-Kriterium für die Bezeichnung Berlin: Beuth Verlag.
GB/T 17670 1999.

Dimension Stone 2004, Přikryl (ed)
© 2004 Taylor & Francis Group, London, ISBN 90 5809 675 0

SRI – A comprehensive web-database for Roman stone monuments

C.F. Uhlir
University of Salzburg, Department Geography, Geology, Mineralogy, Division Regional and Applied Geology, Salzburg, Austria

A. Sartori
FH Salzburg, Department of Information & Network Technologies, Salzburg, Austria

H.W. Müller
University of Natural Resources and Applied Life Sciences, Department of Structural Engineering and Natural Hazards, Institute for Geotechnical Engineering, Vienna, Austria

C. Hemmers & S. Traxler
Universty of Salzburg, Department Classics, Division Classical Archaeology, Salzburg, Austria

ABSTRACT: The web-database SRI is a working database of the interdisciplinary research project Stone-Relief-Inscription and related projects in Austria and its surrounding. It contains general data of selected Roman stone monuments, detailed petrographic information (data, diagrams and photos), detailed relief information (photos and descriptions), detailed inscription information (texts and their translations), and literature. The quarry database contains general information on the quarries and the detailed petrographic information (data, diagrams and photos). A classification of the monuments based on their cultural context and a classification of the used materials (combining petrographic names with local names used by the archeologists) have been developed. The database will be used for the interpretation of social, cultural and economic history of Roman stone monuments of Noricum (Austria) by combining methods of the humanities and natural science.

1   INTRODUCTION

The development of the SRI Database is a result of the interdisciplinary research project Stone-Relief-Inscription (SRI) from both the Division Classical Archeology and the Division Regional and Applied Geology from the University of Salzburg. It aims at the interpretation possibilities of social, cultural and economic history for Roman stone monuments of Northern Noricum by combining methods of both the humanities and natural science (Hemmers et al. in press).

Observation and scientific analyses of reliefs and inscriptions accompanying them, together with the determination of the origin of the rock material, should make this possible.

The result of geochemical and petrographical investigations of Roman stone monuments will be a distribution map of used rock materials. In addition to that, the Roman use of a number of formerly unnoticed quarries in the local surroundings will be proved. Besides, conclusions on transport routes can be drawn by comparing the locations of origin of the stone material with the sites of the stone monuments.

The interpretation of the scientific results will be conducted in combination with stylistic investigations of the reliefs and the assessment of the content of the inscriptions.

It will become apparent, to what extent valuable materials were used for high-quality monuments as regards inscriptions and pictoral messages during Roman Times, and to what extent of regularity the first assessment corresponds to the social status of the donor.

2   THE DATABASE

2.1   Review of existing databases

Several web-based databases on Roman stone monuments focused on inscription and epigraphical interpretation exist. The "Epigraphische Datenbank

Heidelberg" (EDH) aims at integrating Latin inscriptions from all parts of the Roman Empire into an extensive database (Dafferner 2000). At present, the database contains approximatively 35,000 inscriptions. Additionally, photographs of the objects of the Epigraphische Fotothek Heidelberg (EBH) will be supplied with information on the provenance and current location of the inscription and with bibliography. The number of photographs accessible in the web will continuously increase.

Several other web-accessible databases are focused on the Roman monument inscriptions published in Corpus Inscriptionum Latinarum (C.I.L.), Dessau's "Inscriptiones Latinae Selectae", the Année Épigraphique volumes, etc. These are: the Epigraphic Database Eichstätt of the Catholic University Eichstätt, Germany, the Epigraphic Database Heidelberg Clauss/Slaby of the Heidelberg Academy of Sciences, the Epigraphic Database Rome (EDR) and others.

The recently developed database VBI ERAT LVPA of the Viennese Archeological Research society is based on the scientific value of both, the inscription and the relief (Harl 1999). It provides information on currently 5500 Roman stone monuments of Austria and surrounding countries. It is additionally a web-platform for discussions, publications, virtual exhibitions and has a high popular scientific value for educational and touristic matters.

2.2 The concept of the SRI database

Research on Roman stone monuments (Herz 1988, Moens et al. 1992, Müller et al. 1996, Müller & Schwaighofer 1999 and Müller 2002) has shown that the provenance determination of the monument material provides valuable information on technology, trade and transport routes during Roman times. Therefore, a more comprehensive database for selected monuments of Noricum and its surroundings has to be developed, which includes material composition, provenance, location of quarries and a detailed documentation of the relief combined with the traditional epigraphical interpretation of the inscription.

The web-accessibility of the database provides the collaboration of several working groups of the region. During a workshop of the regional scientific community in Graz/Austria 2004 a close cooperation was established, confirming the following classifications of the monuments. The location of the finding place or place of erection and the CSIR reference no. will be used to link to existing databases.

2.3 Classification of the monuments

Before the establishment of the structure of the database several classifications of the monuments and its fragments needed to be adopted for the needs of the research community. Among others, the most important categories of classification are the following:

The material classification of the monuments is based on simplified petrographical investigations, the textural features of the rocks as result of their genesis: igneous, sedimentary, metamorphic, and artificial rocks; a sub-category is the content of the essential minerals (Wright 1981). The petrographical names need to be related to local names of the material as archeologists are more familiar with them.

The monument classification is based on its cultural context (funeral, cult, profane, military, not identified), which is again subdivided into its architectural position and function (altar, portrait, sculpture, inscription, relief, stele, etc.) (see Table 1).

2.4 Structure of the database

The database containing text, consecutive numbers, photos and diagrams is divided into two main sections:

- The monument section provides general information, classification and analysis of the material, information on the relief and inscription as well as their interpretation as well as additional literature.

Table 1. Monument classification based on cultural context and subdivided by its architectural position and function.

Classification of the Roman stone monuments

Funeral context	Cult context	Profane context	Military context	Context not identified
Ash-chest	Architectural elements	Architectural elements	Architectural elements	Architectural elements
Altar	Equipment	Equipment	Equipment	Equipment
Portrait	Sculpture	Building inscription	Inscription	Inscription
Sculpture	Votive altar	Honorific inscription	Portrait	Portrait
Stele	Votive inscription	Milestone	Sculpture	Relief
Inscription stone	Votive relief	Portrait	Function not identified	Sculpture
Sarcophage	Function not identified	Sculpture		
Inscription – Tituli		Function not identified		
Position not identified				

- The quarry section provides general information on the quarry, classification and analysis of the material and additional literature

2.4.1 *The monument database*

The index of monuments and fragments is ordered according to a key with 8 numbers: the first pair of digits indicates the state, the second the country, the third the village or town of erection and the fourth the objects themselves. According to that key all photos, texts, samples, analyses can be associated to a certain object.

The monument database is subdivided into five sections (see Table 2):

- The general section includes the main information on the monument: key, name, photo of the complete object, monument classifications (see Table 1),

Table 2. Content of the monument database.

Structure of the monuments database

General section	Stone section	Relief section	Inscription section	Literature section
Object key	Size (measured or from the literature)	General description of relief and the ornaments	General description of the monuments inscriptions	CSIR
Object name	Material classification	Description of the reliefs and ornaments	Photo of the main inscription	IBR
Photo of the whole object	Material according to the literature	Detailed description of main relief or ornament	Original text of the main inscription	CIL
Monuments classifications (see Table 1)	Photo representative macroscopic surface	Photo of main relief or ornament	Used roman formulas	ILPRON
Place of erection or deposition	Macroscopic description	Detailed description of additional relief or ornament	Used roman abbreviations	Kremer
Kind of erection or deposition	Remarks on macroscopic description	Photo of additional relief or ornament	German transcription incl. literature or name of the translator	Additional literature
Photo of the erection or deposition	Expected material transportation to the place of finding or erection	Detailed description of additional relief or ornament	Interpretation of the main inscription: on buried persons, on honored person, on votive, on donor, military unit	link to ubi-erat-lupa or other databases
Museum inventory No.	No. of sample	Photo of additional relief or ornament	Photos of additional inscriptions	
Web-address of the museum or place of deposition	Sampling from whom and date	etc.	Original text of the additional inscription	
Finding place (roman name, modern name and district)	Thin section analysis		Used roman formulas	
Circumstances of finding	Photo representative area of thin section		Used roman abbreviations	
Condition of the object (weathering, ancient or modern coatings or colors)	Diagram of isotope analysis		German transcription incl. literature or name of the translator	
Dating of the object	Diagram of rare elements			
Background of the dating	Quarry of origin including discussion			
Date and name of the last modification	Date and name of the last modification	Date and name of the last modification	Date and name of the last modification	Date and name of the last modification

165

place of erection or deposition, kind of erection or deposition, photo of the location, museum inventory number, web-address of the museum or place of depository, place of discovery (Roman name, modern name and district), circumstances of find, condition of the object (weathering, ancient or modern coatings or colors), dating of the object, background of the dating, date and name of the last modification.

- The stone section gives all important information on the material: size (measured or from the literature), material classification according to the analyses, if analyses are not available, results according to literature, representative macroscopic surface photo, macroscopic description, including discussion, kind of transportation to the place of finding or erection, key of sample, sampling by whom including date, thin section analysis by whom including date, representative photo of thin section, diagram of isotope analysis, analysis by whom including date, diagram of rare elements, analysis from whom and date, quarry of origin including discussion, date and name of the last modification.
- The relief section describes reliefs and ornaments of the object: general description of relief and ornaments, 10 possible descriptions of the various reliefs and ornaments of one monument including photo and name of photographer, date and name of the last modification.
- The inscription section describes all texts on a monument: general description of the monuments' inscriptions, photo of the main inscription, original text of the main inscription, used Roman formulas, used Roman abbreviations, German transcription including literature or name of the translator, interpretation of the main inscription (on buried person, on honored person, on votive, on donor, military unit), photos of additional inscriptions, original text of the additional inscription, used Roman formulas, used Roman abbreviations, German transcription including literature or name of the translator, date and name of the last modification.
- The literature section is divided into the following references: CSIR, CIL, ILPRON, Kremer, additional literature, link to ubi-erat-lupa or other databases, date and name of the last modification.

2.4.2 *The quarry database*

The quarry database is divided into two main sections (see Table 3):

- The general section includes major information on the quarry: name, overview map of the location, detailed map of the location, photos of the quarry, inventory number of GBA (Geological Survey of Austria), material classification, local name of material, period of Roman mining, background of dating, date and name of the last modification.

Table 3. Content of the quarry database.

General section	Material section
Quarry name	Macroscopic description
Overview map of the location	Representative macroscopic photos
Detailed map of the location	Microscopic description
Photos of the quarry	Representative microscopic photos
Inventory Nr. of GBA	Description of the material's variation
Materials classification	Diagram of isotope analysis
Material's local name	Diagram of rare elements
Period of roman mining	Technical characteristics and nature of weathering
Background of the period's dating	Literature
Date and name of the last modification	Date and name of the last modification

Structure of the quarry database

- The material section provides information on the material composition and characteristics: macroscopic description (grain size, characteristic minerals, characteristic texture, representative macroscopic photos), microscopic description (grain size, grain borders, main minerals and accessories, microscopic texture, representative microscopic photos), description of the material's variation, technical characteristics and nature of weathering.

2.5 *Technical features of the database*

The structure of the database was designed using Sybase Powerdesigner for an Oracle Oracle 9i database. Web-frontend of the database is via php and ADODB class. The photos and graphics are embedded in the database.

The database-function for working groups is managed by user administration with different user rights. There is no limit on the number of users accessing the database simultaneously. The data input forms are designed in a user-friendly way: importing data sets before modification for a similar monument, pull-down menus for places, names, classifications, etc. Extension of classifications by the administrator on user request. The search menus are divided into a simple and complex search as usual.

3 CONCLUSIONS AND PERSPECTIVES

For public access and the handling of this complex type of data sets a web-accessible database is more

166

suitable than a printed catalogue. Other working groups from Austria and surrounding countries are cooperating and continuously submit their data and findings to the SRI-database. The structure can be modified easily for monuments of different cultures. As the number of data-sets is increasing continuously it will be a basic tool for further research, conservation of monuments and for classification of new findings.

ACKNOWLEDGEMENTS

The authors thank the Austrian research community on Roman monuments for a close cooperation and for discussions on the monument classification. Special thanks to Ortolf Harl and the ubi-erat-lupa-database working group.

Financial support from the Stiftungs- und Förderungsgesellschaft of the Paris-Lodron-Universität Salzburg and the Austrian Science Foundation (FWF grant no. P15669) is gratefully acknowledged.

REFERENCES

CSIR Corpus Signorum Imperii Romani D I,1, 1973; Ö I,6, 1979; Ö III,1, 1975; Ö III,2, 1976; Ö III,3, 1981.
Dafferner, A., Feraudi-Gruénais, F. & Niquet, H. 2000. Die Epigraphische Datenbank Heidelberg. In M. Hainzmann & Chr. Schäfer (eds), *Alte Geschichte und Neue Medien. Zum EDV-Einsatz in der Altertumsforschung, Computer und Antike (St. Katharinen 2000)*: 5: 45–65.
Harl, O. 1999. Das Forschungsprojekt "Die Grabsteine von Norikum und Pannonien" – Ein Arbeitsbericht. *Fundort Wien. Berichte zur Archäologie* 2: 194–202.
Hainzmann, M. & Schubert, P. 1985–87. Corpus Inscriptionum Latinarum. Inscriptionum Lapidarium Latinarum Provinciae Norici usque ad annum MCMLXXXIV repertarum indices (ILLPRON Indices), I, II & III.
Hemmers, C., Traxler, S., Uhlir, C.F. & Wohlmayr, W. in press. 'Stein – Relief- Inschrift'. Konturen eines Forschungsprojektes. Proc. of the VIII. *Internationales Colloquium über Probleme des provinzialrömischen Kunstschaffens. Zagreb 2003*: in press.
Herz, N. 1988. Geology of Greece and Turkey: potential marble source regions. In N. Herz & M. Waelkens (eds), *Classical Marble: Geochemistry, Technology, Trade (Nato ASI Series, Ser. E: Applied Sciences* 153: 7–10.
Kremer, G. 2001. Antike Grabbauten in Noricum. *Sonderschriften des Österreichischen Archäologischen Institutes*: 36.
Moens, L., Paepe, P. & Waelkens, M. 1992. Multidisciplinary research and cooperation: keys to a succesful provenance determination of white marble. *Acta Archaeologica Lovaniensia, Monographiae* 4: 247–252.
Mommsen, Th. 1903. Corpus Inscriptionum Latinarum (CIL): III.
Müller, H.W., Schwaighofer, B., Benea, M., Piso, I. & Diaconescu, A. 1996. Greek marbles in the Roman Province of Dacia. *Proc. 3rd Symposium for Archaeometry, 6–9. Nov. 1996*: 22–23, Athens, Greece.
Müller, H.W. & Schwaighofer, B. 1999. Die römischen Marmorsteinbrüche in Kärnten. *Carinthia II*: 549–572, Klagenfurt.
Müller, H.W. 2002. Provenance determination of marble sculptures from Pannonia. *Proc. 31st Archaeometry Symposium Budapest 1998*: 767–775.
Wright, J.B. 1981. Earth materials: minerals and rocks. In D.G. Smith (ed.), *The Cambridge Encyclopedia of Earth Sciences*: 68–91. Cambridge: Cambridge University Press.
Vollmer, F. 1915. Inscriptiones Bavariae Romanae (IBR).

Dimension stone decay: from rock weathering studies to diagnostic and conservation efforts

Dimension Stone 2004, Přikryl (ed)
© 2004 Taylor & Francis Group, London, ISBN 90 5809 675 0

Diagnostic study of the stone surface cleaning at St Lawrence Cathedral (Lugano, Switzerland): comparison of laser, dry mechanical and chemical cleaning

C. Calcagno

Altech srl, Applied Laser Technology and Conservation Artworks, Bassano del Grappa (VI), Italy

ABSTRACT: Effectiveness and quality of laser cleaning in the restoration of stone artworks have been widely investigated, but these results have been so far evaluated by the immediate results of the cleaning, with no assessment on the consequences of this procedure on future conservation of stone.

1 AIM OF THE INTERVENTION

The façade of St Lawrence Cathedral in Lugano is a set of stone decorations (Fig. 1), applied in the 16th century on a previous Gothic façade. The work took almost a century, and was completed at the end of 16th century. The designer of the façade is unknown, but it certainly must have been an important person of the Renaissance, who lived in the Milan-Lombardy area. Some historians

Figure 1. The façade of St Lawrence Cathedral in Lugano.

believe it could be Bramante. The façade is mainly made of Saltrio stone, a limestone from the Pre-Alpine area near Varese, and Carrara marble.

The wonderful decoration and the excellent sculptures must be by the master stone-cutters from Lugano who worked in the area near Como and in the north and centre of Italy during the 15th and 16th centuries. Today it is very altered by extensive adherent black crust and by leaching with exfoliation in the most exposed areas that jeopardize the conservative state of the materials and the elegant original colours.

A pilot project diagnostic study has been made of the cleaning methods used on a representative stone surface. The study identifies and compares the results of the different cleaning methods used, analysing the effects on the surface according to the principle that the method used must not alter the surface with respect to the original patina (Asmus & Lazzarini 1972, Asmus 1976).

2 METHODS

In the pilot project, different cleaning sources were used on similar stone surfaces and with thick layers of black incrustation:

- **Chemical cleaning:** compress AB57 Mora basic complex. Basic complex wrapping .Ph10 made of ammonia bicarbonate, EDTA and paper pulp, 2 hours application.
- **Mechanical cleaning:** precision micro aero abrasive Wulsag type with 0.68 mm nozzle, Al_2O_3 with 440 mash abrasive powder, air pressure from 0.5 to max 3.0 atm.

171

Figure 2. Physical cleaning: Laser Nd:Yag 1064 nm in Q-Swichedt mode in two different energy, "light" and "driven".

- **Physical cleaning (Fig. 2):** Laser Nd:Yag 1064 nm in Q-Swichedt mode: "Light" laser cleaning Nd:Yag 1064 nm in Q-Swiched mode, 9 ns pulse duration, 10 Hz frequency, the fluence has been applied with energies from 260 mJ to 350 mJ, laser beam diameter between 5 mm and 7.2 mm.

"Driven" laser cleaning: Nd:Yag 1064 nm in Q-Swiched mode, 9 ns impulse length, frequency 15 Hz, the fluence has been applied with energies from 360 mJ to 550 mJ, laser beam diameter between 3 mm and 7 mm.

3   DIAGNOSTIC STUDY APPLIED TO THE COMPARISON OF CLEANING METHODS

The purpose of the study is to observe the effects of the different cleaning methods by means of a series of specifications of a chemical and physical nature, in addition to onsite monitoring (Fig. 3, see also Verges-Belmin et al. 1994, Calcagno 1988).

The laboratory studies was focused on the determination of the state of the surfaces both in terms of stratigraphical characterisation and also regarding the

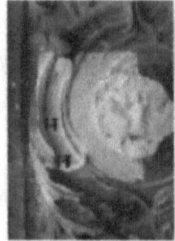 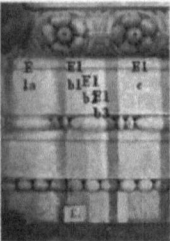

Figure 3. The research has been addressed: to the check of the stone surfaces response after the mechanical, chemical and physical cleaning, by means of measurements on site.

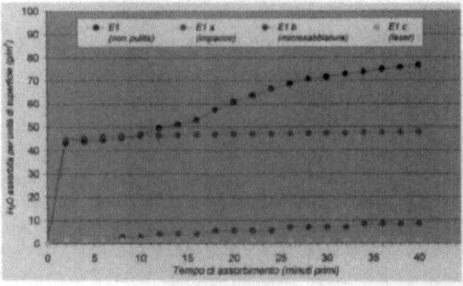

Figure 4. Comparison among cleaning methods used. From the graph it is evident the overwhelming existing difference, in terms of water absorption, between the cleaned surfaces and the not cleaned one. Measurements: trials on site of water absorption at low pressure (E1 = untreated, E2 = wrapping, E3 = microsandblasting, E4 = laser).

determination and extent of phenomena of alteration. At the same time, a series of onsite measurements were made in order to verify the response of the stone surfaces to cleaning carried out both with traditional systems and also using innovative technologies such as laser.

The results were evaluated using following approaches:

- cross-section study;
- SEM electron microscope study;
- EDS microanalysis;
- FTIR spectrophotometry;
- MFO optical microscope study;
- onsite measurements of water absorption according to normal recommendations 44/93 (Fig. 4);
- three-dimensional mapping (roughness) for determining the surface contour (Fig. 5).

E1-Black Layer: The major water absorption is verified on the not cleaned surface, this is due to the major porosity due to the presence of the black layer.

E1a-Wrapping: high absorption values were recorded. This is due to the action of the same wrappings

172

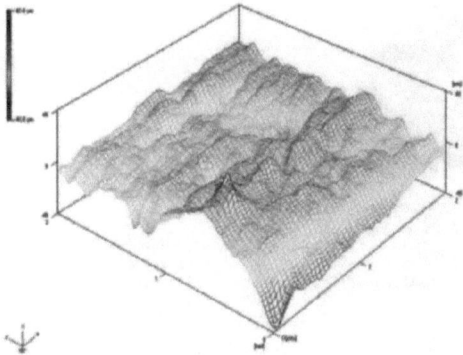

Figure 5a. Three-dimensional mapping of the fragment cleaned using chemical method. Average roughness Ra = 1.78 μm.

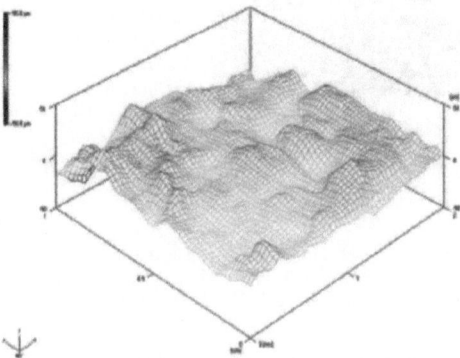

Figure 5b. Three-dimensional mapping of the fragment cleaned using microsandblasting (mechanical cleaning). Average roughness Ra = 1.54 μm.

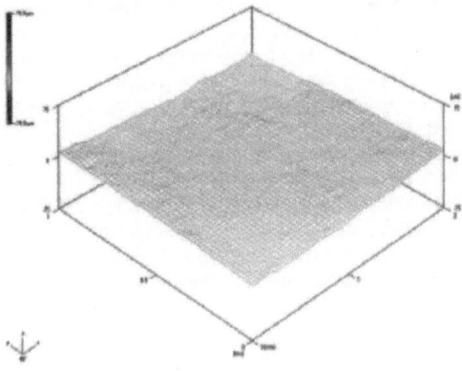

Figure 5c. Three-dimensional mapping of the fragment cleaned using laser (physical cleaning). Average roughness Ra = 0.45 μm.

which determine an increase of the porosity for the removal of the silicate fractions and the solubility of the salts present in the pores.

E1b-micro-sandblasting: the mechanic action removes the first layer of the stone and puts in evidence the whole stone surfaces. The SEM images put in evidence that the areas cleaned by means of microsandblasting are those which show the major surface compactness, but which have a total layer loss.

E1c-Laser: The test has shown more contained absorptions, as a consequence of a surface scarcely waterlogged so that it can exercise a moderate, but effective waterproof action. This is due to the presence of the patina, compared to the original surface and an inferior surface ruggedness.

4 OBJECTIVES OF THE DIAGNOSTIC STUDY

The study criteria are that of an analytical and objective comparison of the treated surfaces, observing the possible appearance of unforeseen phenomena or in any case phenomena generated by the cleaning method used (Calcagno 1988, 1996). The identification of a number of representative samples from specific historic periods favours the understanding of morphological changes over time.

These studies served to determine and confirm if the physical cleaning method by Nd:Yag laser is:

- more effective and efficient than traditional methods;
- more selective and preserves the patina better than traditional methods.

Moreover, the results obtained were in turn compared with similar cleaning methods (chemical, mechanical and physical) used on stone surfaces but carried out over 15–20 years ago (Figs 6–7).

This subsequent comparison serves to appraise and indicate if, compared to traditional methods, the physical cleaning method by means of Nd:Yag laser:

- eventually modifies the stone surface in time;
- slows down the speed of alteration of the stone surface.

A stone surface cleaned in 2002 contains the same morphological and textural information as a stone material treated over 15–20 years ago (Cooper et al. 1993, Calcagno 1988, 1996).

5 CONCLUSIONS

The study led to the identification of several aspects that were not immediately noticeable: the selective and discriminating action of the laser completely removes the carbon deposit, preserving on the surface a layer of a few microns that constitutes the patina,

173

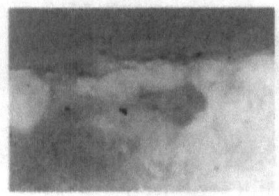

After 15 years the Patina surface is still visible and less rough, the sub-layer appears to be homogeneous and compact, and unaltered.

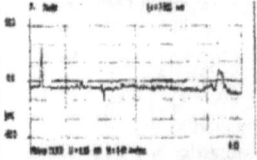

Comparison of the average roughness value Sample cleaned 15 years ago. Average roughness analysis (Ra) Ra = 0.34 µm.

Laser cleaned sample (SEM 800×), 15 years after cleaning and the consolidating intervention there still exists a homogeneous coat of the reinforcing varnish. The surface appears to be unaltered with low surface roughness.

Figure 6. Comparison of the analytic data done on stone surfaces treated using laser cleaning systems in 1987–1988 and monitored 10 years after (in 1998).

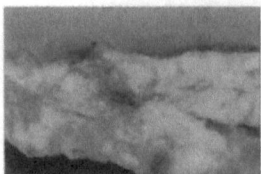

After 15 years, the observation shows a more serious altered stone matrix and the surface is irregular.

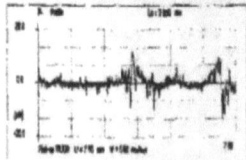

Comparison of the average roughness value Sample cleaned 15 years ago. Average roughness analysis (Ra) Ra = 2.57 µm.

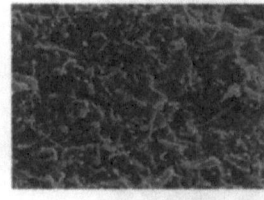

Micro-sandblasting cleaned sample (SEM 800×) 15 years after cleaning and the consolidating intervention, the surface irregular. There are gypsum crystals, a deep alteration and the loss of the consolidating varnish. The average surface roughness is high.

Figure 7. Comparison of the analytic data made on stone surfaces treated using mechanic cleaning systems in 1987–1988 and monitored 10 years after (1998).

i.e. that group of products of calcium oxalate, calcite and gypsum (pseudomorphism of gypsum to the detriment of the calcite) with the total preservation of the shape, volume and solidity of the surface.

Second, it was found that a stone surface becomes more altered if subjected to traditional cleaning methods (chemical and mechanical), with the presence, after cleaning, of more microcracks, microcavities, abrasions or in any case an increase in the area of relative roughness. This fact seems even more significant when compared a long period of time after the treatment. The laser cleaning method retains the conservative parameter in time and therefore the surface becomes more resistant to the pollutant agents in the environment.

Third, the study led to the drawing up of a degradation curve for the surface relative to the cleaning method used. In a medium-long term projection this revealed probable maintenance interventions to be carried out much earlier if the cleaning method used is of the mechanical or chemical type, whereas the maintenance interventions are more distant if the cleaning method used is of the physical type with laser.

The micro-sandblasting appears to be invasive either for the stone surface, or for the patina. It provides little discrimination and selection.

The wrappings highly respect the patina, but they do not completely remove the surface dirt between the little grooves and chinks, and also gypsum often remains in appreciable percentages. They provide very low selectivity.

The laser is the cleaning technique able to put into evidence the stone original surface (in this way preserving the patina). The laser presents very good selection and selectivity capacities that is capacities of calibrating the cleaning action in relation to the type and gravity of the dirt.

A selective cleaning, that is which removes the black layer and maintains the patina as much as possible, helps to have a low ruggedness surface and so a less waterproofing. A low absorption means a strong reduction of the substratum and of the polluting elements responsible for the degradation.

REFERENCES

Asmus, J.F. & Lazzarini, L. 1972. Lasers for the cleaning of statues: initial results and potential. In *1st Int. Symp. on the Deterioration of Building stones*, La Rochelle, Paris.

174

Asmus, J.F. 1976. Lasers Clean Delicate Art Works. *Laser Focus* 12: 56–57.

Calcagno, G. 1988. Verifica della pulitura Laser sugli stemmi del Palazzo del BO' a Padova. Monuments and Fine Arts Service of East Veneto, prot.cant.Log.Ant. Inf.nr.88.

Calcagno, G. 1996. Laser cleaning of Sandstone statue, principal portal of St. Stephen's Church, Vienna. Lab. Bundesdenkmalmat, Tor 4, Vienna.

Cooper, M.I., Emmony, D.C. & Larson, J.H. 1993. The evaluation of Laser cleaning of stone sculpture. In *3rd Int. Conf. Structural Repair and Maintenance of Historical Buildings*.

Verges-Belmin, V., Pichot, C. & Orial, G. 1994. Use of petrography for the comparison of laser-beam and microsandblasting cleaning techniques. Forschungsbericht 13/1994, Eurocare-Euromarble 4. Workshop.

175

Dimension Stone 2004, Přikryl (ed)
© *2004 Taylor & Francis Group, London, ISBN 90 5809 675 0*

Assessment of stone performance 'in use' to inform decision-making during conservation of historic buildings: a case study from Northern Ireland

J.M. Curran
Stone Conservation Services, The Gas Office, Belfast, UK

B.J. Smith
School of Geography, Queen's University Belfast, UK

D. Stelfox & J. Savage
Consarc Conservation, The Gas Office, Belfast, UK

ABSTRACT: Within Northern Ireland there is a severe lack of authoritative data on the performance of building stone, and as climate and pollution patterns are different from Britain, there is no reliable information on which to base decisions for stone replacement and repair. To help with these decisions an interactive inventory of stone performance has been developed. The project is a joint industry/research partnership and it is hoped that this web-based stone inventory will facilitate end-users' choice of repair specifications and selection of new and replacement stone fitted for purpose.

1 INTRODUCTION

There is a tendency for building stone research to focus on studies of prestigious, high cost buildings through schemes such as the Cathedral Research Programme of English Heritage, with the intention of using these as 'show cases' for conservation strategies and technologies. However, the large amounts of money now being spent in countries such as the UK on 'less prestigious' buildings suggests that these also warrant targeted research to allow the development of conservation strategies tailored not only to individual buildings and building stones, but also to local and regional environmental conditions. Furthermore, these 'lesser' buildings are more likely to have undergone inappropriate repairs in the past (e.g. re-pointing with cement-based mortars) that have exacerbated decay and require specific remedies.

Within Northern Ireland, building conservation has grown rapidly in recent years in response to a combination of government support designed to improve the image of the Province, European Structural Funds, funding from the National Lottery and legislation that encourages the restoration of important public and private buildings. However, with this increase in activity there is a danger that many of those who partici-pate in intervention fail to understand the complexities of building restoration and the need for a scientific rationale that underpins decision taking. This applies especially to building owners and others with a duty of care, who are compelled to rely on the veracity of conservation assessments and treatment recommendations from contractors who will ultimately benefit from undertaking the work. Within Northern Ireland, these problems are compounded by a chronic lack of authoritative data on the performance of natural stone (local and imported) used in historic buildings. This is partly because of differences in climatic and pollution conditions between the Province and the rest of the UK (Viles 2002) and partly because of a relatively narrow research base in what is a peripheral region. Both of these factors make it potentially dangerous to simply translate performance data and durability predictions derived from programmes based on and tested in, for example, southern England. As a consequence, there is little reliable information on which to base repair specifications.

These problems were exemplified in Belfast in the 1990s through conditions attached by visiting consultants to conservation funding for sandstone buildings that specified only the physical removal of the salt damaged outer surface of the stonework (Smith et al. 2003).

177

This was despite local research that had shown that in the moist, salt-rich environment of Northern Ireland sandstones often acquire a deep reservoir of stored salts that would be activated and brought to the surface by the dressing back of the stone (Smith et al. 2002, Smith & Curran in press, Warke & Smith 2000). This, and other research carried out on historic buildings in Northern Ireland, has highlighted the fact that due to the numerous factors that control decay and their complex interactions in space and time, stone buildings should not be conserved 'by formula' and in some cases the formulaic application of general rules can lead to even greater problems. However, these and similar projects have also identified that if appropriate questions are asked prior to restoration through structured enquiry it may be possible to formalise conservation strategies (Warke et al. 2003). This clearly results in significant benefits for practitioners, including less time and money spent on the testing and analysis of individual buildings and the specification of repair and replacement based upon a scientific rationale rather than anecdote and a 'gut feeling' as is often the case.

In order to address these issues, Queen's University and Consarc Conservation undertook in 2001 a two year project to develop for Northern Ireland an interactive, web based inventory of stone performance in use based initially upon a survey of key listed buildings. The utilisation of performance in use is seen as an alternative to prediction from standard durability tests that allows performance to be related specifically to preparation and construction techniques, present and past conditions of exposure, previous conservation intervention and the use to which the stone is put within the structure. Once established, the database can also incorporate peripheral information that can facilitate its use and interpretation. This ranges from tutorial information on stone recognition and decay identification to possible sources of replacement stone and data on stone characteristics. The use of this approach is made feasible by: the relatively limited geographical extent of Northern Ireland, the restricted number of local building stones, the concentration of imported stone from lowland Scotland and the pool of expertise and local experience embedded within the two partner organisations.

In the rest of this short paper, the structure of the database is explained and its capabilities are exemplified, both in terms of what it can be used to illustrate and its use in the formulation of conservation strategies.

2 STRUCTURE AND CAPABILITIES OF THE INTERACTIVE INVENTORY

The database includes all condition survey information, chemical and structural data on the building stone used in Northern Ireland. The stone inventory is available as an interactive web-site (www.stone-solutions.net) that can be accessed by all end-users.

There are 3 ways to access the information from the database:

1. **By Stone Type:** information on the main building stones is subdived into: sandstone, limestone, granite, basalt, quartzite and conglomerate. These categories are further subdivided into local and imported stone types and examples provided to show the stone 'in use'.
2. **By County:** buildings are categorized in terms of their county (Antrim, Armagh, Down, Derry, Fermanagh and Tyrone). Under each county there will be key buildings to show the main stone types used and common weathering and decay features under different environmental and pollution conditions. There is a separate category for 'Belfast' due to the range and number of buildings, stone types used in the city and the variations in environmental and pollution characteristics in this urban area.
3. **Individual building search:** particular buildings can be searched by name using the key word search facility.

Both the 'stone types' and 'buildings' are cross-reference so that when information of a particular building is accessed the database provides other examples (from different environments, building types etc.) where the dominant stone type is used.

The website is designed to permit instantaneous access to the database for an appointed administrator. This allows regular updates to the site. It also includes a section on stone identification and other practical information and useful links for those involved in stone conservation.

2.1 Environmental controls

As a geographical database for a Province with variable environmental and pollution conditions, the system is ideally placed to evaluate the performance of the same stone type in relation to these parameters. The examples given in Figures 1 and 2 show a local Triassic sandstone (Scrabo) which, because of its variable bedding and the presence of lenses of smectite clays is particularly prone to damage from the expansion and contraction of precipitated salts. The first photograph in Figure 1 is of St. Matthew's Church in central Belfast (built 1881–1883) and the damage resulting from a combination of pollution derived sulphate salts and chloride salts of principally marine origin. The second photograph (Fig. 2) shows the same stone type used in a church (built 1875–1877) in the small town of Newtownards outside of the Belfast metropolitan area. This is an area of relatively low atmospheric pollution and the stone has largely retained its original integrity.

178

Figure 1. Image showing Scrabo sandstone on St. Matthew's church, Belfast. Decay of sandstone due to exposure in a polluted maritime environment.

Figure 3. Friar's Bush Headstone (1985) showing the onset of contour scaling.

Figure 4. Friar's Bush Headstone (2003) showing damage to sandstone.

Figure 2. Image showing Scrabo sandstone on St. Patrick's church, Newtownards.

2.2 Monitoring change over time

By incorporating photographic data from earlier surveys and studies it is possible to establish patterns of change and decay over time. Figures 3 and 4 show the same quartz sandstone headstone in Friars Bush Graveyard, south Belfast; erected in the mid nineteenth century and photographed in 1983 and 2003. The first photograph (Fig. 3) shows the onset of contour scaling triggered by the subsurface accumulation of gypsum, although much of the original surface

remains intact. The second photograph illustrates the rapid loss of material and hollowing out of the stone over the intervening 20 years once scaling had commenced. In doing so, it demonstrates the episodic nature of sandstone decay in which there may be little if any surface indication of decay for many years. This can be followed by rapid and catastrophic breakdown once the strength/stress threshold for the stone is eventually breached by, for example, the slow build up of salts combined with the a gradual decrease in shear strength (Smith et al. 1994).

2.3 Assessing the effects of interventions

Once the database is established, it is can be easily added to through the re-survey of buildings to record not just further deterioration, but also the effects of subsequent conservation. In the example shown in Figures 5 and 6, it can be seen that intervention need not be positive or even benign.

Figure 5 shows detail of the Locharbriggs sandstone gates to Ormeau Park, Belfast photographed in

179

Figure 5. April 2002 Image showing Locharbriggs sandstone carving on the Gate Pillars of Ormeau Park, Belfast (before cleaning).

Figure 6. June 2002 Image showing destruction of Locharbriggs sandstone carving on the Gate Pillars of Ormeau Park, Belfast (after cleaning).

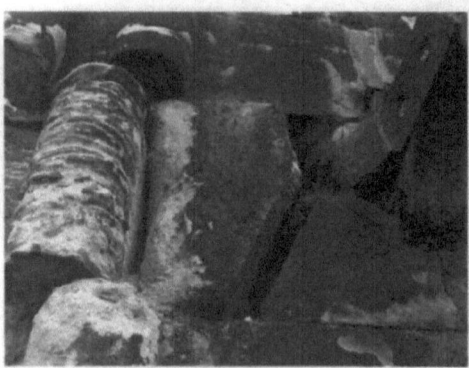

Figure 7. Fracturing of Scrabo sandstone due to rusting of metal fixings. Fortwilliam Gate Posts, North Belfast.

April 2002. The second photograph (Fig. 6) shows the severe erosion and loss of detail from the pillars after they were power washed in June 2002. Clearly this aggressive treatment is totally inappropriate for a relatively soft sandstone. Moreover, a survey of the gate pillars in February 2004 showed that the cleaning, and in particular the opening out of the pore structure, encouraged algal colonisation of the stone and contributed in large measure to the 'greening' of the stone over the intervening period.

2.4 Evaluating performance related to use

The database provides an ideal format for recording examples of how different stone types have been used in structures and how they have performed. Indeed it is particularly effective at recording failure and inappropriate use and can provide object lessons in when and how not to use a particular stone. For example, Figure 7 shows the fracturing and displacement of a Scrabo sandstone gatepost in North Belfast. This is a consequence of the corrosion and expansion of an iron fixing for a gatepost, the rusting of which was facilitated by the open texture and water absorption characteristics of the sandstone.

To demonstrate the capability of the database to demonstrate good practice, Figure 8 shows the same gatepost after successful conservation. The replacement stone used in this case was Stanton Moor sandstone,

Figure 8. Restoration of Fortwilliam Gate Posts, North Belfast.

which was chosen for a combination of its compatability and predicted visual convergence with the original Scrabo and Giffnock sandstone (imported from Scotland) and its proven durability on other structures

180

in the weathering environment of Belfast. This combination of criteria lies at the heart of how the database can be used to inform stone selection for conservation.

3 USE OF INTERACTIVE INVENTORY TO INFORM CONSERVATION STRATEGIES

The most successful stone conservation requires information on:

- Convergence: colours and textures of stone types.
- Compatibility: chemical and structural compatibility of stone types (and mortars used)
- Durability: strength and weathering characteristics of stone types in relation to environmental and pollution conditions.

The natural stone database for Northern Ireland provides this information on a regional scale. The database is designed to provide practical information on building stone to building owners and those with a duty of care for historic buildings in Northern Ireland. Providing this information will also allow them to recognize when to seek professional assistance from conservators who will combine the inventory with their own knowledge and expertise of appropriate and successful approaches to stone conservation.

4 CONCLUSIONS

Hopefully this brief introduction has demonstrated the feasibility and potential value of using performance in use as a tool for stone selection and treatment in conjunction with a web based platform that allows remote access and rapid cross-correlation between components of the database. The primary aim of the research project was to develop and test the structure and philosophy behind the system and it is envisaged that its usefulness will grow exponentially as more buildings are added to the database. It is recognised that contributory factors to its success are the manageable size of Northern Ireland and the relatively limited variety of stone types that have been used in the majority of its buildings. It is felt, however, that it represents a very useful tool for informing conservation

and improving public understanding at a regional scale – where there are often strong traditions of using local stone. It is also considered to be particularly useful in regions that experience distinctive environmental conditions, but which do not have an extensive research base and would otherwise have to rely upon conservation strategies developed initially for other environments and stone types.

ACKNOWLEDGEMENTS

The writers are indebted to the UK Knowledge Transfer Partnership who provided both the stimulus and the funding for this research.

REFERENCES

Smith, B.J. & Curran, J.M. (in press). Urban stone decay: the great weathering experiment. *Geological Society of America, Special Publication.*

Smith, B.J., Magee, R.W. & Whalley, W.B. 1994. Breakdwon patterns of quartz sandstones in a polluted urban environment: Belfast, N. Ireland. In D.A. Robinson & R.B.G. Williams (eds), *Rock Weathering and Landform Evolution*: 131–150. Chichester: Wiley.

Smith, B.J., Warke, P.A., Turkington, A.V., Curran, J.M., Stelfox, D. & Savage, J. 2003. The resolution of conflicting demands in University/Industry collaboration on stone conservation: a UK perspective. In R. Kazlowski (ed.), *Proceedings of 5th EC Conference on Cultural Heritage Research: a pan-European Challenge*: 215–318. Cracow: Polish Academy of Sciences.

Smith, B.J., Turkington, A.V., Warke, P.A., Basheer, P.A.M., McAlister, J.J., Meneely, J. & Curran, J.M. 2002. Modelling the rapid retreat of building sandstones. A case study from a polluted maritime environment. In S. Siegemund, T. Weiss & A. Vollbrecht (eds), *Natural stone, weathering phenomena, conservation strategies and case studies.* Geological Society Special Publication 205: 347–362.

Viles, H.A. 2002. Implications of future climate change for stone deterioration. In S. Siegemund, T. Weiss & A. Vollbrecht (eds), *Natural stone, weathering phenomena, conservation strategies and case studies.* Geological Society Special Publication 205: 407–418.

Warke, P.A., Curran, J.M., Turkington, A.V. & Smith, B.J. 2003. Condition assessment for building stone conservation: a staging system approach. *Building and Environment* 38: 1113–1123.

181

Dimension Stone 2004, Přikryl (ed)
© *2004 Taylor & Francis Group, London, ISBN 90 5809 675 0*

A preliminary investigation into visitor-generated stone deterioration

D. Dragovich & G.A.S. Edwards
School of Geosciences, University of Sydney, Sydney, Australia

ABSTRACT: Rouse Hill Estate near Sydney, Australia includes an historic house which has recently been opened for public visitation. Due to the fragility of the interiors and structure of the house, visits are conducted in the presence of a guide. The potential for accelerated deterioration of sandstone floors and verandahs from pedestrian traffic is of concern to the Historic Houses Trust. Present weathering condition of the sandstone and probable patterns of pedestrian movement are described, and a program for monitoring pedestrian impacts is suggested.

1 INTRODUCTION

1.1 *Heritage management*

Preservation of Australia's European cultural heritage involves a variety of objects, only some of which are located in museum collections. Architectural heritage, because of its exposure to the elements, is often difficult to preserve. As some buildings and structures in Australia are more than 150 years old, interest in such architectural heritage is increasing. Many structures are showing signs of deterioration which leads to concerns about their condition. An added complication is that natural deterioration is non-uniform over the structure, and may not occur at regular and predictable rates over time. This investigation into rates of deterioration of sandstone in an historic house highlights some of the problems of managing cultural heritage under the conflicting demands of preservation and visitation.

In Australia, the management of all heritage buildings is governed by principles outlined in the Burra Charter of 1981 (Stapleton 1985, Aplin 2002). Heritage buildings are available for the education and enjoyment of the public, but natural weathering processes may be augmented by the wear-and-tear accompanying visitation. This was recognised by the Historic Houses Trust of NSW in its Annual Report of 1999–2000 which listed the monitoring of wear-and-tear as a key management objective for one of its less-visited houses at Rouse Hill Estate. It was found that even small numbers of visitors were having an impact on both wall and floor surfaces, and some of these surfaces are sandstone. Within the main house at Rouse Hill Estate there is a large area of sandstone flooring, and a sandstone step at the entrance to the

pantry has been lowered in its centre by several centimetres. This particular weathering feature pre-dates recent visitation but highlights the potential vulnerability of sandstone to abrasion by foot traffic.

2 THE STUDY SITE

2.1 *Rouse Hill Estate*

In this study attention will be focussed on the house at Rouse Hill Estate near Sydney, Australia. The house was built by Richard Rouse between 1813 and 1818 on Vinegar Hill (Thornton 1988), a site located approximately 45 km NNW of Sydney in an area receiving an annual rainfall of around 900 mm. About 500 tons of sandstone were used in the construction of the main house, with the verandahs most likely being constructed in 1856 (Historic Houses Trust of NSW 1997, Thornton 1988). The house was resumed by the Department of Environment and Planning in 1978 and the Historic Houses Trust took over management of the property in 1987 (Historic Houses Trust of NSW 1997). The Trust has a conservation ethic which seeks to preserve the whole fabric of the house as it existed at the time of resumption, requiring that 'All conservation work is to be undertaken strictly with the aim of preserving the artefact. Intervention should aim to be minimal' (Historic Houses Trust of NSW 1994).

2.2 *Site management*

Access to the Rouse Hill Estate is only available in the presence of a guide. In addition to the historic house itself, the gardens, stables and other outbuildings form

183

part of the complex. Currently guides discourage visitors from walking along the verandahs where visitors are required to wear soft overshoes to minimise damage from foot traffic, and carpeting is placed over the stone in trafficked areas during periods of anticipated heavy visitation. However several doorways can only be entered by crossing the verandahs, so pedestrian traffic would have been concentrated in these places throughout the period of occupancy of the house (until 1978), as well as subsequently by visitors. The 'partitioning' of deterioration resulting from previous occupancy and that caused by present visitation can only be estimated by instigating a detailed monitoring process which the Trust is keen to implement. This study describes the current condition of the stones on part of the house's verandah and suggests ways in which monitoring of visitor impact can be undertaken.

As part of establishing the current condition of the blocks and their potential vulnerability to future damage, patterns of present pedestrian traffic were identified on a stone-by-stone basis. In addition to observations on the verandah itself, experimental sandstone blocks have been placed at the entrance to the Estate's visitor centre. The effects of this concentrated pedestrian traffic on the sandstone blocks are being monitored.

2.3 The sandstone

Verandahs extend around the front of the house and about half way along each side. The house and verandahs were constructed from sandstone which underlies the Sydney Basin and outcrops both along the coast and inland in the Blue Mts. The stone, although variable in colour and composition and including some thin shale beds, forms prominent cliffs when exposed and characteristically develops compound hillslopes with structural benches. The nearly-horizontal beds dip gently inland (westwards). The most important formation for dimension stone is the Hawkesbury sandstone, a massive formation up to 200 m thick which provides readily available stone throughout most of the Sydney Basin. This stone is relatively porous, with only moderate strength which decreases further with higher moisture contents (McNally 2000). Nevertheless the stone is relatively durable when used as dimension stone, particularly in sites where salt loads are low. Use of the stone in sea walls requires continued replacement due to rapid weathering by ocean salts (Dragovich 2001) and serious deterioration of sandstone in a cellar at Rouse Hill Estate can probably be linked to salts in the surrounding shales.

2.4 Sandstone composition

Quartz grains comprise between 45% and 75% of Hawkesbury sandstone, with grain size ranging from fine to coarse and shapes varying from well-rounded to angular. In many stones quartz displays overgrowths which cause the stone to glitter. Rock strength is influenced by the bonding between quartz grains which provides a relatively stable framework for the clay minerals present. The proportion of clay in Hawkesbury sandstone is variable, ranging from as low as 10% and as much as 45% (Spry 2000). Although mixed layer minerals are common in some of the Sydney Basin sandstones, the Hawkesbury sandstone is dominated by kaolinite with lesser quantities of mixed layer illite-smectite. The cement is largely siliceous, with minor siderite and sometimes a little calcite present (Spry 1983). The high quartz content and stable clays that dominate the matrix mean that the stone has been widely used as a dimension stone (McNally 2000).

Once cut for dimension stone, the sandstone mellows to a dull yellowish colour within a few years. It appears that some of the iron present migrates through the relatively permeable stone to the surface, creating a slightly case hardened layer to a depth of a millimetre or more. This 'hardening' is sufficient to maintain an intact cut surface which, until subsequently breached by flaking or scaling, retreats only slowly through granular disintegration.

2.5 Sandstone blocks on the verandahs

On the verandahs at Rouse Hill, large blocks of more decorative and partly foliated sandstone were placed at the entrances to doorways. Two of these blocks have dimensions of approximately 1.6 m by 0.9 m. Plain blocks of lesser and variable dimensions make up the remainder; these blocks range in size from 0.4 m to 0.7 m. Most of the originally pale-yellow to pale grey stone has altered to a dull yellow orange colour (around Munsell 10YR 7/2). The sandstone blocks have shifted slightly over time. Since emplacement, settling of individual sandstone blocks has led to the development of gaps up to 36 mm wide between adjacent stones, with some blocks now sloping slightly away from the house, usually at angles of less than 10 degrees. A photograph of the verandah taken in about 1900 (Thornton 1988) shows the comparatively tight arrangement of the stones some 44 years after they were set.

3 METHOD AND RESULTS

3.1 Weathering classification

Various systems have been devised for describing weathering on specific structures (e.g. Fitzner et al. 2003) or largely natural features (e.g. Campbell 1991). In this preliminary investigation, existing weathering was described and classified in terms of type (e.g.

184

disintegration, flaking, fracturing), areal extent, and depth, on a block-by-block basis using a sample area of sandstone blocks comprising part of a covered west-facing verandah. The verandah is 1.85 m wide, allowing for windblown rain and direct sunlight to reach all the blocks. In this preliminary investigation, existing weathering patterns on a total of 13 blocks were described and samples of loose grains were removed for SEM analysis. The small 'study site' included an area where pedestrian traffic would have crossed the verandah to enter the house when it had been occupied. Each of the 13 blocks was categorised using the weathering classification system outlined below. The system involved assessing the general weathering grade (little, minor, moderate or severe – Table 1), describing the weathering features (forms and processes present – Table 2), and categorising expected movements of pedestrian traffic (Table 3).

3.2 Weathering grade

Weathering grades ranged from 0.5 to 3 on the 13 blocks classified. Two blocks were assessed as having

Table 1. Weathering grade.

Grade #0: Little weathering. Original cut surface or <1 mm surface loss.
Grade #1: Minor weathering. <2 mm surface loss, parts of original cut surface may persist.
Grade #2: Moderate weathering or one layer removed, 2–4 mm surface loss.
Grade #3: Severe weathering or two layers removed, >4 mm surface loss.
Intermediate grades were recorded on a sliding scale.

Table 2. Weathering features.

O   : Original cut surface
Fe : Evidence of iron (Fe staining from rusted verandah fix tures or case hardening)
F   : Foliated weathering pattern
GD: Granular disintegration
S   : Scaling/flaking
CR: Concentrated weathering around block corner/s, causing rounding.
Intensity was qualitatively indicated as minor, moderate or severe.

Table 3. Thoroughfare categorization.

Major	–	Everyone using doorway must use this block
Moderate	–	People walking along verandah likely to use this block
Minor	–	People can use this block but are not likely to
Not applicable	–	People physically cannot use this block for thoroughfare

a grade of 0.5 (very little weathered), with six blocks classified at a grade of 2 or above. On two blocks, weathering was very intense but was of limited extent. Most of the blocks – eight of the total – lay in the weathering grade range of 1 to <3.

3.3 Weathering features

Features described included the retention of the original cut surface, evidence of scaling/flaking and disintegration, and the presence of concentrated weathering around block edges especially corners (Table 2). Of the 13 blocks recorded, four retained part of the original cut surface and these stones also showed rounding of corners. Granular disintegration occurred on most stones although only two were categorised as having severe or intense disintegration. Four blocks had moderate or severe flaking/scaling.

The extent of weathering was also described by estimating the proportion of the surface affected by the nominated weathering features. Two of the three intensely weathered blocks had more than 95% of their surface affected. The third block had around 70% of the surface severely affected by granular disintegration and also recorded moderate scaling and corner rounding. Pedestrian traffic on this block was assessed as minor. Apart from these three blocks, the extent of weathering varied between 5% and 40% of the surfaces exposed.

3.4 Thoroughfare categorisation

Blocks were classified according to the likelihood of people traversing them in order to gain entry to the house or to walk along the verandah (Table 3). Smaller blocks which were close to the house walls or blocks where verandah posts had been emplaced would have rendered pedestrian movement impossible or improbable, and the five blocks in these positions were classified as having very minor or minor potential for pedestrian impacts. At the other extreme, all visitors would walk over the two large blocks placed at the entrance way to the house. Both of these blocks also showed scaling, varying degrees of corner rounding, and were weathered over >95% of their surfaces.

3.5 Weathering patterns

A common feature on the majority of blocks was corner rounding. Adjacent blocks are no longer tightly arranged but are separated by gaps up to 36 mm wide, allowing for infiltration and retention of moisture against block edges. This problem is exacerbated by the uneven settling of individual blocks whose now-irregular slopes lead to non-uniform shedding and accumulation of moisture. Although the practice has now ceased, management at one stage experimented

185

with infilling the gaps with clean industrial grade quartz sand. Quartz grains are in any case naturally released from the sandstone by granular distintegration, sometimes in such abundance that loose grains have been removed by vacuuming. The quartz sand used for infilling resulted in additional sand grains being tracked across the blocks and probably augmented the abrasive action of pedestrian traffic.

Corner rounding is likely to have occurred whether or not pedestrian traffic was present as these edges would be exposed to natural weathering processes from both sides. However the uneven settling of blocks may have resulted in some edges being slightly elevated, exposing them to greater pressure from foot traffic, or other edges being in a small depression which may retain moisture for longer periods. Any block adjacent to such a depression would also be subject to greater moisture and to added pressure from foot traffic.

The two most seriously weathered blocks (>95% of the surface affected) were the two largest which had been placed at the entrance to the house. This association between extent of weathering and pedestrian traffic is expected. When information on other blocks was examined, the weathering/traffic linkages were less clear. A third block which had intense weathering was located on the outer edge of the verandah and was assessed as being in a low category of pedestrian trafficability. This example and others suggest that multiple variables are influencing weathering outcomes. In particular, factors such as rock characteristics, stone masons' decisions, past occupation of the house, and non-uniformity in the weathering environment have contributed to a weak general association between weathering and existing pedestrian impacts.

3.5.1 *Rock characteristics*
Kaolinite was the dominant clay identified in the SEM in small samples collected from stone surfaces. However in some samples illite was present in large amounts and kaolinite was only of minor importance. Quartz grain size was also variable, with iron being mainly identifiable as traces only. More detailed petrographic work is needed to establish the nature of individual blocks as, despite the sandstone probably being obtained from the one quarry, the rock varies considerably between different blocks.

In conjunction with mineralogical determinations, permeability assessments would assist in explaining the persistence of a few original cut surfaces and not others, and possibly the distribution of flaking and scaling.

3.5.2 *Stone masons' decisions*
Stone masons will often choose more durable stones for decorative work or for large slabs. In the case of the house at Rouse Hill, the large blocks may have been selected on aesthetic rather than durability grounds, as the foliated appearance creates an attractive ripple effect. Both of these large stones have weathered similarly. The smaller stones vary in their weathering condition, with those nearest the house being least affected. In most cases stone edges appear to show evidence of weathering before the central parts of the blocks, with Grade 2 weathering (moderate, 2–4 mm surface loss) usually appearing around the margins of the stones which recorded generally lower weathering grades overall.

3.5.3 *Past occupation of the house*
Much of the weathering on entrance blocks has historical origins and does not relate to current visitor numbers. It was assumed in assessing pedestrian movements that past patterns of movement would have been similar to present patterns, but that both intensity of use and dispersion of traffic on the verandah would have been greater in the past. This is because current management discourages visitors from walking on the verandah, except to gain entrance to the house.

3.5.4 *Non-uniformity in the weathering environment*
Unless there are special circumstances operating, weathering is likely to preferentially affect block edges. Settling effects and resulting uneven moisture distribution would also lead to non-uniform weathering patterns. On the Rouse Hill house verandah, there is uneven exposure to sunlight and shading, as well as evidence of gutters overflowing at particular points in the past. Fungal growth was identified in the SEM in samples taken from a damp area of probable gutter overflow.

3.6 *Impacts of pedestrian traffic*

Pedestrian impacts involve both pressure where foot traffic occurs, and probable abrasion of surfaces. On some blocks where scaling is evident, recent fractures have developed around the scales. Foot traffic is believed by local staff to have caused the breakage of small scales, leading to an enlargement of the affected areas. It is assumed that pressure from people walking over already-fragile scales hastens the deterioration of these surfaces.

The hollowing of sandstone steps indicates that abrasion is important in contributing to wear. Visitors may bring soil on their shoes which would abrade the stone. However it is likely that the 'tools' for most abrasive action comes from quartz grains released naturally from the sandstone as part of granular disintegration. Once released, these grains will be subject to foot pressure and aid in dislodging further grains.

186

4 CONCLUSIONS AND RECOMMENDATIONS

Initial observations on pedestrian impacts on part of a sandstone verandah suggest that heavily trafficked areas are affected by abrasion. After 150 years most of the blocks have lost some or all of their original cut surface and are showing signs of weathered edges, granular disintegration and flaking. Managers are therefore concerned that the current already-weakened condition of the blocks will deteriorate further if they are subjected to extensive pedestrian traffic.

It is proposed to monitor visitor impacts by first making detailed maps of weathering condition of all blocks on the verandahs. At least two small sites will then be selected for comparative purposes, one with current pedestrian traffic and the other without. In so far as is possible, variables like rock composition, block slope and size, and current weathering condition will be taken into account. Monitoring will involve continuous recording of temperature and humidity; collecting loose quartz grains to estimate weathering loss and for SEM analysis; photographing (microscope) small areas for monitoring the persistence of individual quartz grains on exposed surfaces; and fine-scale mapping of the distribution of flake edges to establish their rate of change in trafficked and non-trafficked areas.

REFERENCES

Aplin, G. 2002. *Heritage: Identification, Conservation and Management*. Melbourne: Oxford University Press.

Campbell, I.A. 1991. Classification of rock weathering at Writing-On-Stone Provincial Park, Alberta, Canada: a study in applied geomorphology. *Earth Surface Processes & Landforms* 16: 701–711.

Dragovich, D. 2001. Condition of the sea wall at Farm Cove, Australia. *Bulletin of the Geological Society of Greece* 34(5): 1749–1754.

Fitzner, B., Heinrichs, K. & LaBouchardiere, D. 2003. Weathering damage on Pharaonic sandstone monuments in Luxor-Egypt. *Building & Environment* 38: 1089–1103.

Historic Houses Trust of New South Wales 1997. *Rouse Hill House: Museum Plan*. Sydney: Historic Houses Trust of NSW.

McNally, G. 2000. Introduction – sandstone and the Sydney environment. In G.H. McNally & B.J. Franklin (eds), *Sandstone City: Sydney's dimension stone and other sandstone geomaterials*: ix–xiv. Springwood, EEHGS of the Geological Society of Australia.

Spry, A.H. 1983. *Australian Building Sandstones*. Frewville, Australian Mineral Development Laboratories.

Spry, A.H. 2000. Sydney sandstone – an overview. In G.H. McNally & B.J. Franklin (eds), *Sandstone City: Sydney's dimension stone and other sandstone geomaterials*: 1–22. Springwood: EEHGS of the Geological Society of Australia.

Stapleton, I. 1985. Conservation philosophy and the Burra Charter. In P. Freeman, E. Martin & J. Dean (eds), *Building conservation in Australia*: 37–38. Red Hill: Royal Australian Institute of Architects.

Thornton, C.R. 1988. *Rouse Hill House and the Rouses*. Nedlands: Caroline Thornton.

Dimension Stone 2004, Přikryl (ed)
© *2004 Taylor & Francis Group, London, ISBN 90 5809 675 0*

Non-destructive IR-spectroscopy measurements at weathered natural stone objects – case studies

A. Ehling
Federal Institute of Geosciences and Natural Resources, Berlin, Germany

J. Stein
Ornamental Stone Investigations, Waldsee, Germany

ABSTRACT: A new method for non-destructive analyses of mineral materials had been tested for its applicability at weathered natural stone surfaces. The Portable Infrared Mineral Analyser (PIMA) is a short wave infrared reflectance spectrometer that operates in the wavelength region from 1300 to 2500 nm. Some weathered surfaces of lime- and sandstones had been analyzed by PIMA. Based on three examples possibilities and limitations of this device are presented.

1 EQUIPMENT AND METHOD

The Portable Infrared Mineral Analyser (PIMA) is a short wave infrared reflectance spectrometer that operates in the wavelength region from 1300 to 2500 nm. In this wavelength region, minerals that contain hydroxyl radicals (such as clays, amphiboles, some sulphates) and carbonate radicals, absorb incident radiation at specific wavelengths and in relative amounts that are diagnostic of the minerals species. A lot of other minerals also produce diagnostic spectra. These are commonly fine-grained minerals and small amounts of them are often difficult to identify.

The PIMA instrument produces reflectance spectra that are saved as individual binary files. The spectra can be interpreted manually by a trained interpreter or by computer-automated interpretation techniques to provide analyses of the minerals. In praxis both methods are often combined. The spectra are measured at an area with a diameter of about 10 mm. Measuring distance can be between 3 (contact with the surface) and 7 mm.

IR spectroscopy itself is not a new method. The advantage is the little size. The PIMA is light, portable and easy to use. The restricted wavelength region is limitation and advantage at the same time. Some minerals can't be obtain in this region but others are better to identify because some dominating minerals like quartz are absent in this region and don't disturb other spectra.

PIMA allows non-destructive high resolution analyses at the object. Even it is not possible to identify all minerals this method delivers first results in situ and considerably decreases sampling.

2 INVESTIGATED STONES

2.1 *Sandstone "Santa Fiora"*

"Santa Fiora" is a Tertiary aged sandstone from Toscana (Italy). It is a massive, auburn, fine- to medium grained, calcareous sandstone. Compounds are quartz, calcite, feldspar, chlorite and rock fragments. Accessory there are only hematite, and biotite. Cementing material consists of authigenic calcite and iron hydroxides. The pore sizes are very small. Surface of the sandstone looks inhomogeneous because of sedimentary textures and irregular distribution of compounds.

The sandstone is used at an external wall of nearly 3 cm thickness, rear ventilated. Exposition time has been 5 years. Already after two years the surface shows irregularly distributed white, rough areas.

2.2 *Limestone "Fränkischer Muschelkalk"*

"Fränkischer Muschelkalk" is a limestone of the Upper Muschelkalk age from Frankonia (Germany). It has a distinctly bedding with beds rich in shells and matrix dominated beds. The fuscous matrix is micritic,

189

irregularly distributed and contains streaks of ochre lime. Parts of fossils and pores are recrystallised. Porosity is 1.85 vol. %, most frequent pore size is less than 0.1 μm (Grimm 1989).

The limestone is situated at an external wall of 2 cm thickness at the base of a building. It is fixed with mortar and surface honed. Exposition time is about 50 years. In order to its site the stone is extremely polluted, especially in the lower part.

2.3 Limestone "Travertin Thüringen"

The travertine is of Pleistocene age and comes from Thuringia (Germany). It is a greyey-white and yellowish, distinctly banded limestone. Massive layers alternate with layers of high porosity. It often contains incrustations of plants, bones and other fossils. Dominating mineral is calcite, aragonite may occur, too.

The limestone had been used at an external wall with a minimum thickness of 10 cm. The stones are fixed with mortar, the surface is stocked. Exposition time is about 100 years. The surface is homogeneously polluted. In the range of the entrance visible pollution at the surface is differently developed due to the protected situation.

3 IR SPECTROSCOPY INVESTIGATIONS

3.1 Sandstone "Santa Fiora"

The visibly most weathered parts of the sandstone had been analyzed in comparison with the apparently less weathered ones. Reference investigations had been made at fresh samples.

In Figure 1 the spectra of these three samples, respectively sample areas are shown. First of all it is apparent that the fresh sandstone shows very few reflectance. This is due to the relative high amount of iron oxides and – hydroxides in the sandstone. Both do not produce spectra in this wavelength region but they hamper reflection of other minerals. As a result only the reflectance of the dominating mineral calcite is to be seen in outlines (quartz doesn't produce spectra in this wavelength region).

The spectra of the weathered stones are very similar. They differ only in the intensity of the reflectance due to the visible intensity of weathering. Peaks of calcite are more clearly visible. Chlorite can be identified, too. It is obviously, that chlorite becomes more dominant in the spectra due to increasing weathering intensity. Interpreting these spectra we can state, that Fe-oxides and hydroxides had been dissolved and led away. This corresponds with the visible bleaching of the sandstone. Without, respectively with less iron, calcite and chlorite could produce clear spectra. In the most weathered area chlorite shows the highest

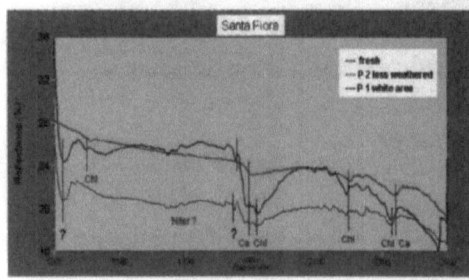

Figure 1. IR-spectra of "Santa Fiora" sandstone: fresh, strong and less weathered (top down).

reflectance. Considering the roughness of this area it could be interpreted as a decrease of calcite amount, caused by dissolution of calcite at the surface.

There are some other peaks in the spectra of the weathered areas, which are obviously but not easily to identify. There is not any known reference spectrum in our data base, which corresponds to them. Identification has to be made by laboratory analysis.

3.2 Limestone "Fränkischer Muschelkalk"

This kind of limestone is very often used in Germany. The exposition we choose is very typical for polluted stones: at the base of the building, in the range of spreading water and debris.

A vertical profile has been analyzed from bottom up to a level of 50 cm (Fig. 2). The results are presented in Table 1 and partly in Figure 3.

The spectra of levels 1 and 4, as well as the spectra of levels 3 and 5 are very similar with few exceptions. Therefore not all spectra are shown in the diagram. The main minerals detected by the IR spectroscopy are calcite, gypsum and kieserite. Intensity of reflectance depends on the amount of iron-bearing pollution and water: the most polluted, dark, and/or wet regions show only little and indistinct peaks.

Kieserite could be identified in most spectra and in the upper parts gypsum becomes more and more evident. Even in these spectra there are some compounds which could not be identified. Some peaks can be assigned to water and organic material but not all of them.

Obviously there are levels with very similar spectra. As a result the determination of sampling points is facilitated and the number of samples for further investigation can be reduced.

3.3 Limestone "Travertin Thüringen"

The travertine had been measured at a protected area sideways to the entrance. In some parts the surface shows light grey discoloration (Fig. 4). Results of the

190

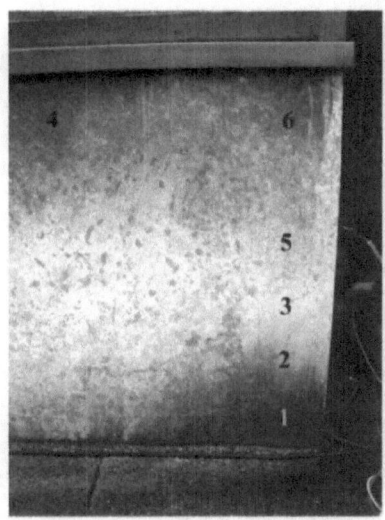

Figure 2. Muschelkalk base at a building with measured points.

Table 1. Detected mineral spectra at a polluted Muschelkalk surface (x – indistinctly; xx – distinctly; xxx – very distinctly).

	Calcite	Gypsum	Kieserite	?
P 6	xxx	xxx	x	–
P 5	xxx	xx	x	–
P 4	xx	–	–	xx
P 3	xxx	–	x	–
P 2	xxx	–	x	xx
P 1	xx	–	–	–

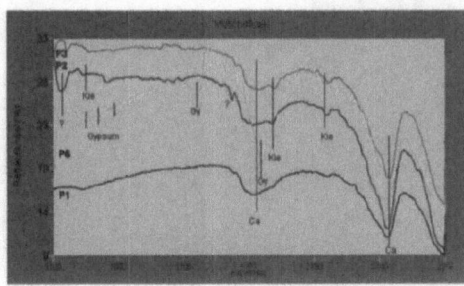

Figure 3. IR spectra of weathered Muschelkalk base at a building.

measurements are presented in Figure 5 together with a reference spectrum of fresh Thuringian Travertine.

The spectra are very clear. As anticipated the spectra of the weathered stone are flattened compared

Figure 4. Travertine area with measuring points.

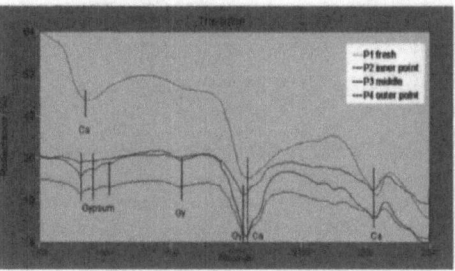

Figure 5. IR spectra of travertine: fresh and weathered.

with the fresh stone because of pollution and humidity. In the spectra of points 3 and 4 gypsum can be identified. There is no hint to other mineral spectra.

4 CONCLUSIONS

Non-destructive measurements with the infrared reflectance spectrometer PIMA at weathered stone surfaces can produce some first results with regard to the mineralogical changes at the surface.

At sandstone surfaces the identification of minerals is facilitated by the fact, that the main compound quartz does not produce spectra. Therefore the other interesting minerals are mostly well identifiable.

At limestone surfaces the spectrum of calcite dominates and often blankets the spectra of other compounds. Nevertheless gypsum and salt could be identified at a calcite surface.

Generally, heavily polluted, iron-bearing, and wet surfaces does not produce any evaluable spectra.

191

Mineral identification is only possible on the base of a comprehensive library with reference spectra. Even if it is not possible to identify all interesting minerals, PIMA can detect areas with similar spectra and therefore reduces sample number for laboratory analyses.

REFERENCE

Grimm, W.-D. 1989. *Bildatlas wichtiger Denkmalgesteine der BRD*. München

Dimension Stone 2004, Přikryl (ed)
© *2004 Taylor & Francis Group, London, ISBN 90 5809 675 0*

Construction material interaction in historical stone bridge structures

M. Gregerová
Masaryk University in Brno, Faculty of Science, Brno, Czech Republic

P. Pospíšil
Brno University of Technology, Faculty of Civil Engineering, Brno, Czech Republic

ABSTRACT: The paper gathers partial results of selected Charles Bridge construction materials study. Potential interactions are studied in relation with outer (climatic) conditions, moisture and pH measured in construction materials in sampling places. It is discussed the mobility of neogenic minerals, their crystallization progress, sources conditioning formation of individual salts. They are discussed within the relation to identified minerals possible variations of mineralised solutions and their migration through material layers in bridge structure together with their influence on degradation of materials.

1 INTRODUCTION

Material investigation of some objects, especially study of heterogeneity, moisture content and temperature variation, content of secondary salts and their time development should precede the reconstruction of every stone historical object. Inseparable part of this study must be determination of external conditions affecting the object. The investigation of secondary salts cannot be reduced only to their identification and determination of their amount. It is very important to know their distribution in material (voids, micro-cracks, contact surfaces of natural and artificial materials etc.).

This problem is usually solved in practice by application of visual investigation. It is resulting in imperfections in reconstruction and formation of failures in construction materials within a short time interval after reconstruction.

Assessment of distribution and amount of salts in construction material is usually complicated. They do not exist right general criteria for assessment salt content and their effect on degradation of construction materials up to now. This situation first of all reflects different properties of salts and their destructive effects together with different properties of individual construction materials and their resistance against to salt action. The behaviour of salts in various climatic conditions is also different. It is depending mainly on temperature and moisture variations of construction materials.

2 METHODS OF STUDY

Construction materials in Charles Bridge (Prague, Czech Republic) are systematically studied since 1994. The methods of investigation and assessment of degradation processes in stone structure were created on the base of detail macro- and microscopical studies of material and failures in the structure during all seasons including verification of effectiveness of individual analytical methods.

Micro- and nanostructure studies of secondary minerals and efflorescences are very specific. Mineralogical and petrologic studies are based on microscopic characteristics (Fitzner 2002). They are focused on assessment of microstructures and mineral composition of construction materials. Polished sections prepared without the application of wetting agents are used for nanostructure study and micro chemical analyses. They are coated by carbon after microscopically analyses. Micro chemical analyses were executed in the laboratory of electron microscopy and microanalysis of the Faculty of Science of Masaryk university and Czech Geological Survey in Brno (analysts R. Čopjaková and R. Škoda) on Cameca SX 100 under following conditions: wave dispersion mode, acceleration voltage 10 and 15 kV, beam current 4 nA, beam diameter 2–5 μm, detection time 10–20 s for one element. Standards: Na K_α – jadeite, Mg K_α – almandine, Al K_α – albite, Si K_α – andradite, Ca K_α – andradite, Fe K_α – andradite, Cl K_α – vanadinite, S K_α – barite, K K_α – orthoclase. Average detection limits are Na – 878 ppm,

Mg – 583 ppm, Al – 486 ppm, Si – 587 ppm, S – 1013 ppm, Cl – 248 ppm, K – 341 ppm, Ca – 746 ppm, Fe – 665 ppm. Gathered data were performed in Excel and Formula. Thaumasite and ettringite were recalculated on 6 apfu Ca and the other sulphates on 4 apfu O. The X-ray maps were created in wave dispersion mode, stage scan, acceleration voltage 15 kV, beam current 15 nA. Efflorescence's – relief sections of salts were coated by gold and documented by images on the same apparatus in BSE and SE modes under following conditions: acceleration voltage 15 kV, beam current 1–4 nA.

3 RESULTS OF ANALYSES

Petrographic analyses proved following materials in the Charles Bridge:

Photo 1. Overview of sandstone structure with calcite matrix. Cameca SX 100, photo R. Čopjaková.

- Carboniferous arkosic sandstone to conglomerates from Kamenné Žehrovice and its vicinity;
- Permian ferrous sandstones red brown in color from Český Brod and Nučice and their vicinity (applied rarely in various places in the structure);
- Cretaceous (Cenomanian) sandstones from Nehvizdy, Vyšehořovice and Brandýs nad Labem;
- Cretaceous (Cenomanian) quartz-kaolinite sandstones from Hořice and its vicinity (pillars V. and VI. arches, in some places in XII.–XVI. arches – reconstruction in 1890–92);
- Cretaceous sandstones from Hloubětín quarries;
- Cretaceous (Cenomanian) quartzose sandstones, arkosic sandstones to arkoses from Božanov (mostly in V., VI. arches, on stairs to Kampa island – reconstruction in 1966–1975, 1996);
- Sandy marlstones (filling material inside of the structure) from Strahov, Bílá hora and Přední Kopanina;
- Lime-mortars – original and from older reconstructions;
- Mixed and cement mortars – from younger reconstructions;
- Lightweight concrete – used within the reconstruction in 1966–75;
- Reinforced concrete – used within the reconstruction in 1966–75;
- Colcrete (cement paste) – used within reconstruction in 1966–75 as binder material within the reconstruction of facial stone ashlars.

The basic finding within the petrographical studies is proving of considerable material heterogeneity within Charles Bridge materials. It is caused primarily by application of various construction materials and consecutively it was increased during the reconstructions.

Facial ashlars (sandstones s.l.) of the Charles Bridge structure differ by carbonate-sulphate matrix

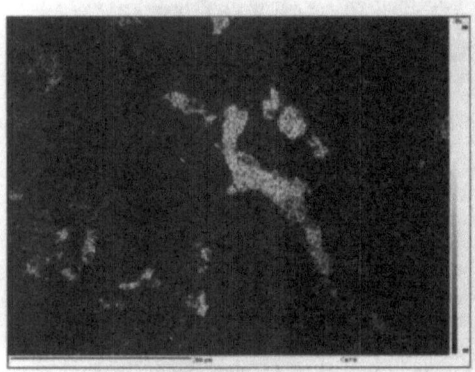

Photo 2. X-ray maps – element CaK$_\alpha$. Cameca SX 100, photo R. Čopjaková.

of primary sources of the sandstones (in quarries) s.l. It was proved the presence of sulphates in depth 1–2 mm, below them up to the depth 5 cm it is identifiable calcite (Photo 1, 2). Microstructure of sandstone ashlars is weakened by crystallization pressures.

Efflorescences on sandstone ashlars (s.l.) were observed during the last ten years (see Table 1). Mineral paragenesis photo documentation is showed in Photos 3–8. Differences in mineral paragenesis relate probably to season, episodic variations of moisture content and pH.

Identification of sulphate degradation in inner concrete elements (Gregerová, in press, Gregerová & Pospíšil 2003) of the structure together with identification of transitional secondary zones between the joint mortars and sandstones (s.l.), between the sandstones (s.l.) and efflorescences is basic for the analysis of joint mortars. They have been studied mortars with cement and sand and joint cement pastes. They

194

Table 1. Efflorescence mineral association of the Charles Bridge structure.

Minerals	Chemical composition	1994	1999	2003
Opal	$SiO_2 \cdot nH_2O$	+		
Bassanite	$2CaSO_4 \cdot H_2O$	+		
Gypsum	$CaSO_4 \cdot 2H_2O$	+	+	+
Anhydrite	$CaSO_4$			
Mirabilite	$Na_2SO_4 \cdot 10H_2O$		+	+
Trona	$Na_3(HCO_3)(CO_3) \cdot 2(H_2O)$		+	+
Sylvine	KCl	+		
Halite	$NaCl$		+	+
Scawtite	$Ca_6(Si_3O_9)_2 \cdot CaCO_3 \cdot 2H_2O$	+		
Jarosite	$KFe_3 [(OH)_6/(SO_4)_2]$	+		+
Amonio jarosite	$(NH_4)Fe_3(SO_4)_2(OH)_6$	+		
Natrojarosite	$NaFe_3^{3+}(SO_4)_2 (OH)_6$	+		+
Nitrokalite	KNO_3		+	+
Nitronatrite	$NaNO_3$		+	+
Syngenite	$K_2Ca[SO_4]_2 \cdot H_2O$	+		+
Giniite	$Fe^{2+} Fe_4^{3+}(PO_4)_4 (OH)_2 \cdot 2H_2O$		+	
Goethite	$FeOOH$	+		
Kutnohorite	$CaMn(CO_3)_2$	+		
Darapskite	$Na_3(SO_4)(NO_3) \cdot (H_2O)$			+
Aphthitalite	$K_{2.25}Na_{1.75}(SO_4)_2$	+	+	+
Amarillite	$NaFe(SO_4)_2 . 6H_2O$	+		
Halotrichite	$Fe^{2+} Al_2(SO_4)_4 \cdot 22(H_2O)$	+		
Pickeringite	$MgAl_2(SO_4)_4 \cdot 22(H_2O)$	+		+
Bílinite	$Fe^{2+} Fe_2^{3+}(SO_4)_4 \cdot 22(H_2O)$	+		
Alunogene	$(Al_2(SO_4)_3 \cdot 17(H_2O))$			+

Photo 4. Sample No. IV-3-D (c). Tabular, irregular confined crystals of alunogene and jarosite owergrowing crystals of gypsum. Cameca SX 100, photo R. Čopjaková.

Photo 5. Sample No. IV-3-D (c). Tabular, irregular confined crystals of alunogene $(Al_2(SO_4)_3 \cdot 17(H_2O))$. Cameca SX 100, photo R. Čopjaková.

Photo 3. Sample No. IV-3-D (c). Corrosion of gypsum crystals. Cameca SX 100, photo R. Čopjaková.

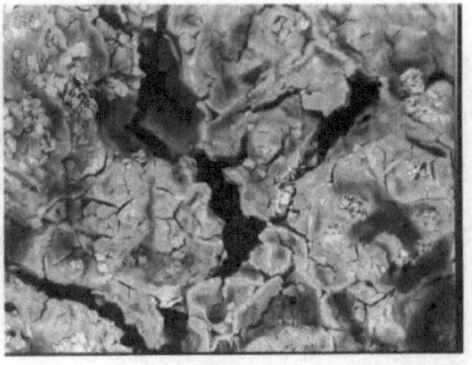

Photo 6. Sample No. IV-3-D-6. Tabular crystals of jarosite, originating by crystallization from gel. Cameca SX 100, photo R. Čopjaková.

195

Photo 7. Sample No. IV-3-D-1. Tabular crystals of jarosite. Cameca SX 100, photo R. Čopjaková.

Photo 8. Sample No. X-2-E. Halite with white coatings of niter. Cameca SX 100, photo R. Čopjaková.

Table 2. Chemical composition and element representation in analysed secondary minerals in the sample points of III. Arch.

Sample	III-1-A	III-1-A	III-1-A	III-1-A
Na_2O	0	0	0.55	0.6
MgO	3.58	5.24	0	0.04
Al_2O_3	1.16	1.14	5.68	0.96
SiO_2	18.65	21.74	6.65	13.15
SO_3	20.02	18.26	18.98	29.77
Cl	0.17	0.07	0.99	0.53
K_2O	0	0.03	0.88	0.45
CaO	40.75	34.1	29.82	36.94
Σ	84.36	80.61	63.55	82.42
Mineral	Ett-th	Ett-th	Ett-th	Ett-th
Element representation in formula				
Na^+	0	0	0.19	0.17
Mg^{2+}	0.65	1.06	0	0.01
Al^{3+}	0.17	0.18	1.18	0.16
Si^{4+}	2.28	2.94	1.17	1.91
S^{6+}	1.84	1.85	2.50	3.24
Cl^-	0.035	0.02	0.30	0.13
K^+	0	0.01	0.20	0.08
Ca^{2+}	5.35	4.94	5.62	5.74
O^{2-}	16.32	17.70	17.27	19.59
Cation Σ	10.29	10.97	10.85	11.31
Anion Σ	16.36	17.71	17.57	19.72
Ca etc. position	6.00	6.00	5.99	6.00
Si + Al	2.451	3.121	2.346	2.071
S + Cl	1.875	1.869	2.799	3.37

were selected three examples of sampled and analyzed points (25) from III. arch (Tables 2–4).

It is apparent from the micro chemical study that with increasing ratio of Mg in ettringite and thaumasite sulphur usually considerably decreases and stoichiometry of mineral is "disturbed". This process was also observed in the border pore parts and mineralized veins.

The mixture of crystals of ettringite and thaumasite in association with Mg secondary minerals (huntite?, magnezite?, dolomite?, event. brucite?) were identified in some places.

Their identification due to their very small size and low concentration was not possible neither by X-ray analysis nor micro analytically. The moisture content of sample was 11.9%, pH 7.

Gypsum and syngenite were identified in the mortar sample III-4-C. Neither Mg^{2+} nor Al^{3+} cations were identified in the secondary filling pore matters.

The increased contents of Cl^- occurred in number of analyses grains. Bond with Ca^{2+} or Mg^{2+} seems to be the most probable. In the given case together with Cl^- the high content of Mg^{2+} were identified in the mixture ettringite-thaumasite. It was found that with the increasing content of Mg^{2+} content of S considerably decreases in the border parts of pores. Neither in this case stoichiometry of sulphates does correspond. The content of Mg, reaching value higher than 15%, is too high for bond in sulphates of ettringite and thaumasite types.

Microstructure of ettringite-thaumasite aggregates is supported by Photos 9 and 10. The moisture content was 12.7%, pH 6.5.

4 DISCUSSION

All studied surface samples (facial parts of ashlars in arch) of construction stones of Charles Bridge structure differ from original ones by carbonate-sulphate matrix. The matrix is formed by crystallization from leaked solutions. The increasing concentration of salt during the evaporation results in gradual filling of pore spaces and cracks. Crystallization pressures

196

Table 3. Chemical composition and element representation in analysed secondary minerals in the sample points of III. Arch.

Sample	III-4-C-dark	III-4-C-dark	III-4-C-light	III-4-C-light
Na_2O	0	0.19	1.27	1.05
MgO	0	0	0	0
Al_2O_3	0	0	0	0
SiO_2	0.23	0.05	0.04	0.35
SO_3	55.36	56.91	47.78	49.72
Cl	0.04	0.03	0.06	0.03
K_2O	0.35	0.24	16.48	15.24
CaO	40.74	40.84	26.95	28.2
Σ	96.79	98.29	92.61	94.6
Mineral	Gypsum	Gypsum	Syngenite	Syngenite
Element representation in formula				
Na^+	0	0.009	0.066	0.053
Mg^{2+}	0	0	0	0
Al^{3+}	0	0	0	0
Si^{4+}	0.005	0.001	0.001	0.01
S^{6+}	0.98	0.99	0.97	0.97
Cl^-	0.002	0.001	0.003	0.001
K^+	0.01	0.01	0.57	0.51
Ca^{2+}	1.03	1.02	0.78	0.79
O^{2-}	4.00	4.00	4.00	4.00
Cation Σ	2.03	2.02	2.38	2.33
Anion Σ	4	4	4	4

Table 4. Chemical composition and element representation in analyzed secondary minerals in the sample points of III. Arch.

Sample	III-2E	III-2E	III-2E	III-2E	III-2E	III-2E
Na_2O	0.08	0.12	0.07	0.16	0.09	0.08
MgO	0.32	0.73	18.05	16.57	16.37	16.98
Al_2O_3	4.65	5.18	13.37	14.27	10.44	14.71
SiO_2	10.54	11.08	1.55	11.59	11.39	11.87
SO_3	24.78	20.47	7.63	5.41	3.35	3.62
Cl	0.18	0.23	1.99	1.64	2.32	2.27
K_2O	0.09	0.12	0.08	0.12	0.12	0.13
CaO	37.83	37.21	16.05	20.21	18.24	7.96
Σ	78.47	75.15	58.78	69.96	62.33	67.72
Mineral	Ett-th Vein	Ett-th Vein		Grain with Cl		
Element representation in formula						
Na^+	11.03	10.72				
Mg^{2+}	18.37	17.23				
Al^{3+}	0.02	0.03				
Si^{4+}	0.07	0.16				
S^{6+}	0.80	0.89				
Cl	1.53	1.61				
K^+	2.70	2.2				
Ca^{2+}	0.04	0.06				
O^{2-}	18.33	17.18				
Cation Σ	11.03	10.72				
Anion Σ	18.37	17.23				
Ca etc. position	6	6				
Si + Al	2.329	2.494				
S + Cl	2.747	2.287				

affect on the surrounding pore walls. The pressure can reach the value of 2–50 MPa (Winkler 1975). The critical situation becomes if the pore space and micro-cracks are completely filled and/or predisposed weakened zones are formed. These whole degraded zones gradually exfoliate by the action of gravity.

The character of efflorescences depends on the reaction kinetics. In the case of highly heterogeneous structures as Charles Bridge is, it is possible to come out from hypothesis based on the interaction between minerals and aqueous solutions. The basis of these degradation processes can be found in the difference of potential between mineral and solution.

This difference depends both on the mineral and the composition of aqueous phase (Helgerson 1983). Water, air oxygen, carbon dioxide and other acid forming oxides of atmosphere represent the essential components of degradation under real conditions. In the case of heterogeneous material environment it is necessary to take into account not only construction materials themselves but it is also important to derive changes gradually. The type of winter season maintenance salts (technical salt and carbamide) utilized in the past period must be taken into account as well.

The sources of salts must be found in the inner structure layers of the Charles Bridge: in concrete elements, sandy marlstone filling material and mortars. This fact must be respected dealing with all

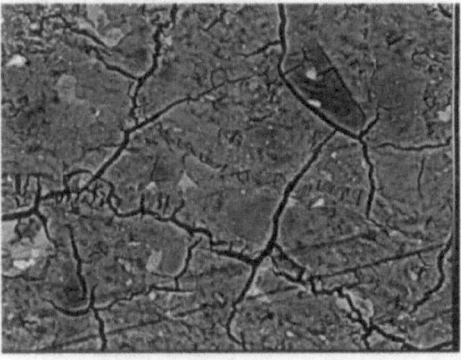

Photo 9. Sample No.: III-2-E Characteristic microstructure ettringite-thaumasite association with noticeable system of fine fissures. Cameca SX 100. Photo R. Škoda.

consequent reactions running on the contact of aqueous solutions and minerals. Distinct heterogeneity of the structure and lost of functionality of insulation layers is the most important precondition for efflorescences formation.

197

Photo 10. Sample No. III-2-E Ettringite-thaumasite veins association with noticeable system of fine fissures. Cameca SX 100. Photo R. Škoda.

Although the study of provenance of utilized sandstone ashlars of Charles Bridge undoubtedly proved that sandstones with calcite matrix were not originally applied, at present times calcite matrix was identified in all samples of sandstone.

In time the efflorescence composition changes the ratio of nitrates relatively rises on the contrary to sulphates. The ratio of anhydrous forms of salts and their hydrates simultaneously varies probably in dependence on climatic conditions. Agriculture and hydrocarbon burning represent the most important producers of nitrates. Nitrates can be washed out by Vltava river and represent the component of atmosphere. Their ratio in atmosphere rises during last years in Prague downtown (Šolc & Hrubý ed. 1997). The increase of nitrate content is in the relation to degradation of carbamide generating nutrient substance for nitrification bacteria.

Based on the study of joint mortars of Charles Bridge the occurrence of thaumasite and ettringite was proved for the first time under further described conditions. Ettringite and thaumasite are formed without gypsum in the range of pH 4.5–6 with moisture content lower or higher than 15%. Under pH higher than 7 and moisture content higher than 15% only gypsum is formed and under the same moisture and pH 7, gypsum together with ettringite and thaumasite are formed.

5 SUMMARY

The most important knowledge concluded is the verification of considerable material heterogeneity in Charles Bridge structure. The least damaged and relatively the most resistant are the sandstones from Božanov, youngest in the structure. The sandstones (s.l.) from Kamenné Žehrovice are within the original materials the least degraded and the most resistant ones. Degradation of all studied materials is very advanced. On the basis of realized study it is apparent that newly formed mineral phases are destructive for materials of Charles Bridge structure. At first it is necessary to remove the main reason of salt migration i.e. critical variations of moisture content in the bridge caused by infiltration of precipitation water (reconstruction of insulation). The salt migration will not stop immediately but there will be long-term decreasing trend. The advanced sulphate degradation of all concrete structure elements (concrete, reinforced concrete, lightweight concrete and colcrete) is alarming (Gregerová & Pospíšil 2003). The facial surface of vaults must be treated by desalination and in some cases the ashlars with higher degree of degradation have to be replaced. The new pointing by appropriate type of mortar has to be completed. Efflorescence formation is attendant process of material degradation and simultaneously noticeable and alerting phenomenon in the structure.

ACKNOWLEDGEMENT

The research was supported by Czech Grant Agency project No.103/02/0990.

REFERENCES

Fitzner, B. 2002. Damage diagnosis on stone monuments – in situ investigation and laboratory studies (English). Proceedings of the International Symposium of the Conservation of the Bangudae Petroglyph, 15.07.2002, Ulsan City/Korea: 29–71, Stone Conservation Laboratory, Seoul National University, Seoul, Korea.
Gregerová, M. in press. Ettringite and Thaumasite Formation Conditions in Concrete Structures. *Acta Geologica Universitatis Comenianae.*
Gregerová, M. & Pospíšil, P. 2003. Thaumasite and ettringite formation affected by aggregate and cement composition in concrete. 9th Euroseminar on Microscopy Applied to Building Materials. Extended Abstracts & CD-ROM. 8 pp. 9, 12.9.2003 Trondheim, Norway.
Helgerson, H. 1983. Calculation of mass transfer among minerals and aqueous solution as a function of time and surface area in geochemical processes. 1 Computational approach. *Math. Geol.* 15: 109–130.
Šolc, J. & Hrubý, O. ed. 1997. *Zdroje znečišťování ovzduší v Praze.* Praha: Institut městské informatiky hl.m. Prahy.
Winkler, E.M. 1975. *Stone. Properties, durability in man's environment.* 2nd ed., Vienna and New York: Springer Verlag.

Dimension Stone 2004, Přikryl (ed)
© 2004 Taylor & Francis Group, London, ISBN 90 5809 675 0

Damage examination for a conservation concept at the Jewish cemetery, Hamburg Altona, Germany

S. Guse & A. Gervais

Norddeutsches Zentrum für Materialkunde von Kulturgut e.V., Hannover, Germany

ABSTRACT: The natural stones of the Jewish cemetery in Hamburg Altona are strongly affected by environmental influences and other deterioration factors. Especially the marble and limestone show severl damage. The precious inscriptions on the tombs have almost vanished. The damage examination focused on about 60 monuments. The multiple damage symptoms were intensely investigated. The diagnostic investigations were carried out with regard to conservation.

1 INTRODUCTION

The Jewish cemetery is one of the oldest cemeteries in northern Europe. Sephardic Jews from the Iberian Peninsula and Ashkenazic Jews from Eastern Europe came to Hamburg Altona at the end of the 16th century and found their last resting-place there. Circa 6.400 tombstones of the cemetery consist partially of marble and limestone. 53 marble and 2 limestone tombs were selected for our investigations as the basis of the planned conservation measures. Different environmental influences, in particular in the last ten years, have caused an increasing loss of original material and matter and as a consequence a loss of the Hebrew and Portuguese inscriptions.

The project started with the observation of the tombstones most severely damaged by environmental influences and other factors. In the interdisciplinary project the first aim was the investigation of the specific reasons for deterioration. After the detailed investigation of the reasons for and extent of deterioration and after testing possible conservation measures, conservation will follow as a final step.

2 HISTORY OF THE CEMETERY

In 1611 the Jewish cemetery in Hamburg Altona was founded. Two parts can be distinguished: the Portuguese/Sephardic part with flatness brasses and insular pyramidal tombs (Fig. 1) and the Ashkenaszic part with upstanding steles. Both cover an area of

Figure 1.   Marble pyramidal tomb (Ohel).

about 19.000 m^2 and originally contained more than 8000 graves.

The graveyard was mainly used until 1869. During World War II the bombing damaged the cemetery but fortunately most of the graves survived this period undamaged.

3 METHODS

The project activities concentrated on selected marble and limestone tombs. To explain the deterioration a great variety of analytical work was done mainly on the marble. First of all the characteristic phenomena of deterioration and also the visible conservation measures were photographed, mapped and described. Based on macroscopic, microscopic and

199

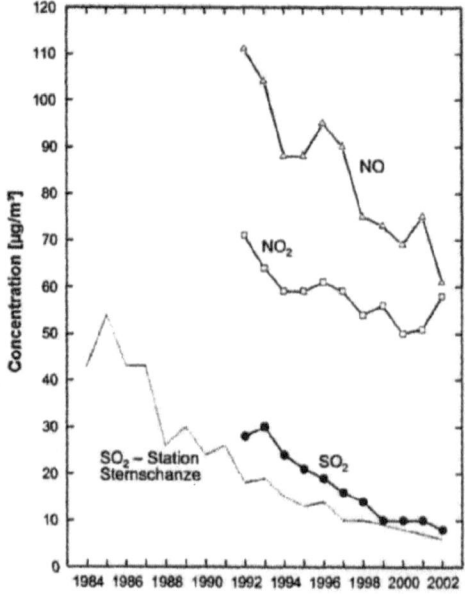

Figure 2. Annual average of the concentration of SO_2, NO and NO_2 at the Stresemannstrasse gauging station since 1992, as well as reference values of the Sternschanze gauging station (Data: HaLM 2003) (Steiger 2003).

Table 1. Ion concentration in rain samples, Jewish cemetery in mg/L (except pH) (Steiger 2003).

a) Year 2003

From–until	Vol. [mL]	Na^+	K^+	Ca^{2+}	Mg^{2+}	NH^{4+}
23.09–30.09	220	2.00	0.25	0.56	0.29	0.96
30.09–07.10	135	1.25	1.31	0.58	0.18	1.52
07.10–14.10	190	4.44	0.46	0.54	0.55	0.99
14.10–28.10	130	2.32	2.77	2.71	1.33	>0.1
28.10–18.11	180	2.68	5.43	2.10	0.62	1.19
18.11–25.11	130	1.51	1.82	1.30	0.37	3.36

b) Year 2003

From–until	Vol. [mL]	Cl^-	NO_3^-	SO_4^{2-}	pH
23.09–30.09	220	3.87	7.46	2.74	5.33
30.09–07.10	135	2.62	2.64	4.42	4.82
07.10–14.10	190	7.86	3.15	3.41	4.88
14.10–28.10	130	6.97	1.62	4.50	4.92
28.10–18.11	180	8.12	2.67	7.97	5.10
18.11–25.11	130	3.77	3.37	8.74	5.17

isotopic-geochemical analysis, the marble fits best to Carrara marble. In a second step petrographic and petrophysical investigations were carried out to characterise the state of preservation. In-situ ultrasonic-measurements were made as well as investigations on drilling cores. The in-situ velocities of the marble extend from 1.5 km/s to 5.5 km/s, those of the limestone from 5.9 km/s to 6.3 km/s. To characterise the impact of atmospheric gases on the tomb – surfaces the amount of carbon dioxide, sulphur dioxide and nitrogen oxide in the atmosphere during the last decades was assessed (Steiger 2003, Fig. 2).

Recent rainwater samples of the cemetery were analysed looking at their acidity as a damaging factor especially for carbonaceous materials (Table 1).

By means of polarising microscopy, REM with EDX the effect of microorganisms was investigated.

4 RESULTS AND CONCLUSIONS

As a very important point when dealing with Jewish cemeteries, the compatibility of all activities and possible planned measures was discussed with the authorities responsible. The results of the different non-destructive diagnostic investigations have shown the need for conservation measures to prevent further collapse.

The highly reduced ultrasonic velocities of the marbles point to an alarming situation. The velocities of the limestone exclude structural deterioration. The chemical weathering process on the tombs is affected by the emission load. Acid deposition has damaged the tombs too. The investigations of the last 20 years prove deterioration by corrosive gases. On all surfaces of the tombs there is an intense natural covering of biogenous material. The stone material itself, as well as the inclination of the tomb surfaces and the location of the individual tomb, control the growth of microbiology. Microscopically the microbiological settling is detectable to a depth of up to 0.4 cm. Larger crystals show round forms by dissolving at the surface. Marble and limestone show an analogous settlement.

The marble shows manifold damage images with extreme back weathering, intensive crack pattern and distinctive biological impact, additionally the is visible obvious damage such as opening of grain boundaries, twinning lamellae and cleavage plains. Together this will lead to a total material loss at the surface. To reduce the progressive damaging process it is necessary to clean the stone surfaces carefully by means of proved techniques (including the laser technique). In order to maintain the historic character of the monuments it is necessary to save the tombs with traditional and new conservation skills.

ACKNOWLEDGMENTS

This project was generously supported by the Deutsche Bundestiftung Umwelt (DBU) and the Denkmalamt Hamburg.

REFERENCES

Neumann, H.-H. 2003. *Schadensdiagnose von 4 Grabsteinen im Hinblick auf spätere Reinigung und Konservierung*. Interner Untersuchungsbericht (unpublished report).

Siegesmund, S. Weiß, T. & Rüdrich, J. 2002. *Schadensuntersuchungen von Marmortumben des 17. Jahrhunderts auf dem Jüdischen Friedhof Königstraße in Hamburg Altona*. Interner Untersuchungsbericht (unpublished report).

Steiger, M. & Willers, U. 2003. *Belastungssituation am Standort jüdischer Friedhof Königstraße in Hamburg bezüglich atmosphärischer Schadstoffe*. Interner Untersuchungsbericht (unpublished report).

Steiger, M. 2003. *Untersuchungen von Niederschlagsproben auf dem Jüdischen Friedhof Königstraße in Hamburg*. Interner Untersuchungsbericht (unpublished report).

Studemund-Halévy, M. & Zürn, G. 2002. *Zerstört die Erinnerung nicht. Der jüdische Friedhof Königstraße in Hamburg*. Hamburg: Dölling und Galitz Verlag.

Von Plehwe-Leisen, E. 2003. *Ultraschallmessungen an Marmorgrabmalen auf dem Jüdischen Friedhof in Hamburg Altona*. Interner Untersuchungsbericht (unpublished report).

201

Dimension Stone 2004, Přikryl (ed)
© 2004 Taylor & Francis Group, London, ISBN 90 5809 675 0

Dimension stone durability: evaluation of climatic data for several European and North American cities

M.J. Scheffler
Wiss, Janney, Elstner Associates. Inc., Northbrook, Illinois, USA

K.C. Normandin
Wiss, Janney, Elstner Associates, Inc., New York, New York, USA

ABSTRACT: This paper presents a review and analysis of climatic conditions in several North American and European cities. Weather data reviewed includes temperature extremes, precipitation, acid rain, freeze-thaw cycles, and freezing after a saturating rain. The weathering data reviewed is useful when evaluating stone for both building facades and monuments. The information presented may also be useful when comparing the weathering characteristics of the same stone in different environments or for predicting the weathering effects of a previously used stone proposed for use in a new environment. The number of freeze thaw cycles found by others to cause damage to stone in laboratory accelerated weathering testing is shown to correlate to certain environmental conditions. A weathering index for stone is proposed as a way to quickly compare certain environmental factors between cities.

1 INTRODUCTION

Evaluation of future stone durability is performed in several ways. History of use of a stone can in certain circumstances provide useful but limited information. Stone used in the past is often not from the same quarry or location and may not be representative of that which is proposed to be used. Also, for certain cities, environments may be significantly different and past use may not be a good indicator or predictor of future satisfactory performance. Evaluation and study of the historic performance of a particular stone type in service is the most commonly used approach, and is used on almost all projects where stone is being evaluated for use in an exterior environment. A benefit of this approach is that often for widely used stone a significant amount of material may be available for review, and that the material available for review may represent a significant range of the material's variability. This approach is of limited use for many stones, particularly those that have not been used extensively or those which have a limited history of use. This paper reviews the issue and significant factors that affect stone durability and the variations in environmental factors in North America and Europe that may affect the value of using in place examples for judging historic durability of stone. These factors are discussed in more detail below.

More important than reviewing past performance is review of the physical and mechanical properties of a particular stone. The stone's tested properties are often compared to minimum standards or to the physical and mechanical properties of other durable stones, or historic to data for that stone. Often past test data is readily available for review for common stone types, however it is often old and not representative of new stone. Testing should usually be performed on a large enough sampling of the new material being used so that it is representative of the stone that will be used and takes into account the variability which may be present.

Petrographic evaluation is also commonly used to evaluate stone in an effort to identify the mineral composition and structure of a stone, and based on these observations and past knowledge of those characteristics, to predict future performance. Mineral size and interlocking, as well as the presence of flaws or veins may be observable. Evaluation of a stone using a hand held field microscope is shown in Figure 1. Again the amount of material that can be reviewed petrographically is usually only a small portion of that actually used on a project. Although this can be mitigated by having the petrographer visit the quarry to review the local geology and exposed stone there, and review prior uses of the stone.

Figure 1. Examination of dimension stone panel using a field microscope.

Figure 2. A photo of a limestone specimen exhibiting severe deterioration after 100 cycles of accelerated weathering.

Another method of evaluation is to expose samples of stone to an accelerated weathering procedure, and evaluate the stones physical and mechanical properties for changes. Bortz and Wonneberger (2000) have shown that the strength loss of stone exposed to a specific accelerated weathering regimen correlates favorably to strength loss observed for that stone when exposed to real time in service weathering. The laboratory accelerated weathering regimen that has been found to be predictive of in service deterioration includes the following: submersion of flexural strength test prisms, 1/4 inch (6.3 mm) to 3/8 inch (9.5 mm) deep, in a 4 pH sulfurous acid solution, to simulate acid rain. Specimens are subjected to 100 cycles of heating

and cooling between −10°F to 170°F (−23°C to 77°C). This accelerated weathering test procedure has been performed on many different stone, including granites, marbles, limestone, and sandstones. Figure 2 is a photograph of a limestone specimen exhibiting severe deterioration after 100 cycles of accelerated weathering to the above described procedure. The procedure has shown that the strength loss for certain stone exposed to 12 to 16 cycles of accelerated weathering is approximately equal to one year of actual weathering for stone exposed to a temperate climate such as the Midwest United States.

When studying the historic use of a particular stone it is beneficial to understand the environment in which the stone has previously been used, and compare it to the proposed environment. The relevancy of a stones observed durability to its environmental expose is important to evaluate. Several key environmental factors are critical to properly understating environmental weathering exposure of stone and stone durability.

2 KEY ENVIRONMENTAL FACTORS AFFECTING STONE DURABILITY

Deterioration of stone is related to a number of environmental factors, chief among these is temperature, temperature change, freezing and thawing cycles, exposure to moisture, acid rain, air pollution, acidic gases and deicing salts and other chemicals, and exposure to the sun. In addition to temperature affects and freezing and thawing, stone subjected to wet or saturated conditions while undergoing freezing and thawing is also known to deteriorate stone. It is known that moisture is often needed for bowing distortion to occur in thin marble panels. Heating and cooling cycles are usually needed for granite deterioration to occur as the mineral crystals expand and contract relative to each other. Environmental climatic data is summarized and compared for temperature, precipitation, rain pH, and freeze-thaw, and freezing after rain, for several different cities in the United States and Europe.

2.1 Comparison of environmental conditions for various cities

2.1.1 Temperature extremes
Stone which experiences wider temperature variations is known to experience more deterioration than stone that is in more temperature stable environments. The average annual extreme high and low air temperatures for various cities were reviewed. The annual average is obtained from averaging the highest, lowest, temperatures reported for each year, from daily readings, over several decades. For comparison in

204

Table 1a. Extreme annual average temperatures °F and differences for various cities. From ASRAE 2001 Book of Fundamentals.

Location	High °F	Low °F	Diff. °F
Minneapolis, Minnesota	97	−22	119
Denver, Colorado	97	−11	108
Chicago, Illinois	96	−12	108
New York, New York	97	6	91
Atlanta, Georgia	96	9	87
Houston, Texas	98	24	74
Helsinki, Finland	83	12	71
Munich, Germany	90	−1	91
London, England	85	16	69
Rome, Italy	94	26	68

Table 1b. Extreme annual average temperatures °C and differences for various cities. From ASRAE 2001 Book of Fundamentals.

Location	High °C	Low °C	Diff. °C
Minneapolis, Minnesota	36.1	−30.0	48.3
Denver, Colorado	36.1	−23.9	42.2
Chicago, Illinois	35.6	−24.4	42.2
New York, New York	36.1	−14.4	32.8
Atlanta, Georgia	35.6	−12.8	30.6
Houston, Texas	36.7	−4.4	23.3
Helsinki, Finland	28.3	−11.1	21.7
Munich, Germany	32.2	−18.3	32.8
London, England	29.4	−8.9	20.6
Rome, Italy	34.4	−3.3	20.0

Figure 3. Bowing marble panels on building in Houston, Texas, USA.

Figure 4. Bowing marble panels on building in Houston, Texas, USA.

Minneapolis the extreme average high air temperature is 97°F (36°C), while the extreme average daily low air temperature is −22°F (−30°C). Temperatures of stone will be higher than where the stone is directly exposed to the sun. A table listing the extreme temperature conditions for several North American and European cities is provided below:

The cities represent those in northerly as well as middle latitudes. The data shows that the annual average extreme temperatures for Minneapolis, Minnesota, and Denver, Colorado, and Chicago, Illinois, are very similar and more than significant than other cities. Temperature extreme differences for Atlanta, New York, Munich, and Helsinki are slightly less severe. London and Rome do not experience high or low temperature extremes that are comparable to those experienced in Minneapolis, Denver and Chicago. Houston has high temperatures but not as significant variation between extremes. The temperatures shown in Table 1 will be less than those experienced by stone directly subjected to the sun, especially dark colored stone which could have surface temperatures of as much as 170°F (77°C).

Interestingly, thin white marble panels used on buildings in Helsinki, Chicago, and Houston have experienced significant weathering related deterioration. Figures 3 and 4 are photographs of a building in Houston with significantly bowed panels. Bowing of marble has been shown to be indicative of weathering deterioration and the result of marble hysteresis. Hysteresis, when described for stone, is a breakdown of the mineral components and structure of the stone usually associated with cyclic heating and cooling of certain calcitic marbles. It may be that for locations as warm as Houston the difference between extreme temperatures is sufficient to cause marble deterioration even if freezing temperatures are not often experienced.

205

2.1.2 Precipitation

Increases in moisture and humidity are known to increase the rate of stone deterioration. Stone in areas receiving significant rainfall are more likely to experience deterioration than those in drier climates. Also stone minerals undergo expansion when exposed to moisture. The effects of cyclic and wetting drying deterioration are not well understood, but may contribute to weathering deterioration. Moisture has been found to be needed for marble bowing to occur. The average annual precipitation for various cities is listed below. The precipitation totals include rain and snow.

From this data one can see that the southern portions of the United States receive significantly more rain than the European cities listed. Denver is particularly dry and based on rainfall would not seem to provide a particularly damaging environment for stone as a result of moisture.

2.1.3 Acid rain

Sulfates and nitrates, produced from burning fossil fuels, are present in the atmosphere and have been shown to elevate the acidity of rain above normal levels; the result is referred to as acid rain. Acid rain includes deposition of acids from rain, fog, snow, and airborne particles. Acids, even weak acids applied over time, will significantly dissolve stone, certain stone types particularly those made of calcium carbonate such as limestone, and certain marbles, are susceptible to deterioration from acid rain. The effect of acid rain is to dissolve the surface of certain stones, particularly marbles and limestones, with less of an affect on granites. The loss of surface can diminish the mechanical properties and strength of stone, particularly in thin stone panels.

Stone in Architecture (Winkler 1994) provides a contour map of rain pH for Europe and North America. Based on this map the rain pH for various cities is shown below. The lower the rain pH the more deterioration will result to stone exposed to rain over extended periods.

While water has a neutral pH of 7 (neither acidic nor alkaline) rain is normally considered to have a pH of approximately 5.6 (Winkler 1994). A pH of 4 is ten times more acidic than a pH of 5. By comparison, lemon juice has a pH of between 2 and 3.

The data shows that Munich and New York have more acidic rain than the other cities listed. All of the cities except Denver have pH levels lower (more severe) than "normal" rain. However, since a pH of 4, is ten times more acidic than a pH of 5, and 100 times more than a pH of 6, the acidity of rain in Munich and New York could be as much as 100 times that of rain in Denver. From this data, the effects of acid rain may be more similar for Munich and New York, but stone which has been successfully used in Rome may not

Table 2. Average annual precipitation for various cities. From United States National Oceanic and Atmospheric Administration.

Location	Inches	mm
Minneapolis, Minnesota	29.4	747
Denver, Colorado	15.8	401
Chicago, Illinois	36.2	919
New York, New York	44.4	1,128
Atlanta, Georgia	50.2	1,275
Houston, Texas	47.8	1,214
Helsinki, Finland	25.6	650
Munich, Germany	35.7	907
London, England	32.0	813
Rome, Italy	31.6	802

Table 3. Approximate rain pH for various cities (from Winkler 1994).

Location	pH
Minneapolis, Minnesota	5
Denver, Colorado	5.5–6
Chicago, Illinois	4.5
New York, New York	4–4.5
Atlanta, Georgia	4.5–5
Houston, Texas	5
Helsinki, Finland	4.5
Munich, Germany	4–4.5
London, England	4.5
Rome, Italy	5

fair as well in Munich or New York, as a result of increased effects of acid rain.

It is interesting that the solution used in the accelerated weathering test procedure shown to be predictive of stone durability described earlier is a 4 pH sulfurous acid solution. The pH of the solution is comparable to that of rain in New York and Munich.

2.1.4 Freeze-thaw cycles

Heating and cooling as well as freeze-thaw cycles are known to be a significant contributor to the deterioration of stone. The cyclic thermal expansion and contraction of stone minerals can also result in a breakdown of the stone's structure over time. Moisture present in the stone which freezes, expands, and also causes deterioration. Review of temperature data for freeze-thaw days and freezing periods recorded for a particular location provides a degree of understanding of the number of temperature cycles occurring. Obviously, for very warm environments the number of freezing periods will be minimal although temperature variations could be significantly high.

2.1.4.1 Freeze-thaw days

Freeze-thaw days, and therefore freeze thaw cycles, can be estimated by determining when the maximum

Table 4. Estimated annual freezing periods and freeze-thaw cycles for various cities based on review of NOAA data.

Location	Freezing periods	Freeze-thaw cycles	From years
Minneapolis, Minnesota	12	76	1995 to 2002
Denver, Colorado	21	141	1995 to 2002
Chicago, Illinois	23	96	1994 to 1999
New York, New York	15	50	1998 to 2003
Atlanta, Georgia	16	37	1998 to 2003
Houston, Texas	7	7	1997 to 2002
Helsinki, Finland	30	115	1994 to 1999
Munich, Germany	31	79	1994 to 1999
London, United Kingdom	15	50	1994 to 1996
Rome, Italy	6	12	1994 to 1997

temperature for a location is above 32°F (0°C) and the minimum temperature is at or below 32°F (0°C), daily. Review of average daily weather data from 1995 through 2002 shows that Minneapolis, on average, experiences approximately 76 freeze-thaw days annually. For the same period, Denver on average, experiences approximately 141 freeze-thaw days annually. This indicates that Denver experiences almost twice as many freeze-thaw cycles as Minneapolis; however, Denver is a much drier environment. Using only daily minimum and maximum temperatures likely overestimates the actual number of freeze thaw cycles for a stone in service, as the actual time the temperature of a particular stone is sustained below freezing likely depends on the duration of below freezing temperatures. For example, a stone unit exposed to a single day of below freezing temperature will likely not freeze, but one that is exposed to several days of continuous freezing temperatures likely will freeze. Using the number of periods of freezing temperatures is likely a better way of representing freeze thaw cycles experienced by stone.

2.1.4.2 Freeze-thaw periods

For this study one freezing period is considered the duration in which the temperature for a day or period of days remains below the freezing point of water, 32°F (0°C), based on either of the recorded maximum or minimum daily temperatures. Review of daily weather data for 1995 to 2002, this way, shows that Minneapolis on average, experiences approximately 12 freezing periods each year, while Denver experiences approximately 21 freezing periods each year. This shows a consistent relationship between freezing cycles and freezing periods between these two cities. Weather data summaries obtained show that from 1994 through 1999, Munich experienced on average 31 freezing periods each year, more than Minneapolis and Denver.

For comparison, from 1997 through 2002 Houston experienced only 7 freezing periods on average

annually. Since the duration of each freezing period was almost always not more than one day, the average number of freeze thaw cycles annually was the same, 7. The number of average annual freezing periods and annual freeze-thaw cycles for several cities are listed in Table 4.

2.1.4.3 Freezing periods and accelerated weathering

As described earlier, the accelerated weathering procedure conducted by Bortz and Wonneberger (2000) has shown that the strength loss for certain stone exposed to 12 to 16 cycles (12 to 16 freezing periods) of accelerated weathering is approximately equal to that which occurs in one year of actual weathering for stone exposed to Midwest United States climates. From review of climatic data, the number of freezing periods in Chicago a Midwest United Stases climate is estimated to be approximately 23 each year. This is approximately twice the number of freeze thaw cycles Bortz and Wonneberger (2000) have shown to occur in stone each year based on the rate of stone deterioration. It is expected that the actual number of freezing periods or cycles each year experienced by a stone saturated by rain, conditions similar to that experienced by stone in the accelerated testing method, would be less than the reported number of periods measured by air temperature alone. In service the stone is not always wet when freezing occurs, sun radiation may warm the stone, and building heat loss may also protect the stone from freezing. It is reasonable to estimate based on this that approximately half of the freezing periods result in stone experiencing freezing while also being saturated. Whether the stone is saturated when freezing occurs will also be affected by the amount of precipitation in the environment.

The number of periods in which the daily maximum temperature remained below freezing was also evaluated for Chicago from 1994 to 1999. This review showed that there were on average 10 periods of varying duration annually in which the temperature remained below freezing. This is slightly lower than but close to the 12 to 16 freeze-thaw cycles obtained by Bortz and Wonneberger (2000) based on their accelerated weathering testing.

The results of accelerated weathering testing strongly indicate that subjecting stone to the accelerated weathering procedure for 100 cycles is comparable to approximately 7 to 8 years of actual weathering exposure when the number of freezing periods each year is 12 for that locality. The review of data indicates that 12 to 16 cycles of freezing, while a stone is saturated is reasonable to expect, for a stone in service in the Midwest United States climate, or similar climate.

2.1.5 *Weathering index*

It can be useful to compare the environment and climatic conditions in Chicago to that in other parts of the US and Europe. The freeze-thaw data and precipitation data reviewed above indicate that the chances of freezing temperatures occurring shortly after a significant, saturating rainfall is likely similar for Denver, Munich, and Minneapolis. This effect is described in ASTM C216-02, "Standard Specification for Facing Brick," which provides a method for rating weathering severity based on location in the United States, called the Weathering Index. The weathering index for brick is defined in ASTM C216 as follows:

The effect of weathering on brick is related to the weathering index, which for any locality is the product of the average number of freezing cycle days and the average annual winter rainfall in inches (millimeters), defined as follows.

A Freezing Cycle Day is any day during which the air temperature passes either above or below 32°F (0°C). The average number of freezing cycle days in a year may be taken to equal the difference between the mean number of days during which the minimum temperature was 32°F (0 °C or below, and the mean number of days during which the maximum temperature was 32°F (0°C) or below.

Winter Rainfall is the sum, in inches (millimeters), of the mean monthly corrected precipitation (rainfall) occurring during the period between and including the normal date of the first killing frost in the fall and the normal date of the last killing frost in the spring. The winter rainfall for any period is equal to the total precipitation less one tenth of the total fall of snow, sleet, and hail. Rainfall for a portion of a month is prorated.

The contour map in ASTM C216 provides weathering indices throughout the United States. From this map the weathering index for Minneapolis is 500 or greater and for Denver is 500, both of which are considered "severe" weathering environments by the standard, while 100 to 500 is considered moderate and below 100 negligible. The weathering index value from ASTM C216 for various US cities are listed below as well as the product of average rainfall (prorated for months with freezing temperatures) and number of freezing cycles, for comparison. Using the table below, it is estimated based on the limited data reviewed that Munich would likely have a similar weathering index as Minneapolis, severe while Houston and Rome would have a similar weathering index, negligible.

In Table 5 and Table 6 annual average precipitation is adjusted for winter by dividing the estimated freezing months by 12 months and multiplying by one-half, based on the study above and on that basis that most precipitation occurs in non-winter months.

Table 6 presents the product of average annual rainfall and number of average annual freezing periods, for comparison to the weathering index given in Table 5

Table 5. Average Annual: Freeze-thaw cycles (F-T), precipitation (inches) (P), product of F-T × P*, compared to weathering index from ASTM C216 (W.I.), for various cities.

City (Est. Months of Winter Pre-cip.*)	F-T	P	F-T × P*	W.I.
Minneapolis, Minnesota (7)	76	29	643	>500
Denver, Colorado (6)	141	16	564	500
Chicago, Illinois (6)	96	36	864	>500
New York, New York (6)	50	44	550	500
Atlanta, Georgia (5)	37	50	385	500
Houston, Texas (3)	7	48	42	<50
Helsinki, Finland (7)	115	27	906	n/a
Munich, Germany (6)	79	36	711	n/a
London, England (2)	50	32	133	n/a
Rome, Italy (2)	12	32	32	n/a

*For US cities the estimated number of months having freezing temperatures is based on published NOAA data from 1951–1980 for freeze free months per year. European cities are estimated. Precipitation is adjusted by multiplying by winter months of freezing divided by 12 and divided again by 2 for low rainfall during winter months.

Table 6. Average Annual: Freezing periods (FP), precipitation (inches) (P), and product of FP × P or proposed Weathering Index for Stone (S.W.I), for various cities.

City (Est. Months of Winter Precip.)	FP	P	FP × P* (SWI)
Minneapolis, Minnesota (7)	12	29	102
Denver, Colorado (6)	21	16	84
Chicago, Illinois (6)	23	36	207
New York, New York (6)	15	44	165
Atlanta, Georgia (5)	16	50	167
Houston, Texas (3)	7	48	28
Helsinki, Finland (7)	30	27	236
Munich, Germany (6)	31	36	279
London, England (2)	15	32	40
Rome, Italy (2)	6	32	16

for brick. The value could be used as a Weathering Index for stone.

It may be more appropriate to use this weathering index for stone than the weathering index used in ASTM C216 for brick. Based on the calculated weathering index for stone above, Munich may have the most severe weathering environment for stone given its rainfall and number of freezing periods. Also Munich has rain with a relatively high pH, which also will increase the rate of stone weathering. It seems likely that a weathering index for stone above approximately 150 should be considered a severe weathering environment for stone.

There are limitations to the above index because marble panels in Houston have bowed; likely mainly from heating and cooling cycles and moisture, even though the index is above is low. Therefore appropriateness and use of the weathering index needs to be carefully considered and needs to be further developed for certain types of stone, such as marble.

3 DISCUSSION AND CONCLUSIONS

Deterioration of stone is related to a number of environmental factors, chief among which are heating and cooling, and freezing and thawing. In addition to freezing and thawing alone, stone subjected to wet or saturated conditions while undergoing freezing and thawing has also been shown to deteriorate stone. Exposure to acid rain will over time also lead to stone deterioration. The environmental data provided above shows that environmental conditions for temperature, precipitation, and freeze and thawing can vary significantly between cities and are not similar, for example between Rome and Minneapolis. For certain other cities, such as Munich and Minneapolis, weathering exposure and historic weathering performance of that stone should provide a reasonable comparison.

There are several cautions about using experience in stone from one location to predict its performance in another location. For example, one can never be certain, unless detailed records exist, whether a stone used in any one of these environments is representative of that which will be used elsewhere, because material quarried from different layers or areas of the geologic formation or quarry may have different characteristics and perform differently. In addition, the environment continues to change over time. For example acidity of rain has increased significantly over what it was 50 to 100 years ago, for certain regions. A stone that has performed well over hundreds of years when exposed to a rain pH of 5 may not perform well when exposed to a rain pH of 4.

History of use of a stone can in certain circumstances provide useful information, but as has been shown for certain cities, environments may be significantly different and past use may not be a good indicator or predictor of future satisfactory performance. With careful consideration historic performance of a stone may be useful in understanding past stone durability, as well as anticipated performance in a new and different environment. Review of past history of stone usage along with testing of physical mechanical properties, petrographic evaluation of the stone, and accelerated weathering can aid in understanding and predicting future stone durability for a particular environment. The proposed weathering index for stone could be a useful tool for comparing and assessing exposure environments in which stone is used and for projecting future stone durability

ACKNOWLEDGEMENTS

The authors gratefully acknowledge the information and guidance provided by Mr. S.A Bortz in preparing this paper.

REFERENCES

Bortz, S. & Wonneberger, B. 2000. Predicting the Durability of Building Stone Using Accelerated Weathering. ASTM STP 1385 Durability 2000.
Winkler, E.M. 1994. *Stone in Architecture. Properties, Durability.* 3rd ed., Berlin: Springer-Verlag.

Dimension Stone 2004, Přikryl (ed)
© 2004 Taylor & Francis Group, London, ISBN 90 5809 675 0

Marble deterioration

S. Siegesmund, J. Ruedrich & T. Weiss
Geoscience Center, Georg-August-University, Goettingen, Germany

ABSTRACT: Marbles as ornamental and dimensional stones as well as in their natural environments show complex weathering phenomena. The physical weathering of marbles due to thermal treatment is often discussed as the initial stage of deterioration. Experimental data on moisture content and freeze thaw successions also demonstrate a pronounced loss in strength and most spectacularly an ongoing amount in bowing. Moreover, it could be clearly documented that the rock fabric is a major control on the physical weathering of marbles. Finally, the assessment of the intensity of marble degradation is quantified by ultrasonic wave velocities.

1 INTRODUCTION

Numerous cases of damage on sculptures (Figs. 1–2), architectural heritage or façade stones (Fig. 3) made from marble indicate that the deterioration of building stones depends mainly on climate. Chemical mechanisms have received much attention in recent years with a special emphasis on the effects of acid rain or biofilms. It has been shown that the initial reaction of calcite surfaces to incident rainfall produces clear morphological alteration even within a short term of exposure (e.g. Grimm 1999). Recently, the physical weathering is discussed to be the initial stage of deterioration of marbles (see Siegesmund et al. 2000).

Durability is an important issue to consider when specifying stones as cladding material for exterior exposure. The use of stone panels as cladding materials for façade has undergone a considerable increase in the last decades. The observed durability problem, for example the most spectacular deterioration feature of some marble slabs is their bowing behaviour, has given a negative image to these material. The

Figure 1. Sculpture in the Orangerie of Potsdam-Sanssouci.

Figure 2. Total damage of the sculpture from Figure 1 after 110 years of exposure.

211

Figure 3. Bowing of marble panels at the façade from the library building in Goettingen (Germany).

complete replacement of façade panels of some prestigious buildings like the Amoco building in Chicago (Logan et al. 1993), the Finlandia Hall in Helsinki or the Grand Arche de la Defense in Paris all made of some varieties of marble coming from the Carrara area are often cited examples for the existing concerns on the durability of those materials. However, bowing is already frequently reported from ancient gravestones (e.g. Grimm 1999). Detailed knowledge of the mechanisms and rates of decay is essential to protect historic stone monuments or in the case of marble panels an economical quantity is important. The replacement of all 43000 marble panels of the Amoco building with granite required US $ 65 million (Cohen & Monteiro 1991).

The reasons for the observed deformation are still in discussion. Kessler (1919) found that repeated heating may lead to permanent dilatations due to microfracturing. Logan et al. (1993) explained the bowing of marble slabs on the Amoco building as being due to the anomalous expansion–contraction behaviour of calcite combined with the release of locked residual stresses based on laboratory testing. The hypotheses of other researchers have required the presence of moisture or gravity variation (e.g. Winkler 1994).

2 MARBLE AND ITS FABRICS

Although marbles have a very simple mineralogical composition, the physical weathering due to its extremely anisotopic physical properties seems to be essential. The rock fabric, which includes grain size, grain aspect ratio, grain-shape preferred orientation, lattice preferred orientation (texture) and the microcrack populations control the materials behaviour.

Geological fabric analyses are usually applied in order to reconstruct the rock-forming processes (mineralization, deformation) and to define the environments in which they are formed. Often, rock fabrics are the consequence of polyphase events, which occurred over a long time span under varying conditions (pressure, temperature, type of deformation, chemical environment) leading to complex final rock fabrics. In conclusion, all above mentioned fabric elements are responsible for the physical anisotropy of rocks, i.e. for the directional dependence of the bulk physical properties. Hence, the quantification of rock fabrics and their correlation with physical and mechanical properties may contribute significantly to the understanding of rock weathering (e.g. Siegesmund et al. 2000).

3 THERMAL EXPANSION BEHAVIOUR

The thermal expansion expresses the relative length change of a sample. The connection to the temperature is non-linear, i.e. the thermal expansion coefficient α which describes the specific length change ($10^{-6} K^{-1}$) depends on the considered temperature interval. More than 24 different marble types worldwide were investigated (Zeisig et al. 2002). In sum, up to now all experimentally determined polycrystalline thermal expansion behaviours of marbles as a function of a heating and subsequent cooling cycle can be classified in four overall categories: (a) isotropic thermal expansion without residual strain; (b) anisotropic thermal expansion without residual strain; (c) isotropic thermal expansion with residual strain; and (d) anisotropic thermal expansion with residual strain (Fig. 4).

Thermally treated marble samples, which do not return to the initial length after cooling (i.e. the length before heat treatment), can show a residual strain even as a result of very small temperature changes, as shown for the temperature range between 20 and 50°C by Battaglia et al. (1993). In general, the measured relative expansion exhibits a significant increase of ε_{max} and ε_{min} with increasing temperature, whereas the slope for each sample may be different. To describe these phenomena in more detail not only a single direction of a rock is investigated since most physical properties are extremely anisotropic. Many samples show a large difference between ε_{max} and ε_{min} as a function of temperature and, thus, a strong directional dependence. Moreover, a residual dilatation can be observed after subsequent cooling down to room temperature for many samples (see Fig. 4). It ranges from 0.2 to 0.4 mm/m seldom higher and are usually smaller in the direction of ε_{min}. The thermal expansion coefficient α of calcite is extremely anisotropic (Kleber 1959): $\alpha_{11} = 26 \times 10^{-6} K^{-1}$

212

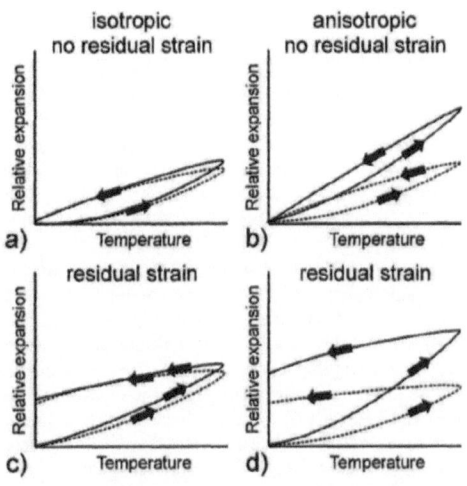

a) isotropic no residual strain

b) anisotropic no residual strain

c) residual strain

d) residual strain

Figure 4. Schematic illustration of the thermal expansion behaviour of marbles. The arrows indicate direction of temperature change (heating and subsequent cooling).

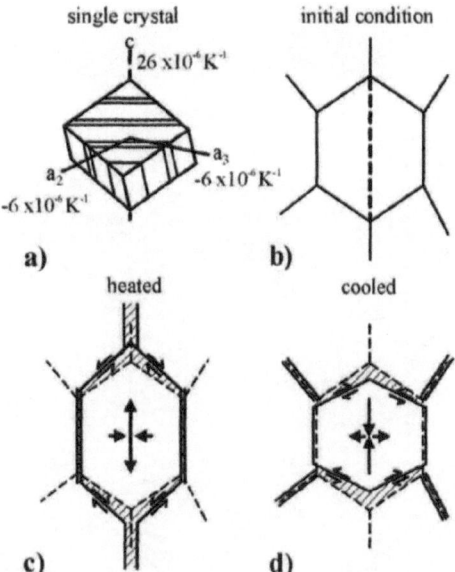

a) single crystal

b) initial condition

c) heated

d) cooled

Figure 5. Thermal dilatation of calcite. (a) Thermal dilatation coefficients of the calcite single crystal with respect to the crystallographic orientations, (b) schematic illustration of a calcite crystal at initial conditions, (c) volume change by heating and (d) by cooling.

parallel and $\alpha_{22} = -6 \times 10^{-6}$ K^{-1} perpendicular to the crystallographic c-axis (Fig. 5), i.e. calcite contracts normal to the c-axis and expands parallel to the c-axis during heating.

The total anisotropy for a monomineralic rock has to be between an isotropic situation (random orientation of all crystals) and a situation of maximum anisotropy where all crystals have the same crystallographic orientation which corresponds to the single crystal anisotropy. Consequently, the thermal expansion should be mainly controlled by the degree of a crystallographic preferred orientation (texture) of the calcite crystals. A significant amount of the residual strain can also be observed for marbles after heating up to 90°C which also seems to be controlled by the texture.

3.1 Thermal expansion and modal composition

The modal composition is an important factor for the thermal properties of a marble, since the thermal expansion behaviour is at least partially controlled by the single crystal properties. Both calcite and dolomite show an extreme directional dependence of α at different crystallographic directions. Parallel to the c-axis, both minerals show an α value of about 26×10^{-6}K^{-1}. However, parallel to the a-axis dolomite shows a positive α value about 6×10^{-6}K^{-1}. Thus, even strongly anisotropic dolomite marbles will not show any contraction with increasing temperature. Since dolomite and calcite marbles may show a comparable thermal degradation, the residual strain is likely not to be controlled exclusively by the composition.

3.2 Effect of the grain size

Marbles with large grain size exhibit the same magnitude of residual strain as marbles with a small grain size. Therefore the grain size cannot be the most important factor for marble degradation as was already discussed by Tschegg et al. (1999).

3.3 Grain shape and grain boundaries

The grain shape anisotropy significantly triggers the thermal degradation, as shown for several samples by Siegesmund et al. (2000) or Ruedrich et al. (2002).

Thus, grain boundary cracking is the most prominent factor for marble degradation, a substantial part of the observed directional dependence of residual strain must be attributed to shape fabrics. The irregularity of grain boundaries (Fig. 6) does not play such an important role (e.g. Zeisig et al. 2002) which is in part contradict the observations by Royer-Carfagni (1999). Marbles with interlobate fabrics as well as marbles with polygonal fabrics may show a residual strain after thermal treatment.

3.4 Texture

The texture clearly determines magnitude and directional dependence of α, since there is a general

213

Figure 6. Two distinct types of grain boundaries ranging from (a) straight to (b) more irregular or lobate ones.

agreement between calculated (texture-based) and experimentally determined anisotropies. Thermal degradation changes this relation, i.e. the anisotropies increase or decrease according to a coincidence or contrariness of thermal degradation and intrinsic dilatation, respectively. A general observation is that the maximum of thermal degradation is closely linked to the c-axis maximum. A deviation from this behaviour is correlated with the shape preferred orientation. However, the individual grain to grain orientation, i.e. the misorientation may produce internal stresses leading to microcracking (e.g. Tschegg et al. 1999 or Weiss et al. 2003). This could occur by an almost random orientation of the grains or even in the case of a strong preferred orientation. However, the total amount of thermal-induced microcracking must be validated in the future by single grain texture measurements and modal calculations.

3.5 Microcracks

Pre-existing microcrack systems may be of importance as well. Siegesmund et al. (2000) have shown that a change in the anisotropy patterns between modelled and experimentally determined values may be explained by pre-existing microcracks, resulting from a complex geological history.

4 EVIDENCE FROM FINITE-ELEMENT MODELLING

Microstructure-based finite element simulations were used by Weiss et al. (2002, 2003) providing an excellent insight in the magnitude and mechanisms of thermal degradation. The basic observation is that the thermal expansion behaviour of marbles can be modelled with a good coincidence to real experiments. Onset and magnitude of thermal microcracking vary for calcite and dolomite marbles, when assumed to have exactly the same microstructure and texture, being greater for calcite. Thus, finite element modelling indicates that dolomite marbles may be more resistant against thermal weathering than calcite marbles. The variations in the texture may significantly affect the distribution of thermal stresses within the marble. There is a strong inverse correlation between thermal stresses and degree of texture, since higher elastic strain energies are associated with weakly textured marbles, and vice versa.

However, grain to grain orientation relationships, frequently called misorientations, and their distributions are also important parameters. The same bulk texture may be achieved when the grains have small or high angles in between from a statistical point of view. Different misorientations lead to variations in the elastic energy density which are in the same order of magnitude than those related to the texture itself (Weiss et al. 2003). Thus, more information on misorientations is required in the future in order to constrain thermal stresses in marble.

In summary, it can be stated that there is a clear fabric dependence of residual strain after thermal treatment and, thus of thermal degradation. Thermally induced microcracks lead to a residual strain after heat treatment and, thus, to a deterioration of the rocks quality. The degradation is controlled by an interaction of all fabric patterns.

Moreover, many authors (e.g. Sage 1988 or Koch & Siegesmund 2004) could demonstrate that the increase of residual strain stops after few heating cycles if moisture is absent. Therefore Bucher (1956) or Winkler (1994) point out the importance of moisture in the bowing process.

5 THE EFFECT OF MOISTURE

To discuss these effects more detailed, measurements of progressive residual strain on a calcite marble were performed. Eight dry cycles up to 90°C were followed by 25 additional wet cycle's altogether, whereby the first six wet cycles were carried out in a way that at the end of a heating cycles the samples in the climate chamber were run until totally dry.

The findings from this approach are: (1) The progressive residual strain indeed proceeds continuously;

214

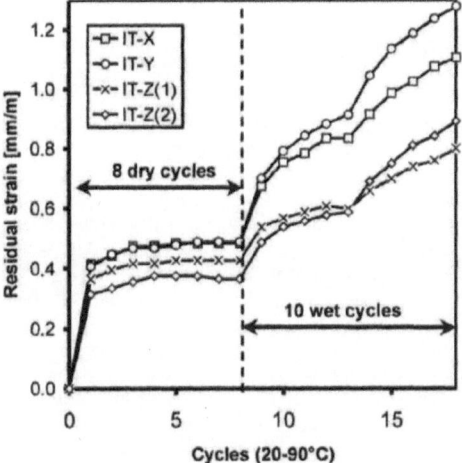

Figure 7. Progressive increase of residual strain of a calcite marble as a function of 8 dry cycles and following wet cycles (8-18) taken from Koch & Siegesmund (2004).

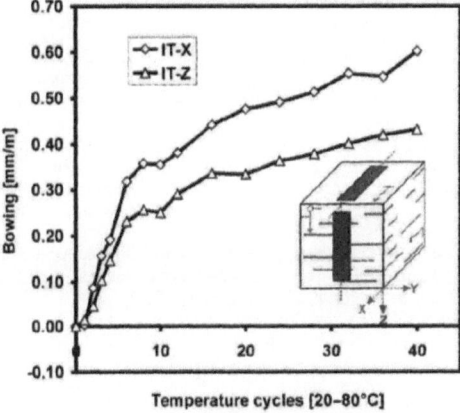

Figure 8. Bowing of marble slabs versus number of heating cycles in two different directions (Koch & Siegesmund 2004).

the increase is constant even after 25 cycles under wet conditions. (2) The moisture content after heating cycles apparently influences the intensity of marble degradation (Fig. 7). The strain versus cycles curve (Fig. 7) tends to get flat as long as water is still available after cooling down. As soon as samples totally dry up after each heating cycle the durable marble expansion accelerates again. This observation cannot be explained sufficiently by the theory of Winkler (1994) that oriented molecule layers (thickness of

2–3 nm) in capillaries <0.1 μm may cause swelling by elongation and stone disruption.

As a result he concludes that panels start to bow outwards if the sun and high humidity exposes the panel from outside only, and inwards if moisture is available behind the panel in a closed cavity where the relative humidity can remain near 100%. In addition, the bowing potential of the same marble was tested in the laboratory. The test was performed in such a way that the marble specimen (slab of 400 m × 100 mm × 30 mm) was exposed to moisture on one side and infrared heating on the reverse side. The applied temperature ranged between 20°C and 80°C and a total of 40 cycles was performed. The permanent length change and the effect of anisotropic bowing with the applied cycles are clearly demonstrated in Figure 8.

6 FREEZE-THAW CYCLES

Thermal dilatation experiments were carried out on a variety of marbles in the temperature range between −40°C and 60°C. From single crystal properties it is clear that when cooled, calcite crystals contract along the c-axis, but expand along the a-axes while the opposite behaviour occurs when heated. For the Sterzing marble the strong directional dependence of heating and cooling is given in Figure 5. A residual strain is observed along the X- and Z-direction. For a long term freeze-thaw cycle (204 cycles) over a period of 14 months the effect of strength loss is illustrated in Figure 9 for some selected marbles. In addition to fresh ones also artificially weathered ones of the same stone types were used. To characterize the state of deterioration the Young's modulus was determined. A pronounced difference is observable.

The highest decrease of the Young's modulus from the fresh to the artificial ones can be observed for the Carrara marble, where the values changes from 55 GPa to less than 10 GPa, while for the dolomitic Palissandro marble the reduction is less pronounced. After five to seven cycles a first remarkable loss occurs in the Young's modulus. A second pronounced decrease in the Young's modulus of around 5–10 GPa is seen after 100–115 cycles especially for the fresh samples.

The Young's modulus decreases continuously during the experimental run and is at a maximum for Carrara where a reduction in strength of up to 50% must be recognized. Microstructural observations give evidence that open grain boundaries, but also twins and cleavage planes are open due to frost action. In the highly weathered examples the open grain boundaries are interconnected with intergranular microcracks, i.e. the formation of a progressive network is being developed.

215

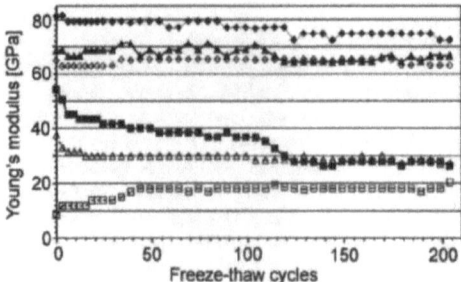

Figure 9. The Young's modulus of Palissandro (rhombohedrons), Sterzing (triangles) and Carrara (squares) marbles as a function of 204 freeze-thaw cycles. The filled symbols represent the fresh marbles, whereas the open symbols represent the artificially weathered equivalents.

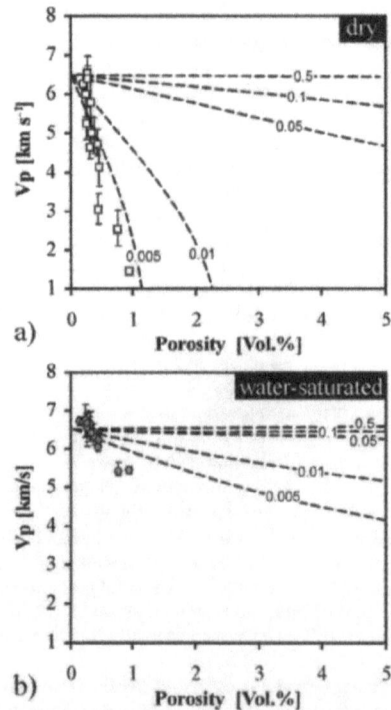

Figure 10. Average compressional wave velocities for different marble samples as a function of porosity and sample condition: (a) dry samples and (b) water-saturated samples. The anisotropy of the respective marble is given as error bars. The theoretical predictions according to the models of O'Connel & Budiansky (1974; hatched lines) are added for comparison. For the model calculations, the aspect ratio of the cracks assumed in the computations are given.

7 QUANTIFICATION OF MARBLE DEGRADATION BY ULTRASONIC VELOCITIES

Ultrasonic wave velocity measurements are a powerful and sensitive tool for the damage assessment of marble. Experimental data reveal that the state of preservation of a marble is clearly documented by compressional wave velocities. For a maximum porosity of up to 1%, velocities determined on dry samples range from about 1 km/s to over 6 km/s (Fig. 10a). Modal calculations reveal that the velocity reduction is caused by cracks with an extremely aspect ratio of about 0.005 or even less.

Water saturation has an important influence on the magnitude and directional dependence of ultrasonic velocities. Hence, it is essential to gather sufficient information on the state of water saturation of an object made from marble to quantify the state of deterioration since water as a pore fluid significantly increase the velocities (Fig. 10b). This observation is besides the anisotropy a key observation when on-site inspections are performed.

REFERENCES

Battaglia, S., Franzini, M. & Mango, F. 1993. High sensitivity apparatus for measuring linear thermal expansion: preliminary results on the response of marbles. *Il Nuovo Cimento* 16: 453–461.

Bucher, W.H. 1956. Role of gravity in orogenesis. *Bulletin of the Geological Society of America* 67: 1295–1318.

Cohen, J.M. & Monteiro, P.J.M. 1991. Durability and integrity of marble claddings: a state-of-the-art review. *J. Perf. Constr. Facil. ASCE* 5: 113–124.

Grimm, W.D. 1999. Beobachtungen und Überlegungen zur Verformung von Marmorobjekten durch Gefügeauflockerung. *Z. dt. geol. Ges.* 150/2: 195–236.

Kessler, D.W. 1919. Physical and chemical tests on the commercial marbles of the United States. Technology papers of the Bureau of Standards, No. 123.

Kleber, W. 1959. *Einführung in die Kristallographie*. Berlin: VEB Verlag Technik.

Koch, A. & Siegesmund, S. 2004. The combined effect of moisture and temperature on the anomalous expansion behaviour of marble. In: S. Siegesmund, H. Viles & T. Weiss (eds), *Stone decay hazards; Environmental Geology*.

Logan, J.M., Hadedt, M., Lehnert, D. & Denton, M. 1993. A case study of the properties of marble as building veneer. *Int. J. Rock Mech. & Min. Sci.* 30: 1531–1537.

O'Connell, R.J. & Budiansky, B. 1974. Seismic velocities in dry and saturated cracked solids. *J. Geophys. Res.* 79(35): 5412–5426.

Royer-Carfagni, G. 1999. On the thermal degradation of marble. *Int. J. Rock Mech. Min. & Sci.* 36: 119–126.

Ruedrich, J., Weiss, T. & Siegesmund, S. 2002. Thermal behaviour of weathered and consolidated marbles. In S. Siegesmund, T. Weiss & A. Vollbrecht (eds), *Natural* stones, weathering phenomena, conservation strategies and case studies. Geol. Soc. Spec. Pub. 205: 255–271.

Sage, J.D. 1988. Thermal microfracturing of marble. In P. Marinos & G. Koukies (eds), *Engineering Geology of Ancient Works*: 1013-1018. Rotterdam: Balkema.

Siegesmund, S., Ullemeyer, K., Weiss, T. & Tschegg, E. 2000. Physical weathering of marbles caused by thermal anisotropic expansion. *International Journal of Earth Science* 89: 170–182.

Tschegg, E.K., Widhalm, C. & Eppensteiner, W. 1999. Ursachen mangelnder Formbeständigkeit von Marmorplatten. *Z. dt. Geol. Ges.* 150/2, 283–297.

Weiss, T., Fuller, E. & Siegesmund, S. 2002. Thermal stresses in calcite and dolomite marbles quantified by finite element modelling. *Geol. Soc. Spec. Pub.* 205: 89–102.

Weiss, T., Fuller, E. & Siegesmund, S. 2003. Directional dependence of thermal degradation in marble: A finite element approach. Buildings and Environment, Elsevier, 1251–1260.

Winkler, E.M. 1994. *Stone in architecture.* 3rd ed., Berlin: Springer Verlag.

Zeisig, A., Weiss, T. & Siegesmund, S. 2002. Thermal expansion and its control on the durability of marbles. *Geol. Soc. Spec. Pub.* 205: 65–80.

Dimension Stone 2004, Přikryl (ed)
© 2004 Taylor & Francis Group, London, ISBN 90 5809 675 0

Assessment of building stone degradation by ultrasonic measurements

R.J.G. Sobott

Labor für Baudenkmalpflege Naumburg, Germany

ABSTRACT: Visual inspection of building stone surfaces alone is in most cases insufficient to get a proper notion about the intensity and depth of weathering. While polarized light microscopy of thin sections from drill cores would give conclusive answers, the necessarily destructive character of sample taking excludes this method of investigation when decorative stones are the objects under consideration. Here ultrasonic measurements can be helpful to assess the degree of stone damage and were applied to a pinnacle shaft at the northern portal of St Vaclav Church in Naumburg. The pinnacle was worked from an oomoldic limestone ("Schaumkalk") which is the predominant building stone of the region.

1 INTRODUCTION

The St Vaclav Church in Naumburg was built between 1426 and 1511 in the late Gothic architectural style. The church has several portals of which one on the northern side opening to the market-place is richly decorated and framed by two pinnacles on each side (Fig. 1).

The building material was a local oomoldic limestone called "Schaumkalk" (foam-limestone) because of the vesicular appearance of the oomolds. Weathering processes caused by the reaction with acid rain containing dissolved carbon and sulfur dioxide damaged the limestone considerably, especially in areas sheltered from rainwash. Calcite dissolution and the migration of Ca^{2+} ions to the surface resulted in the formation of a sub-surface layer depleted in calcite with low mechanical strength and the growth of a hard gypsum crust on the surface. Different thermic, hygric, and mechanical properties of the two layers led to surface losses by flaking and scaling. Once the hard crust has been separated from the stone surface a powdery limestone appears which is prone to rapid further damage. In order to interrupt and stop this process of stone decay it is necessary to locate the areas with low mechanical strength and give them proper conservational treatment. As the propagation of ultrasonic waves depends on the mechanical properties of the matter traversed non-destructive ultrasonic measurements were tentatively carried out at the shaft of a pinnacle.

2 PETROGRAPHY OF THE LIMESTONE

The limestone under discussion belongs stratigraphically to the Lower Muschelkalk which comprises the

Figure 1. Northern portal of St Vaclav Church (the lower right pinnacle was investigated).

facies types biogenic/peloidal ("Wellenkalk") and oosparitic ("Schaumkalk") limestone. The lower part of the "Schaumkalk" section contains the building stone "Naumburger Schaumkalk" which derives its name

219

from the town of Naumburg in the vicinity of the historic and modern quarries. The lower "Schaumkalk" bed can be subdivided into five lithological units of which only the lower four are relevant to the quarrying of building stones. The description of the three main facies types in the "Schaumkalk" is given according to Koch et al. (1999).

Type 1 is an oosparite in which most of the ooliths have been dissolved leaving behind a meshed fabric of sparitic cement made up of hypidiomorphic calcite crystals which are 50–100 μm in size. It forms the greater part of the lithological unit II which consists of a homogeneous, fine-grained, and cross-bedded oolite. This is the material mostly favoured by the stonemasons because it can be easily shaped into decorative stones.

Type 2 is a biosparite which occurs either as beds with a thickness of a few centimetres to about 1 dm or more frequently as cemented layers with shell debris in alternating oosparite/biosparite sequences. The dissolution of the aragonite component of the shells gave rise to the formation of biomolds which were partially filled with later block cements.

Type 3 occur as thin micritic and sometimes biomicritic limestone intercalations in all lithological units, especially as hardgrounds in unit II.

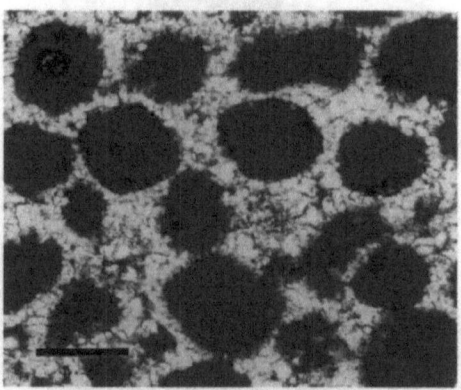

Figure 2. Oosparite (facies type 1) (scale length 1 mm).

Table 1. Petrophysical properties of main "Schaumkalk" facies types.

Facies type	Total porosity [%]	Permeability [mD]	Capillary water uptake [wt. %]
Oosparite	34.7	0.5–2.6	5.3
Oo/biosparite	28.5	0.1–3.5	4.9
Oobiosparite, cemented	22.7	0.04	3.9

Other facies types are combinations of the three main facies types which distinguish themselves also by distinctly different petrophysical parameters.

3 ULTRASONIC MEASUREMENTS

3.1 *Instrumentation*

The instrumentation used for the measurements which were performed in transmission consisted of an ultrasonic pulse generator USG 20, a magnetostrictive transmitter UPE-T and a piezoelectric broadband receiver UPE-T, all produced by GEOTRON Electronic, Pirna/Germany. The input pulse frequency was 46 kHz, corresponding to a wavelength of 6.5 cm. The output signal was amplified, recorded by an oscilloscope with a digital storage capacity for 20 measurements (FLUKE Scopemeter PM99), and transferred by FLUKE software to a PC. The oscillograms were evaluated with respect to the onset and amplitude of the longitudinal wave (P-wave).

3.2 *Measurements and results*

The transmitter and receiver were pressed manually against the surface of the pinnacle shaft. Clay was used as coupling medium to ensure a good signal transfer from and to the probes. Two sets of measurements orientated at a right angle to each other were carried out through the shaft in horizontal planes with a distance of 10 cm. The results are given in Table 2.

4 INTERPRETATION OF RESULTS

The time average equation by Wyllie et al. (1956) defines the total rock velocity in terms of the uniform

Table 2. Results of ultrasonic measurements.

Plane	Measuring direction	v_p [km/s]	Depth of weathering	Measuring direction	v_p [km/s]
100	NW–SE	3.1	9 mm	NE–SW	3.6
90	NW–SE	3.8	2 mm	NE–SW	3.8
80	NW–SE	3.4	5 mm	NE–SW	3.5
70	NW–SE	2.8	13 mm	NE–SW	2.8
60	NW–SE	2.1	29 mm	NE–SW	2.2
50	NW–SE	2.3	22 mm	NE–SW	2.0
40	NW–SE	2.2	25 mm	NE–SW	2.4
30	NW–SE	2.2	27 mm	NE–SW	2.3
20	NW–SE	2.6	18 mm	NE–SW	2.6
10	NW–SE	3.3	7 mm	NE–SW	2.6
0	NW–SE	3.6	6 mm	NE–SW	3.5

220

intergranular porosity, the rock matrix velocity, and the velocity of the pore filling.

$$\frac{1}{v} = \frac{\phi}{v_f} + \frac{(1-\phi)}{v_{ma}}$$

where v = total rock velocity, v_f = velocity of pore filling, and v_{ma} = velocity of rock matrix.

For example, in a pure calcite marble (calcite $v_{p\perp c}$ 7.7 km/s, $v_{p\ //c}$ 5.9 km/s) with a very low porosity <0.5% the longitudinal wave travels at a speed of 6.2 km/s, while it takes about three times longer to traverse a sandstone (quartz 5.5 km/s) with 25% porosity. An increase of porosity and loss of mechanical strength due to weathering processes dissolving cements and instable mineral grains and weakening the grain/crystal bonds is therefore indicated by a decrease of the total rock velocity. If type material which is well defined with respect to mechanical properties is at hand total rock velocity intervals can be assigned to weathering states. In this study the upper limit of the total rock velocity (4 km/s) was attributed to fresh "Schaumkalk" of facies type 1, while the lower limit (1 km/s) was set by a soft and powdery limestone which could be easily scratched by a finger nail. With a simplistic model in which sound core is surrounded by a weathered shell (Fig. 3) and the thin gypsum crust on the surface is neglected the depth of weathering can be calculated by the following equation:

$$s_1 = \frac{v_1}{2} * \frac{tv_2 - s}{v_2 - v_1}$$

where v_1 = velocity of weathered shell, v_2 = velocity of sound core, t = total travel time, and s = total length of travel path.

As input data for v_1 and v_2 the velocities for powdery (v_1) and fresh (v_2) "Schaumkalk", respectively, were taken. The calculated weathering depths range between 2 and 29 mm which agrees well will on-site observations (Fig. 3).

The amplitude of the P-wave is attenuated by a number of factors such as ultrasonic frequency, density, porosity, grain/crystal size, etc. If the measuring conditions are strictly kept identical (especially the coupling of transmitter and receiver to the stone surface) and if the measurements are performed on the same stone material then the amplitude can be described as a function of the length of the travel path by an exponential law equation.

$$A = A_0 e^{-bs}$$

where A_0 = P-wave amplitude for a short length of travel path, say 10 cm, b = attenuation coefficient, s = length of travel path.

The evaluation of amplitude data requires a calibration curve in which the amplitude is plotted against the length of the travel path. A distinct negative deviation of the amplitude for a given length of the travel path from the calibration curve can be regarded as an indication for the disintegration of the fabric due to the opening of grain/crystal boundaries or fracturing. A conspicuous attenuation of the amplitude is observed in the section of the pinnacle where the velocity reduction points out to weathering depths of more than 10 mm (Fig. 4).

So by combining the information about the travel time (velocity) and the amplitude read from the oscillogram it is possible to get a very good idea about the extension and intensity of weathering. Simple, mathematically defined geometries of the plane under investigation such as squares, rectangles, circles or ellipses favour the data evaluation in a tomographic manner.

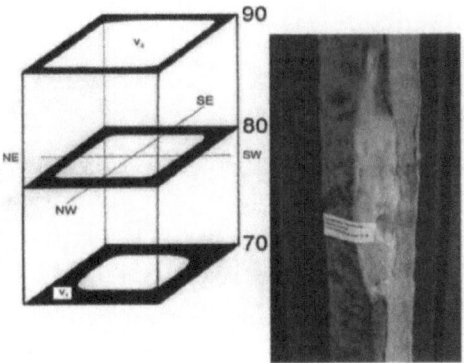

Figure 3. Model (black: weathered limestone zone, white: sound core) and edge of weathered pinnacle.

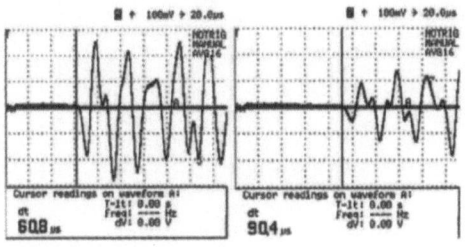

Figure 4. Oscillogram of fairly sound (left) and strongly weathered (right) "Schaumkalk" (the travel path in both measurements was 187 mm).

221

REFERENCES

Koch, R., Sobott, R. & Lorenz, H. 1999. Der Schaumkalk (Trias, Unterer Muschelkalk) am Naumburger Dom als Baustein: Einfluß von Fazies und Diagenese auf die Gesteinsqualität. In N. Hauschke & V. Wilde (eds), *Trias – Eine ganz andere Welt. Mitteleuropa im frühen Erdmittelalter*: 449–471. München: Dr. Friedrich Pfeil.

Wyllie, M.J.R., Gregory, A.R. & Gardner, L.W. 1956. Elastic wave velocities in heterogeneous and porous media. *Geophysics* 21: 41–70.

222

Dimension Stone 2004, Přikryl (ed)
© 2004 Taylor & Francis Group, London, ISBN 90 5809 675 0

Staining of building stones by mortar components: prevention and elimination

J. Twilley

Miller Druck International Stone Ltd., London, UK; Miller Druck Specialty Contracting, New York, USA

ABSTRACT: Stains in stone pavers and cladding, especially pale colored limestones, are a common problem in new construction and renovation work. In general, they take the form of darkened areas resembling damp spots or "mustaches" along joints. However, they may become so extensive as to involve most of the surface – especially in floors. In some cases they develop a color distinct from that of the stone but more commonly they consist of a more saturated shade of the existing color. Laboratory investigation holds the key to determining the root cause of such blemishes and the development of effective solutions for their removal. Scientific analysis can thereby avoid the potential for new damage due to misapplication of products designed for conditions other than those actually responsible for the specific case. The scientific findings from several case studies are described along with their implications for the prevention and removal of stains.

1 INTRODUCTION

It has been found through testing that stains with similar appearance may arise from a surprisingly wide variety of different causes. Very often blemishes that are prematurely attributed to a particular stone type or to installation errors have been found to derive from associated construction materials. In other cases they are provoked by a specific combination of job site conditions acting in consort with installation details in a manner that was not anticipated by the designers.

2 THE ROLE OF WATER

It is often apparent to even the casual observer that stains relate in some way to water used in the installation or to some source of dampness. Sometimes stains are localized over the placement of a spot of setting mix. In indoor wall installations these may be plaster spots while outdoor installations may stain over anchors or tie wires set in portland cement mortar. Wall installations incorporating an air gap may exhibit staining up to the level of a back-fill consisting of plaster or mortar intended to reinforce the lowest course against impacts. Floors may exhibit stain patterns spanning pavers that correspond to the position of materials or equipment stationed on them after installation while adjacent construction proceeded. Perhaps most frustrating of all, stains may be found after removal of temporary protective layers that mimic the placement pattern of the protective material itself, as if flaunting the very intent of its use.

Often the initial presumption is that, when the last of the moisture evaporates, the stains will disappear. Only when this fails to happen on its own, and when forced drying fails to improve the situation, does it become apparent that a more serious problem exists. Water played an initial role, but the stain itself has another component.

3 THE NATURE OF STAINING

What constitutes a stain and where does it reside? In the simplest cases stains are due to some colored substance contaminating the stone surface. However such cases make up only a fraction of the ones encountered. It is more often the case that stains result from colorless materials. To understand the causes of staining it is important to recognize that the appearance of stone is entirely a result of how it reflects light to the observer. This, in turn, is determined by how light is both absorbed and scattered from the stone. To a greater degree than is often recognized, a stone's appearance is determined by some degree of translucency that exists near its surface. A stone lacking translucency which is effective in scattering light off its top surface appears bright, while one that allows a greater portion of the light to penetrate and be absorbed in the interior

223

looks darker. The principal causes of light scattering in a pale stone are surface roughness and the pores between translucent grains just beneath the surface that scatter light back out to the observer.

When water fills these subsurface pores, scattering is reduced and absorption increased, darkening the stone. Reduced light scattering is not only a function of moisture, however. Non-volatile materials which fill the pores may similarly reduce scattering and result in darkening. Thus, a stain which starts as a consequence of moisture-filled pores may persist long after the moisture is gone as a consequence of pore-filling by water-soluble substances left behind when the water evaporates.

A reasonable expectation would be, therefore, that dissolution of this residual matter by water could solve the problem. The real situation is more complicated, however, due to the properties of the typical staining material. The prevailing conditions inside the stone at the time of setting, in which the material was able to migrate, are not those present at the surface from which the stain must later be removed. Furthermore, reintroducing the quantities of water required to extract the stain often reactivates the conditions that first created it and leads to bleeding of the stain over a wider area. The latter is a principal cause of the failure of commercial poultice formulations which attempt to treat a broad spectrum of different conditions with a single material.

4 STAIN TYPES

Trace levels of naturally occurring materials that are capable of forming stains are an intrinsic part of most stones, especially limestones, and in no way constitute a defect in the material. These often include traces of soluble minerals and organic matter derived from ancient plant, animal and microbial sources. Since stain formation is a surface process, it is easy to calculate how a soluble natural component, present in the bulk stone at levels of a few hundred parts-per-million, is sufficient to entirely fill the pores at its surface that are responsible for its appearance if this material is effectively transported there by conditions during the installation.

Associated construction materials may serve as sources for soluble matter that ultimately migrates to the stone surface, staining it. Casting plaster (plaster of Paris) can serve as a source of the slightly soluble mineral gypsum, as can manufactured wall board made of this material ("sheet rock"). Paper products that serve as facings on wall board may contain soluble organic matter that is extractable by water.

One of the principal sources of soluble matter is portland cement and the myriad proprietary mixtures based upon it. Modern portland cement mortars are highly sophisticated materials including multiple agents to control both working and setting properties. While specifications control its main performance properties, considerable variation exists among ingredients that control these properties. Trace additives such as grinding aids, agents to reduce moisture absorption during storage, agents to limit the water proportions in the mix, agents to increase adhesion, and agents to control the setting rate and ultimate strength under varying temperatures are a few of the most common types.

A usable mix contains sand and water as well, of course. Mix formulators interested in securing an increasing share of the stone setting market will impose quality controls on the sand and exclude any unstable minerals and contaminants prone to staining. However sands that might be acceptable in general concrete work can occasionally be unsuitable for stone setting. The stone installer who would create a mix from locally available material would be well-advised to scrutinize the sources carefully, preparing installation mockups as an added precaution. This is particularly advisable in international projects that may pair stone from distant sources with local mix materials with which there has been no prior stone installation experience.

It has become increasingly common to use prepackaged setting mixes including "thinset" that include a liquid component that is intended to be added in fixed proportion to the pre-mixed dry solids. These can have substantial benefits for stone installation but they also open new, and often unforeseen, possibilities of stain development. To understand, and thereby avoid this complication, it is useful to know a bit about the main classes of portland cement admixtures, how they work, and what they are intended to achieve.

Many mixtures make use of a liquid component including a polymer, or latex, emulsion (a suspension of fine droplets in water with a milk-like appearance). These are not, as some perceive them to be, water-free. All hydraulic cements require water to set. Therefore, they do not eliminate the potential for water-borne solubles to cause staining. They do, however, offer a potential advantage in the form of minimizing the water content of the mix through the incorporation of water-reducing (WR) admixtures. Without such an admixture a portland cement mix will typically require water in excess of that actually consumed in its setting reaction in order to achieve the necessary "slump", "suction" and workability. Water that is not consumed in the setting reaction must ultimately be lost through evaporation. Therein lies the potential for stain formation, as soluble matter is carried along with the water to its point of evaporation. One goal of using a WR admixture is to keep the water content to a practical minimum while maintaining workability. Further reductions in water content are possible through the use of sand that has been sieved for a selected size distribution that minimizes fluid-filled spaces between the grains.

224

In stone setting it can be an error to extend a polymer emulsion component by diluting it with water. Often the temptation arises from a desire for cost reduction and the assumption that sufficient admix will be imparted even in a diluted form. However, in so doing, the proportion of the admixture to the water is reduced. A corresponding increase in the proportion of total water to portland cement will result (because workability will require it in the absence of the intended admixture level) and the portion of water remaining to be lost by evaporation will increase.

Polymer emulsions require stabilization aids (to resist separation of the emulsion ingredients) to ensure a reasonable shelf life under a wide range of storage and job conditions. These ingredients open some new potential for unexpected staining problems since these ingredients typically remain soluble after setting of the mix and may be transported to the stone surface if there is an unexpected source of continuing moisture migration.

Apart from additives, portland cement has intrinsic sources of stain-forming materials as well. Free alkali that is released upon mixing with water can be particularly effective in mobilizing organic matter from limestone that would remain evenly distributed throughout the stone in its absence. This problem affects many polishable stones commercially termed "marbles" that are actually limestones in the geological sense of being sedimentary in origin. The use of portland cement meeting the requirements of a "Low Alkali" designation is not a solution to this problem, as the requirements of the designation seek to minimize the overt effects of alkali, such as efflorescence formation and aggregate attack but are not sufficiently low to eliminate limestone staining phenomena. "Masonry Mortar" compositions that incorporate hydrated lime can be helpful in reducing this specific problem, as can the use of a white portland cement, some types of which can be found to significantly exceed the baseline requirements of the "Low Alkali" designation.

One of the most intractable staining problems can be the result of migration of the free lime that is released from the setting of portland cement or subsequently leached from it by migrating moisture. The disparity between the environment within the stone pores and that of the outside air is particularly problematic in this case, although it can be a factor in the difficulty of stain removal in other cases as well. In the case of free lime, the slightly soluble species calcium hydroxide is converted (carbonated) to insoluble calcium carbonate upon exposure to carbon dioxide in the air. This typically occurs at the same point, or just beneath, the stone surface where evaporation of the transporting moisture takes place. Translucent, light colored granites in exterior installations like those of Figure 1 have been particularly affected by this phenomenon when drainage from the setting bed

Figure 1. Stained margins due to subsurface calcium carbonate deposition in granite installation with a saturated bed.

was neglected by the designers, leading to long-term migration.

This reaction puts into focus the reason why staining problems with apparently simple mechanisms often lack correspondingly simple solutions. The stain formation occurred under persistent migration of water carrying the soluble species gathered from a large volume of setting material to a tiny zone near the surface. There it occluded the pores, reducing their light scattering role and resulting in darkening of the regions most favorable to moisture egress. In its carbonated form the solubility that the lime once had is lost. Some granites will afford treatment to dissolve this calcium carbonate (with difficulty, since it resides in a protected zone beneath the surface) but limestones or marbles that are similarly affected cannot be treated, since no discrimination is possible between the calcium carbonate of the stone and that in the stain.

5   CONVERGING FACTORS

Staining problems are seldom the result of a single property of the stone, setting materials, substrate or job conditions. They more often result from unforeseen combinations of these. Much of the apparent capriciousness with which they occur can be explained by the interaction of the factors listed above.

A variant of the problem of alkali extraction from the setting bed of an outdoor installation serves as an example. A large plaza surrounding a fountain situated over an underground parking structure was paved with a radiating pattern of grey-veined Cararra marble in travertine. The structural concrete beneath was given

225

a heavy bitumen-based coating as part of the water proofing design. The marble surface was equipped with drains and additional drains were provided for seepage within the setting bed. Over a period of years the marble developed intense orange-brown stains that originated along the grey markings and spread across the surface from them. The travertine was unaffected. The appearance suggested a serious problem of rusting pyrite in the marble by the way that the stains bore a clear relation to the marble markings (Fig. 2). In fact pyrites are often encountered within the grey markings of white marbles.

However, laboratory study coupled with site investigations and a review of the construction drawings revealed an entirely different mechanism that had nothing to do with iron. Tests showed that the stain contained no more iron than the unstained areas and consisted of dark organic material derived from the bitumen water proofing underlayment. Samples of the bitumen released nothing into fresh water when extracted in the laboratory but when the composition of the water was changed to that typical of the water percolating through the setting bed, bearing the extracted soluble components of the mortar, it produced a solution resembling strong tea in color. Upon contact with the air, this colored matter gradually became insoluble once again.

Examination of the bed drainage system showed that loose sand had obstructed parts of it, creating a situation where the inclination of the plaza led to parts of the marble surface being below the level of water entry up-slope. This created a positive pressure and resulting upward water migration, analogous to thousands of artesian wells on a tiny scale. Water bearing the extractives emerged through the more porous regions of the marble pavers. These pores tended to be larger and more numerous along the grey markings.

Unfortunately, the effects of sunlight and air on the brown matter resulted in it becoming permanently insoluble within the marble pores. However, the issue of stain removal was moot since an understanding of the process and evaluation of the construction made it clear that staining would recur unless the water proofing and drainage system was entirely revamped.

Stains that occur shortly after installation in indoor settings can often be avoided by attention to the problem of excess moisture evaporation from the setting materials. An interior limestone wall installation using casting plaster over old, cast-in-place concrete developed stains corresponding to the plaster "spots" only near inside corners. Outside corners were unaffected unless the joints were caulked shortly after setting. In this case, better air circulation behind the limestone near the outside corners eliminated the excess moisture before it had time to migrate through the limestone face carrying solubles with it. Where the joints were caulked soon after installation, air circulation behind was reduced, alternate routes for moisture evaporation were eliminated, and its one remaining route, through the stone, became stained. Elsewhere in the same job, stones installed over recently cast concrete used to fill an unused doorway stained even though the joints were left uncaulked for a time. In this case the recently cast, and not fully dry, concrete served as a reservoir of moisture which was coupled to the stone through the setting plaster, resulting in prolonged moisture migration in excess of that which could escape into the airspace behind the stone.

An interesting example of limestone floor staining involving a pre-packaged two-pack setting mortar, incorporating a polymer emulsion and portland cement-based dry mix, illustrates the potential pitfalls of such a system. The pale limestone developed widespread darkening that bore no relation to any apparent variations in the stone itself (Fig. 3). Removal of one paver revealed that areas free of staining corresponded to

Figure 2. Cararra marble paver with stains due to alkali-transported bitumen extracts. Pores are associated with grey veining, leading to the erroneous impression that the stain is derived from minerals associated with the markings.

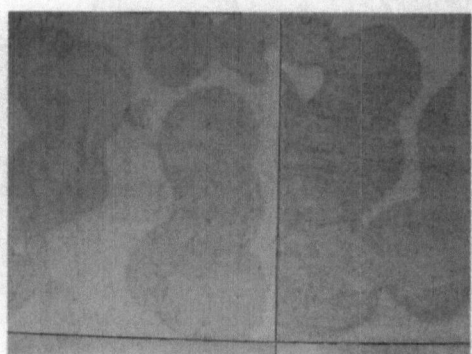

Figure 3. Light beige limestone interior floor stained by a set accelerating mortar admixture.

226

small regions of air trapped between the stone and setting mix and, therefore, suggested that staining involved transport of some soluble species from the mix to the top of the stone. Under the microscope the fossil matter of which the stone was comprised remained completely visible and no specific cause of the stain could be seen.

Washing of the stone had no effect, nor did mildly alkaline stone cleaners drawn from commercial sources. Samples collected from the stone surface for analysis by Fourier Transform Infrared Spectroscopy (FTIR), a method of chemical micro-analysis that has broad range applicability to both minerals and organic compounds, showed substances that are not normal components of the stone. In fact, the stained area was found to be covered by a minute, transparent layer of calcium formate with a little synthetic polymer derived from the setting mix (Fig. 4).

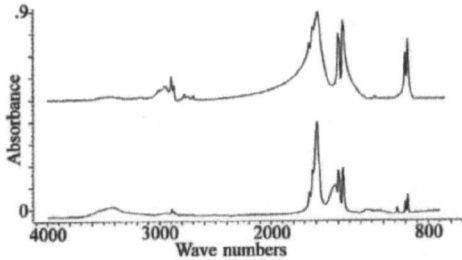

Figure 4.   FTIR spectrum of stain residue scrapings (lower curve) compared with the spectrum of calcium formate.

Figure 5.   Transition between stained (left) and cleaned (right) surface of limestone shown in Figure 3 as viewed under oblique illumination at 30× magnification. Note that the fossil matter of which the stone is comprised remains visible through the crust that is responsible for the stain. In the absence of the step transition produced by cleaning, the layer was indistinguishable from the honed surface of the stone.

An absorbtive cleaning material, based on clay and using minimal water (so as not to reactivate the migration of more staining matter), was tailored to the specific problem based on this information and successfully removed the stain. Comparison under the microscope of the cleaned and adjacent uncleaned areas reveals a "step" where the uniform layer of deposited material ends (Fig. 5). In the pre-cleaned state, the layer was invisible under all forms of lighting but was nonetheless responsible for the diminished brightness of the stone.

The precise objective of the manufacturer's use of the calcium formate is unknown. However, the use of this compound was patented in 1965 as a set accelerator for portland cement as an alternative to calcium chloride and could be expected to improve its cold weather performance (Dodson et al. 1965). Sodium formate has also been used in the role of an accelerator and would be expected to produce calcium formate from available calcium ions released during setting. Review of the actual installation revealed that the emulsion component had been extended by mixing 1:1 with water, a factor that may have exacerbated the transport of the formate to the surface.

6   DEVELOPING METHODS OF STAIN REMOVAL

As noted above, a wide variety of staining problems result from changes in the translucency of the stone surface – a property that is strongly influenced by the deposition of water-borne compounds on the sub-surface pores. Over twenty years of investigating stone problems with the aid of chemical microanalysis to verify the compounds involved, this writer finds that a diagnosis of the materials involved by visual assessment alone is accurate less than half the time. Specific conditions on the job site during and after construction are often impossible to verify when attempting to determine the contributory factors. "Obvious" causes, like the apparent corrosion of pyrite in the first case described above, are often found not to be responsible. Requests for technical assistance with stains often arise as a consequence of failed, trial-and-error attempts to eliminate the stain without a knowledge of the underlying cause. It is not unusual for the outcome of such assistance to be an effective method for the stain removal that succeeds everywhere except where previous attempts at experimentation have resulted in "setting" the stain or creating a new class of problems.

The diverse materials that have been found to cause stains require treatment with a knowledge of what is to be removed in order to prevent the possible fixing of a stain through the inadvertent conversion of the staining matter to a less soluble product. Subsequent attempts with other methods often are

227

made impossible, resulting in re-finishing or replacement operations with all their attendant disruptions to the occupants. This has been widely observed as a consequence of the use of commercial cleaners and "poultices" which, in order to be economically viable as off-the-shelf products, cannot be formulated to deal optimally with individual problems and, instead, must attempt a moderate level of success and broad applicability to widely different cases.

An effective resolution of the stain problems on a high-value project is often possible through the application of chemical analysis and the subsequent formulation of a treatment material that addresses the specific cause without undesirable side effects. The technical information required is not only provided by standardized testing of the stone and installation materials but, most importantly, by an assessment of the installation details, job conditions and analysis of the stains themselves through instrumental means. This analysis should seek not merely evidence of a feasible explanation for their occurrence, but to find evidence that excludes alternative causes.

Many agents implicated in staining have multiple possible sources in the same installation. Identification of the actual cause may have important financial ramifications in terms of the complexity and duration of stain removal and the avoidance of new problems. Such an approach allows an informed decision to be reached that includes considerations such as whether the problem will recur and whether it is likely to occur in future installations.

The tools of analytical chemistry: Fourier Transform Infrared Spectroscopy (FTIR), Scanning Electron Microscopy – X-ray Spectrometery (SEM – ES), X-ray Diffraction (XRD) and Polarized Light Microscopy (PLM), all serve complementary roles in such an effort. Each can provide key evidence for the underlying stain mechanism but real understanding requires synthesis of the different results with an eye toward the strengths and limitations of each method. Input from a scientific advisor with experience in the application of these techniques to stone problems in advance of an important project often yields savings both in terms of immediate costs and in terms of the performance reputations of the firms involved.

REFERENCE

Dodson, V.H., Farkas, E. & Rosenberg, A.M. 1965. *Non-Corrosive Accelerator for Setting of Cements*. U.S. Patent No. 3,210,207, October 5, 1965.

Dimension Stone 2004, Přikryl (ed)
© 2004 Taylor & Francis Group, London, ISBN 90 5809 675 0

Complex weathering effects on the durability characteristics of building sandstone

P.A. Warke, B.J. Smith & J. McKinley
School of Geography, Queen's University Belfast, UK

ABSTRACT: Durability characteristics of Stanton Moor sandstone samples were assessed by a standard salt crystallization test and a modified laboratory weathering simulation using a combination of salt weathering and freeze/thaw cycles. Data indicated significant differences in durability characteristics between fine – medium-grained and medium – coarse-grained Stanton Moor sandstone with the latter being more durable than the former under both test conditions. Inclusion of salt weathering and freeze/thaw cycles in the main experimental run, introduced complexity into the decay process that more accurately reflected 'real world' conditions. Data indicated that relatively minor structural and mineralogical differences between samples of the same stone type can significantly influence weathering behaviour, resulting in distinct rates and patterns of breakdown.

1 INTRODUCTION

The weathering of building stone involves an often, complex progression from 'fresh' to 'failed' stone. Typically, however, this change of state does not follow a simple linear route but proceeds episodically, often in response to the breaching of intrinsic strength/stress thresholds and/or extreme weathering events, with intervening periods of apparent quiescence (Smith et al. 2002). Frequently, at any one point in time different stages of the decay sequence may be evident on a single block or adjoining blocks of stone, reflecting the complexity of intrinsic controls such as structural/mineralogical properties and extrinsic factors such as the temporal and spatial variability of microenvironmental conditions at the rock/air interface (Smith 1996, Warke 2000). Feedback conditions, which arise from interactions between these intrinsic and extrinsic controls, introduce further complexity making the weathering behaviour or the durability of stone difficult to predict in all but the most general of terms.

Durability characteristics of different stone types may vary greatly reflecting differences in the ability of stone to resolve stresses imposed by the cumulative and sequential effects of weathering processes under a particular set of environmental conditions. Even within a single stone type relatively minor structural and mineralogical variations can significantly alter durability and, as stone weathers, original durability may alter as stone properties such as porosity change due, for example, to enlarging of pores through chemical dissolution or blocking of pores through infilling by salt (Smith & Kennedy 1999).

Despite the dynamic nature of stone properties, the suitability of stone for use on different parts of buildings is typically determined by a number of standard industry tests which aim to predict durability of fresh, unweathered stone (Smith et al. this volume). Unfortunately, most standard durability tests only register the two end extremes in this progression i.e. stones are inserted as fresh blocks and only revisited upon disintegration. Durability tests (e.g. the sodium sulphate test) are therefore not designed for, or capable of, identifying the subtle, non-linear and invariably spatially specific changes that characterise stone decay. This dichotomy between standard tests that assume constancy of stone properties, and 'real world' conditions where some stone properties can change significantly over time, may lead to a difference between predicted and actual behaviour of stone (Smith 1999, Turkington 1996).

The ability to predict 'change during lifetime' will greatly help in the selection of stone for specific purposes and environments and in the design of structures that will retain their architectural integrity as they age. Preliminary data are reported from laboratory simulation experiments where the effects of incorporating high magnitude, low frequency weathering cycles (freeze/thaw) into a regime of low magnitude, high frequency weathering (salt weathering) on Stanton Moor sandstone samples are examined. Rates and patterns of decay for both stone types are reported with

229

specific emphasis given to identification of extrinsic controls and lithologically-specific and non-specific intrinsic factors that determine the ability of this sandstone to resolve both high and low magnitude stress events. Attention is also paid to the early stages of the decay sequence when degradation of structural and mineralogical properties is just beginning. These are the stages in stone weathering about which much is assumed but little is actually known.

2 METHODOLOGY

2.1 Materials

Stanton Moor sandstone is a member of the Carboniferous Millstone Grit series and is described as fine – medium-grained quartz sandstone. However, grain size can be variable with beds of medium – coarse-grained material reflecting the original conditions of fluvial deposition. In this study, experimental data from two different grain-size representatives of Stanton Moor are reported separately and, for the sake of brevity, are subsequently referred to as coarse-grained Stanton for the medium – coarse-grained member and fine-grained Stanton for the fine – medium-grained variety. Summaries of their respective structural and mineralogical properties are shown in Table 1.

2.2 Sample preparation

Sixty-six blocks ($750 \times 750 \times 750$ mm) of fresh Stanton Moor were cut (33 each of fine-grained and coarse-grained sandstone). Each block was washed and air-dried at c. 20°C until block weight remained stable over three consecutive days. These blocks were then divided into eleven subsets each comprising six blocks with two blocks in each subset used as control samples. Average permeability data were collected from six of the eleven subsets of blocks (total of 36 blocks) using a portable probe gas permeameter to give an indication of any permeability differences between the two varieties of sandstone (Table 1).

Thin-section examination of coarse and fine-grained sandstone samples identified the abundance of quartz with the presence of extensive interlocking quartz overgrowths, confirmed by scanning electron microscopy (Fig. 1). Interlocking overgrowths were also evident in the feldspars and these structural properties may account for the relatively low permeability and porosity characteristics of the fine-grained and especially the coarse-grained Stanton sandstone (Table 1).

2.3 Experimental design

Each of the sixty-six blocks of Stanton sandstone was placed on a separate tray inside an environmental

Table 1. Summary of Stanton Moor structural and mineralogical properties.

Stone type	Porosity	Permeability (mD)	Mineralogy*
Fine – medium Stanton	17%	Mean 61 Range 7.7–206	62% Quartz 7.5% Feldspar 0.75% Mica 11.25% Clays
Medium – coarse Stanton	13.5%	Mean 58 Range 4.8–114	62.3% Quartz 12% Feldspar 2% Mica 7.5% Clays

* Preliminary analysis identified clays as chlorite.

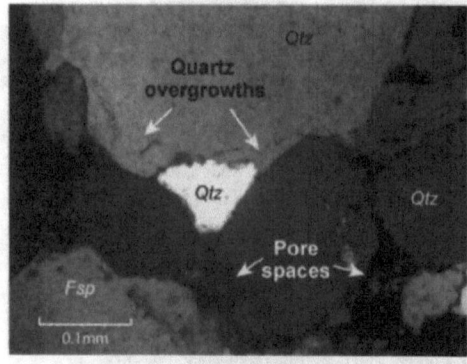

Figure 1. Thin-section showing detail of grain boundaries with well-developed quartz overgrowths in coarse-grained Stanton sandstone (*Qtz* = Quartz; *Fsp* = Feldspar).

chamber. Each day four blocks from each of the eleven subsets were fully immersed in a 2.5% solution of Na_2SO_4. Any debris released in the tray whilst in the chamber and during daily wetting was collected and weighed. The temperature regime used for salt weathering cycles is shown in Figure 2a and was selected to provide temperature parameters representative of 'real-world' conditions. The total daily experimental run lasted for 20 hours comprising two consecutive cycles of 10 hours each (one 'wet' cycle following immersion and one 'dry' cycle). This combination of 'wet' and 'dry' cycles more accurately reflects conditions experienced by stone on a building when it dries between wetting events but is still exposed to the same fluctuations in temperature. The salt weathering conditions described comprised the high frequency, low magnitude element of the experimental regime. The choice of a 2.5% solution of Na_2SO_4 was dictated by the need to avoid conditions so extreme that the subtleties of different stages of the decay sequence would be lost in rapid breakdown. A pilot

Copyrighted Material

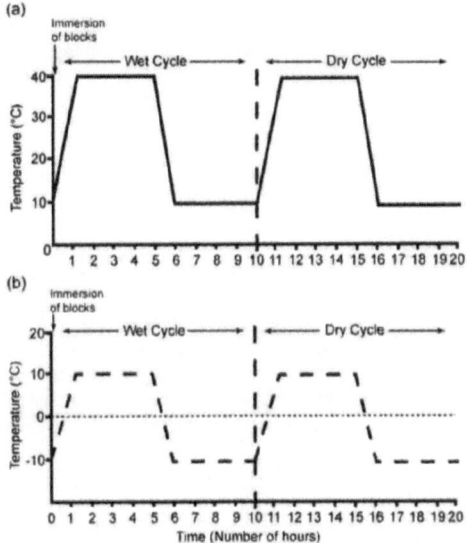

(a)

(b)

Figure 2. Temperature regimes and cycle profiles used in the main experimental run.

study had shown that a 10% solution of Na_2SO_4 resulted in complete and rapid sample breakdown before a representative number of experimental cycles could be completed.

The low frequency, high magnitude component of the experimental run comprised the episodic exposure of selected block sets to an incremental number of freeze/thaw cycles. After every 20 salt weathering cycles pre-selected sample subsets were fully immersed in deionised water and then exposed to two consecutive freeze/thaw cycles. The temperature regime used for the freeze/thaw weathering cycles is shown in Figure 2b. In order to assess the specific effect of freezing on the nature and rate of breakdown not all sample sets experienced the same number of cycles. The weathering histories accumulated by each subset of blocks at the end of the experimental run span a spectrum of conditions from subset one with 220 salt weathering cycles only, subset six with 210 salt weathering and 10 freeze/thaw cycles through to subset eleven which was exposed to a combination of 20 freeze/thaw and 200 salt weathering cycles. The various combinations of freeze/thaw and salt weathering cycles created a complex set of known 'weathering histories' thereby facilitating identification of the effects of different combinations of high and low magnitude weathering events on decay pathways.

In each of the eleven subsets two blocks provided the experimental control samples with one remaining dry throughout the experimental run while the other

underwent daily wetting with deionised water. The two control blocks in each subset experienced the same combination of freeze/thaw and/or salt weathering cycles as their subset counterparts.

2.4 Sample analysis

With regard to results reported here, the main method of assessing block response to different weathering regimes was through analysing weightloss data. Following each weathering cycle (frost and salt) the amount of material released during wetting was washed to remove as much salt as possible, dried and weighed. At the end of the experimental run the cumulative amount of debris lost from each block was expressed as a percentage of the original block weight. In addition to weightloss data, selected blocks were removed at preset intervals during the experimental run for analysis of substrate material by ion chromatography (IC) and atomic absorption spectroscopy (AAS) to identify the distribution of sodium sulphate. However, preliminary data only are currently available from these analyses.

2.5 Standard durability testing

As part of the general characterisation of both the fine-grained and coarse-grained Stanton Moor sandstones in preparation for the main simulation experiment, additional samples were tested using the British Standards Institution and European Standard prEN12370 procedure for 'Determination of Resistance by Crystallisation Cycles' (for methodology see Ross & Butlin 1989). This was done to provide a standard baseline assessment of durability against which to compare data derived from this study.

3 RESULTS AND INTERPRETATION

3.1 Standard durability test

Results from the standard durability salt crystallisation test showed marked differences in the durability characteristics of the two types of Stanton Moor sandstone. Coarse-grained samples proved to be more durable under the test conditions losing, on average, 6.9% of initial block weight after fifteen cycles whereas fine-grained samples showed an average loss of 67% of initial block weight. Differences in durability characteristics of the two types of Stanton Moor may reflect the influence of porosity and mineralogy with the lower porosity and permeability of the coarse-grained Stanton samples acting to retard penetration of salt solution. Results from the standard salt crystallisation test provided a reasonably accurate, if somewhat generalised, guide to the behaviour of both Stanton types

231

in the main experimental study but its simplicity masked subtleties in the decay sequences of both types of Stanton Moor sandstone.

3.2 Weightloss data

Weightloss data were recorded after each cycle for each block during the experimental run and expressed as cumulative percentage weightloss. In the interests of clarity, only one curve from each fine-grained and coarse-grained sample blocks in alternate (1, 3, 5, 7, 9 and 11) subsets are shown in Figure 3. It is important to note that no debris was released from any of the control blocks during the experimental run. Data indicate that the fine-grained Stanton blocks released a consistently higher proportion of material in comparison to their coarse-grained counterparts with several decay stages identified.

3.2.1 Stage 1: preparation

Stage one involves preparation for the onset of decay – a stage characterised by the absence of obvious evidence of block deterioration or release of debris. During this stage the blocks (both coarse and fine-grained) exhibited a prolonged period of apparent quiescence during the first 100–120 cycles of the experimental run. Initiation of breakdown occurred within a narrow span of cycles from 104–126 with no significant correlation identified between the beginning of breakdown as defined by actual debris loss and the total number of weathering cycles experienced. So for example, blocks in subset eleven started to breakdown at approximately the same stage in the experimental run as samples from subset one. These data indicate that some internal strength/stress threshold was being breached at this stage irrespective of the type or combination of weathering cycles applied

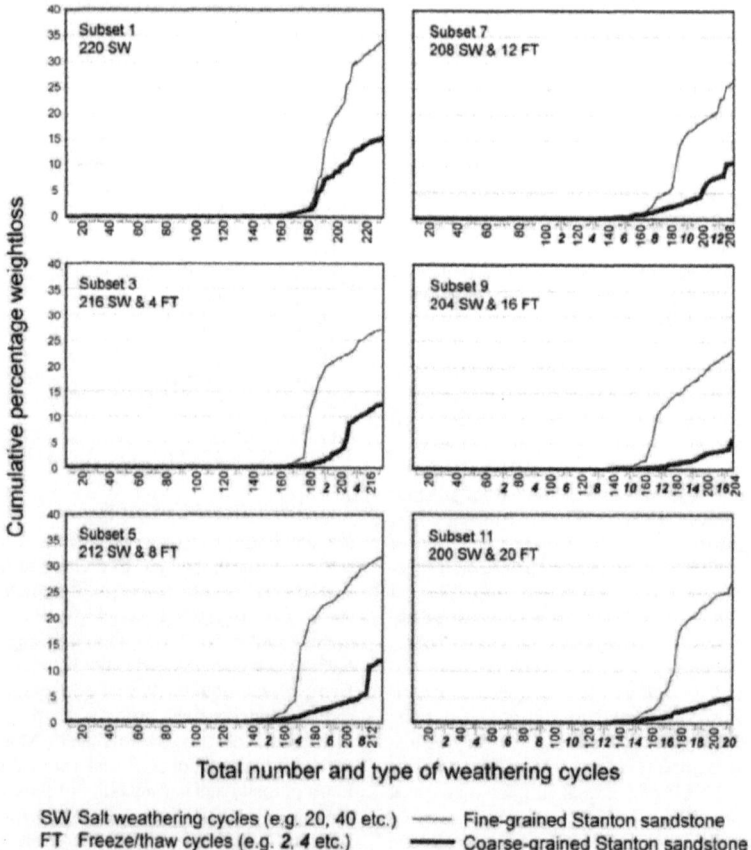

Figure 3. Selected cumulative percentage weightloss curves.

(i.e. whether salt only or salt and freeze/thaw). This threshold may be structurally determined, related to the constraining effect of well-developed interlocking quartz and feldspar overgrowths evident in both the fine and coarse-grained samples.

It was observed that the fine-grained Stanton samples exhibited some surface deformation immediately prior to initiation of failure, manifest as a slight 'bowing' of the block surface. Preliminary analysis of salt distribution in substrate material of blocks removed at predetermined intervals during the first 120 cycles of the experimental run indicated that Na_2SO_4 tended to concentrate in pore spaces some 5–10 mm below the block surface. However, it is important to note that salt penetration was not uniform and, particularly in the fine-grained Stanton samples, may have been associated with more permeable points on block surfaces. From these points of initial localised penetration and concentration it would appear that salts further penetrated the substrate aided by repeated cycles of dissolution and recrystallisation and in some subsets by the additional disruptive effects of repeated freezing and thawing of moisture, all of which appear to have facilitated the gradual disruption of interlocking grain boundaries and eventual lateral extension of microfracture networks. The initial depth of salt penetration was related to maximum wetting depth, which was restricted to the upper c.5 mm of stone as determined by porosity and permeability characteristics.

Fine-grained Stanton blocks exhibited the greatest range of permeability values with possible 'hot-spots' of greater permeability that may have provided loci for salt/moisture penetration providing points of access to substrate material. Such spatial variation may help to explain the behaviour of sandstone under 'real world' conditions where initiation and development of decay features can be spatially variable over the surface of one block.

Coarse-grained Stanton samples also exhibited spatial variability in surface permeability characteristics but the range of readings was much less than their fine-grained counterparts. These data indicate that penetration of salt in solution would have been less effective in the coarse-grained Stanton with a resultant accumulation of salts in the upper few millimetres of stone. In comparison, the fine-grained Stanton would have provided more points of access to substrate material for salts in solution and may have facilitated their deeper penetration and gradual accumulation.

3.2.2 Stage 2: initiation and nature of early decay

Stage two of the decay sequence covers the period from time of first debris release to establishment of stone breakdown. This is best described as a transition phase, representing the period in the experimental run when blocks were undergoing the most dramatic changes in form through rapid release of debris.

The onset of block deterioration did not appear to be related to the nature of the weathering history i.e. whether this comprised exposure to salt weathering cycles only or a combination of both salt and freeze/thaw cycles. The nature and rate of early debris release differed for each of the two types of Stanton Moor. The combination of low porosity and low permeability appeared to retard moisture penetration during wetting in the coarse-grained Stanton samples. Consequently, salts tended to concentrate in the upper few millimetres of stone with the result that surface and near-surface grain boundaries became foci for salt accumulation. Initiation of breakdown in coarse-grained samples was, typically, heralded by the release of small amounts of debris through granular disintegration with the amount released increasing gradually as the experimental run progressed.

In comparison, the fine-grained samples had higher porosity values and although mean permeability values were similar to those of the coarse-grained samples, they exhibited a much greater within, and between block, range of values. In addition, fine-grained Stanton sandstone showed a higher percentage content of clay minerals (11.25%) and clay minerals have been linked to increased efficacy of salt weathering because they provide points of ingress for moisture and can act as foci for salt accumulation (McGreevy & Smith 1984, Rodriguez-Narvarro et al. 1997, Warke & Smith 2000). Initiation of decay in fine-grained samples was heralded by deformation of block surfaces followed by the development of surface fractures. As the fractures enlarged small amounts of debris was released and typically this was followed by catastrophic breakdown through extensive scaling and granular disintegration in the following 2–4 cycles.

3.2.3 Stage 3: reinforcing decay

Once established, decay of fine and coarse-grained Stanton appeared to follow different pathways. Stage 3 for fine-grained Stanton samples was characterized by a relatively consistent loss of material primarily through granular disintegration which followed the initial scaling failures in Stage 2. Preliminary examination of samples removed just prior to and following initiation of decay, indicated that during Stage 1 substrate material was weakened primarily through disruption of inter-granular bonds. Evidence indicated that following initial rapid loss of material through extensive surface scaling in Stage 2, the weakened substrate facilitated salt and moisture ingress with decay progressing steadily through granular disintegration and localised structural disaggregation.

Data indicated that if the experimental run had continued, the fine-grained Stanton samples would have continued to diminish in size through conditions of

233

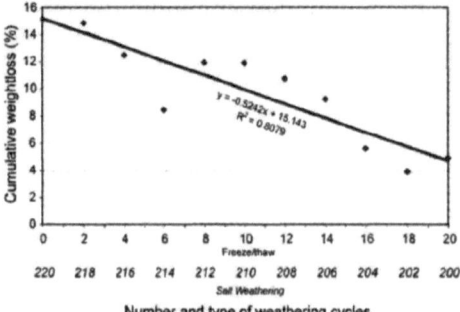

Figure 4. Correlation between weathering cycles and weightloss for coarse-grained Stanton Moor sandstone samples.

positive feedback. Patterns and rates of decay exhibited by the fine-grained Stanton samples throughout the experimental run appeared to be controlled by a combination of intrinsic structural controls and the number of weathering cycles regardless of whether these comprised a combination of freeze/thaw and salt weathering or only the latter.

Defining the end of Stage 2 and the beginning of Stage 3 in the decay sequence of coarse-grained Stanton samples was difficult because of the influence of freeze/thaw cycles. Nevertheless the nature of debris loss remained consistent throughout, proceeding almost entirely through granular disintegration. In coarse-grained samples, preliminary IC and AAS analyses showed that salt accumulation was restricted to the outer few millimetres of stone. This was attributed to the combined effects of a lower porosity and a slightly lower but less variable permeability that prevented uniform penetration of moisture, development of organised subsurface micro-fractures and thus scaling. Consequently, debris loss progressed through the release of individual grains with an uneven lowering of the block surface.

The precise nature of the interaction between breakdown patterns of coarse-grained samples and freeze/thaw cycles is unclear at present but preliminary analysis of data indicated the existence of a significant negative correlation between cumulative percentage weightloss and the total number of freeze/thaw cycles experienced during the experimental run (Fig. 4). The association between these variables may be partly explained by the experimental procedure itself because of the use of deionised water for block immersion prior to freeze/thaw cycles which removed some salts in surface and near-surface material with a subsequent reduction in the efficacy of subsequent salt weathering cycles.

4 CONCLUSIONS

Although these data are still in the preliminary stages of analysis, they highlight several features of the decay sequence for Stanton Moor sandstone.

Decay sequences for fine and coarse-grained Stanton samples comprised different stages. Both Stanton types initially followed similar decay pathways with a preliminary stage of apparent quiescence followed by a transitional stage when breakdown began. At this point the two decay pathways diverged with fine-grained samples exhibiting rapid and dramatic initial breakdown while coarse-grained samples broke down at a much slower rate. The final stage was difficult to define but represented the establishment of positive feedback conditions for both Stanton types and the progressive loss of material primarily through granular disintegration.

In the transition from Stage 1 to Stage 2, some intrinsic strength/stress threshold was breached in both Stanton types. This threshold appeared to be directly related to the structural characteristics of Stanton and not to the type of weathering cycles each subset of blocks was exposed to (i.e. freeze/thaw and salt or salt weathering cycles only).

Data indicated a relationship between the number of freeze/thaw cycles and breakdown of the coarse-grained Stanton but as yet the precise nature of this remains unclear.

ACKNOWLEDGEMENTS

Financial support for this work was provided by EPSRC grant GR/R54491/01. Special thanks must also go to Gill Alexander in the Cartographic Department of the School of Geography.

REFERENCES

McGreevy, J.P. & Smith, B.J. 1984. The possible role of clay minerals in salt weathering. *Catena* 11: 169–175.

Rodriguez-Navarro, C., Hansen, E., Sebastian, E. & Ginell, W.S. 1997. The role of clays in the decay of ancient Egyptian limestone sculptures. *Journal of the Institute of American Conservators* 36: 151–163.

Ross, K.D. & Butlin, R.N. 1989. *Durability tests for building stone.* BRE Report BR 141. Watford: Building Research Establishment.

Smith, B.J. 1996. Scale problems in the interpretation of urban stone decay. In B.J. Smith & P.A. Warke (eds), *Processes of urban stone decay*: 3–18. London: Donhead.

Smith, B.J. & Kennedy, E. 1999. Moisture loss from stone influenced by salt accumulation. In M.S. Jones & R.D. Wakefield (eds), *Aspects of stone weathering, decay and conservation*: 55–64. London: Imperial College Press.

Smith, B.J., Turkington, A.V., Warke, P.A., Basheer, P.A.M., McAlister, J.J., Meneely, J. & Curran, J.M. 2002. Modelling the rapid retreat of building sandstones: a case study from a polluted maritime environment. *Geological Society of London Special Publication* 205: 339–354.

Smith, B.J., Warke, P.A. & Curran, J.M. (this volume). Implications of climate change and increased 'time-of-wetness' for the soiling and decay of sandstone structures in Belfast, Northern Ireland. Rotterdam: Balkema.

Smith, M.R. 1999. Stone: building stone, rock fill and armourstone in construction. *Geological Society of London Engineering Geology Special Publication* 16, London.

Turkington, A.V. 1996. Stone durability. In B.J. Smith & P.A. Warke (eds), *Processes of urban stone decay*: 19–31. London: Donhead.

Warke, P.A. 2000. Micro-environmental conditions and rock weathering in hot, arid regions. *Zeitschrift für Geomorphologie, Suppl.* 120: 83–95.

Warke, P.A. & Smith, B.J. 2000. Salt distribution in clay-rich weathered sandstone. *Earth Surface Processes and Landforms* 25: 1333–1342.

Dimension Stone 2004, Přikryl (ed)
© 2004 Taylor & Francis Group, London, ISBN 90 5809 675 0

Porosimetric studies in rocks: methods and application for weathered building stones

Z. Weishauptová

Institute of Rock Structure and Mechanics, Academy of Sciences of the Czech Republic, Prague, Czech Republic

R. Přikryl

New Institute, Gouda, Netherlands

ABSTRACT: Porosity is the inevitable part of rock fabric. It influences many rock physical properties. Type and extent of porosity provokes and affects deterioration processes in rocks. Detection and quantification of porosity presents one of the most difficult tasks of rock study. In this paper, overview of methods is presented together with application of selected methods for dimension stone weathering study.

1 INTRODUCTION

1.1 *General*

The durability of porous stones is vitally connected with its porous structure, which can be modified by natural or artificial weathering. In order to characterize the materials themselves and to understand the mechanism of their alteration it is necessary to measure the pore structure of the materials used.

Structure of stone porosity is often described as 'chaotic' (e.g., Fitzner & Basten 1992), although being directly influenced by rock genesis (primary porosity), rock setting after genesis (secondary porosity), and deterioration processes (weathering porosity) (Simmons & Richter 1976, Kranz 1983, Fitzner 1993). The deterioration of porous building materials is caused by chemical attacks combined with mechanical stresses. Porous structure of the stone is the place where main weathering processes occur. Frost attack or salt crystallization is directly related to the presence of pores whereas mechanical stresses producing cracks and fractures are related indirectly. In this process, the pores and/or microcracks behave as fracture nuclei (e.g., Gramberg 1989).

Natural stones possess complex porous structure. The primary porosity of a rock is a function of its genesis, fabric, and mineralogical composition. Rocks generally show extreme variability of both total porosity and pore-size distribution. According to literature the value of total porosity of sandstones varies from 5 to 42%, of hard limestones from 0.8 to

27%, of soft limestones from 4 to 42%, of granites from 0.1 to 2.8% and of marbles from 0.1 to 6% and even the stones of the same class can be very different in structure.

1.2 *Methods of porosity evaluation*

Pore-structural examinations can be carried out using both direct and indirect methods. Direct methods involve standard optical microscopy (using either stained samples, e.g., Nishiyama & Kusuda 1994, or electron microscopy techniques, e.g., Sprunt & Brace 1974, Fitzner 1993). Among the indirect method strictly devoted to the measurement both of the porosity and of the pore size distribution high-pressure porosimetry. The use of this method is stimulated by the favourable results obtained in many fields of applications concerning porous materials.

2 DETERMINATION OF THE PORE VOLUME

2.1 *Densities of porous materials*

The pore volume of a porous stone can be determined as a difference of the two specific volumes (apparent and real), that is a difference between the reciprocal values of the apparent and real density of the substance.

The real density is defined as a mass to volume ratio of solid phase and is also a material constant. For the value of porosity, however, the apparent density is

237

the determining quantity defined as a ratio of mass to the apparent volume, i.e., the solid phase volume including pores.

There are several methods for determination of both real and apparent density; their choice depends on textural character of material investigated and on the accuracy of determination required (Clark 1966, Goodman 1989, Weishauptová & Medek 1990).

2.2 *Real density determination*

Determination of the real density is based on two assumptions: (i) the absence of closed pores, (ii) perfect penetration of the entire pore volume. The principle of determination is measuring of the solid phase volume by pycnometry with gaseous or liquid penetration medium. Entirety of penetration depends on size of a molecule and its interaction with the solid phase surface. Helium is considered as the most suitable gaseous pycnometric medium and the 'helium density' is regarded as standard. More frequently than the gas pycnometry, however, liquid pycnometry is used which requires simple equipment and is not time consuming. Pycnometric liquids are used depending on character of material and its porous system, e.g., methanol is quite common. Values determined with various pycnometric liquids can differ and are thus denoted as the effective ones. Samples with very fine grains are always used for determination of the real density.

2.3 *Apparent density determination*

The apparent density can be determined in two ways, first by independent determination of the solid phase volume and the pore volume separately with the apparent volume expressed as their sum, second by simultaneous determination of the solid phase volume and the pore volume. The second approach is more frequent, and there are several methods whose choice depends on textural character of material studied. For example, the stereometric method is based on measuring dimensions of a geometrically regular body.

The impregnation method does not require the body evenness, pores are closed with a suitable impregnation agent and the pycnometric determination follows. Mercury pycnometry is quite common as a part of porosimetric measurement. However, accuracy is restricted by the principle of the method as the pores with a diameter >7.5 μm are not included into the porous system. Thus, determination of the apparent 'mercury density' is not suitable for course porous substances and can be realized for samples with size limited by dimensions of a pycnometer vessel.

3 DETERMINATION OF THE PORE VOLUME AND DISTRIBUTION

3.1 *High Pressure Mercury Porosimetry*

Simultaneous determination of volume, surface, distribution of pores and determination of the apparent mercury density is enabled by the method of the *High Pressure Mercury Porosimetry*. This method is based on the phenomenon of the mercury capillary depression, when the wetting angle is >90° and mercury thus penetrates into pores only by pressure action. Volume of mercury forced down into the porous system is generally interpreted as the total volume of pores in a measured sample, with the relation between the actual pressure P and radius of a cylindrical pore r given by Washburn's equation (Drake 1949):

$$P = -\frac{2\sigma \cos \phi}{r}$$

where σ = surface tension of mercury and φ = wetting angle.

For irregularly shaped pores the ratio between the pore cross-section and the pore circumference is not proportional to the radius and depending on the pore shape, Washburn's equation will give slightly lower values.

The minimum measured radius is given by the maximum generated pressure, and the maximum radius corresponds to the initial pressure 0.1 MPa. Using the maximum pressure for example 200 MPa, pores with the radii ranging from 7500 to 3.7 nm can be indicated, which represents identification of meso- and macropores regarding the standard pore classification according to their radii as micropores ($r < 2$ nm), mesopores ($r = 2$–50 nm), macropores ($r = 50$–7500 nm) and course pores ($r > 7500$ nm) (Brunauer et al. 1973). In rock weathering studies, the pores are often classified as micropores ($r < 0.1$ μm), capillary pores ($r = 0.1$–1000 μm) and macropores ($r > 1000$ μm) (Klopfer 1985 in Fitzner 1993).

The dependence of the mercury volume forced down on increasing pressure P represents the porometric intrusion curve. Recalculation of the pressure P for the radius r according to Washburn's equation gives the cumulative curve whose derivative is used to calculate the pore distribution according to radii. The dependence of the volume on the decreasing pressure is recorded by the extrusion curve whose position over the intrusion curve indicates retention of a certain amount of mercury by the porous substance after the pressure release.

Based on the pore volume obtained and the cylindrical model, the pore surface can be calculated. Calculation of the pore surface is sometimes made more accurate assuming other shapes as cone, slit or bottle.

238

Geometric properties of the sample measured whose size limited by the size of dilatometer determines the measurement reproducibility; provide a representative description of the porous structure. Choice of the size of the sample depends on the structural character of the material studied.

3.2 Experimental

Samples with the size approximately 5×5 mm were dried at 105°C until a constant weight and evacuated (0.27 Pa) in a dilatometer before its filling by mercury on a Mercury Filling Device. For porosimetric measurement, the automatic device POROSIMETER 2000 Carlo Erba with the working pressure range up to 200 MPa was used. Correction for the mercury compressibility and expansion of the glass dilatometer was provided by a blank run.

Evaluation of the measurement was done using the code MILESTONE 200 Carlo Erba at the following conditions: wetting angle 141.3°, surface tension of mercury 480 mN.m^{-1}, and the cylindrical pore model. The pump rate providing pressure changes in the system and controlling duration of analysis and a failsafe course of the intrusion curve was 3, and the time required to equilibrate penetration of mercury into the pores at a given pressure determined by its stepwise increase was 3 s.

This resulted in determination of the volume of meso- and macropores V_{mm}, their surface S_{mm} for the model cylindrical shape of pores, the mean weighted pore radius r_{mean}, and the apparent mercury density d_{Hg}, the apparent mercury density corrected for the pore volume V_{mm} which at the absence of micro- and macropores represents an effective real density, and porosity Por. The progress of measurement has been documented with a numerical printout and graphically processed in form of two records, (i) mercury intrusion/extrusion vs. pressure (corrected for the blank run and hydrostatic pressure), and (ii) cumulative volume vs. pore radius together with a histogram of the relative distribution of the sample porosity.

4 DIRECT OBSERVATION OF PORES

The limits of indirect methods – the impossibility to observe real geometry of pores – can be overcome using microscopic techniques. Although the scanning electron microscopy provides finer resolution in comparison to conventional optical microscopy, the optical methods are cheaper, quicker and more accessible. There are several sample treatments available before its microscopic observation. These involve staining of thin section after its preparation (Ruzyla & Jezek 1987) or impregnation of sample with colour dyes (e.g., Hishiyama & Kusuda 1994, Anselmetti et al. 1998).

For this study, the pore impregnation technique before thin section preparation has been employed (Hishiyama & Kusuda 1994). The rock treatment before thin section preparation depends on the state of the rock. For low porosity non-weathered rocks, the rock chips of firstly diamond sawn. The flat surface of the rock is impregnated using mixture of low viscosity epoxy resin and fluorescent dye.

For weathered or soft rocks, the samples are impregnated 2 times. First impregnation is done before any sawing to increase cohesion of the specimen. Such treated sample is diamond sawn and the second impregnation is similar to that of non-weathered samples.

After impregnation and hardening of the mixture, the flat surface is carefully polished to remove any excessive amounts of dye on the sample's surface. After that ordinary, but uncovered, thin section is prepared. Impregnated thin sections are observed under ordinary petrographic microscope equipped with the source of fluorescent light.

5 LEITHAKALK LIMESTONE CASE STUDY

Combination of above mentioned methods – high pressure mercury porosimetry and thin section impregnation technique have been applied during the study of Leithakalk limestone. Leithakalk limestone is organodetritic limestone (calcarenite) composed almost entirely of fragments of *Lithothamnium* algae. Beds containing these limestones have formed during Tertiary (Miocene). Leithakalk limestone can be found on numerous architectural and sculptural projects in Central Europe (Hungary, Austria and partly also in SE Czech republic) for more than two millennia (Török in print).

The studied samples were collected during pre-restoration material research of the Colonnade (Přikryl

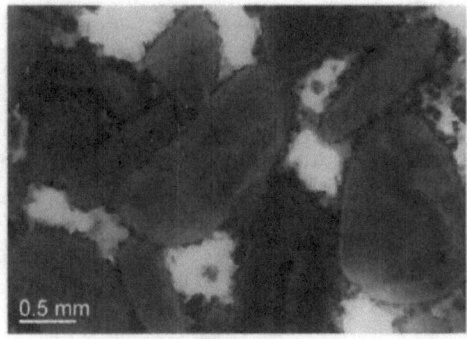

Figure 1. Microscopic character of porosity in studied Leithakalk limestone. The pores (bright areas) were visualised using fluorescent dye.

239

et al. 2002). The Colonnade is one of the numerous architectural objects built in the Lednice-Valtice area, a large man-cultivated landscape in south Moravia (SE Czech Republic). The whole area is cared as UNESCO heritage site. The use of local Leithakalk limestone resources for the Colonnade was documented during previous research (Příkryl et al. 2002).

The Leithakalk limestone from the Colonnade was categorized into two groups based on the degree of weathering. These groups were clearly distinguishable from the rock mechanical properties study (Příkryl et al. 2002) but not fully visible from microscopic observation (Fig. 1).

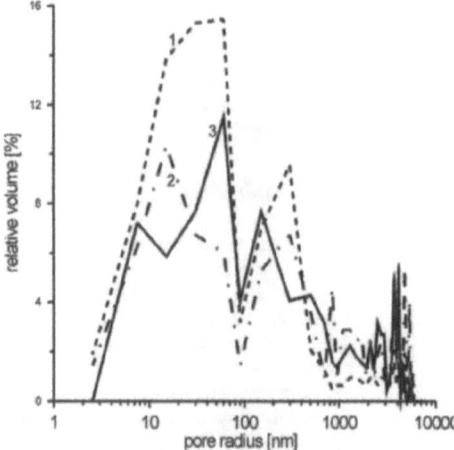

Figure 2. Pore size distribution of gradually weathered organodetritic limestone (Leithakalk limestone) from the Colonnade monument in Lednice-Valtice area (southern Moravia, Czech Republic). 1 – seriously weathered limestone, 2 – almost non-weathered limestone, 3 – non-weathered limestone collected in the original quarry.

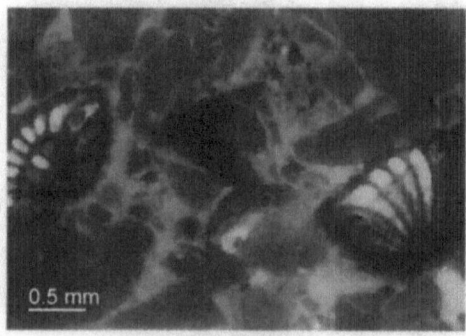

Figure 3. Intrafossil pores (left and right margins) in studied Leithakalk limestone. The pores (bright areas) were visualised using fluorescent dye.

The differences in the degree of deterioration are, however, markedly shown from high pressure porosimetry study as shown on Figure 2. Three groups shown in this figure include material from the Colonnade (almost non-weathered and severely weathered limestone) and one non-weathered material sampled in the abandoned quarry in the area

ACKNOWLEDGEMENTS

This research is conducted in the frame of the project IAA3046401 (provided by the Grant Agency of the Academy of Sciences of the Czech Republic). The experimental material was studied under the projects MSM 520000001 and Cz5/2002.

REFERENCES

Anselmetti, F.S., Luthi, S. & Eberli, G.P. 1998. Quantitative characterization of carbonate pore system by digital image analysis system. *AAPG Bull.* 82(10): 1815–1836.

Brunauer, S., Skalny, J. & Odler, I. 1973.Complete pore structure analysis. In S. Modry (ed.), *Pore structure and properties of materials*: C3–C26. Proc. of the Int. Symposium RILEM/IUPAC, Prague: Academia.

Clark, S.P., ed. 1966. *Handbook of Physical Constants*. Geological Society of America, Memoir 97.

Drake, L.C. 1949. Pore-size distribution in porous materials. *Ind. and Eng. Chem.* 780–785.

Fitzner, B. 1993. Porosity properties and weathering behaviour of natural stones – methodology and examples. In F. Zezza (ed.), *Stone material in monuments: diagnosis and conservation*: 43–54. 2nd course, Heraklion, Crete, C.U.M. University School of Monument Conservation.

Fitzner, B. & Basten, U. 1992. Gesteinsporosität – Klasifizierung, messtechnische Erfassung und Bewertung ihrer Verwitterungsrelevanz. In *Jahresberichte aus dem Forschungsprogramm Steinzerfall – Steinkonservierung*, Band 4: 19–32. Berlin: Ernst & Sohn.

Goodman, R.E. 1989. *Introduction to Rock Mechanics*. New York: John Wiley & Sons.

Gramberg, J. 1989. *A non-conventional view on rock mechanics and fracture mechanics*. Rotterdam: Balkema.

Gregg, S.J. & Singh, K.S. 1982. *Adsorption, surface area and porosity*. London: Academic Press.

Nishiyama, T. & Kusuda, H. 1994. Identification of spaces and microcracks using fluorescent resins. *Int. J. Rock. Mech. Min. Sci. & Geomech. Abstr.* 31: 369–375.

Příkryl R., Svobodová J., Siegl P., Chvátal M., Novotná M., Andrade Sanchez R., Mézlová M., Myšková K., Faltus J. & Korecký J. (2002): Weathering of limestone cladding above the waterproofing layer: salt action due to previous restoration of the Colonnade (Valtice-Lednice area, Czech Republic). In R. Příkryl & H.A. Viles (eds), *Understanding and managing stone decay*: 209–221. Prague: The Karolinum Press.

Ruzyla, K. & Jezek, D.I. 1987. Staining method for recognition of pore space in thin and polished sections. *J. Sed. Petrol.* 57(4): 777–778.

Simmons, G. & Richter, D. 1976. Microcracks in rocks. In R.G.J. Strens (ed.), *The Physics and Chemistry of Minerals and Rocks*: 105–137. London: John Wiley & Sons.

Sprunt, E.S. & Brace, W.F. 1974. Direct observation of microcavities in crystalline rocks. *Int. J. Rock. Mech. Min. Sci. & Geomech. Abstr.* 11: 139–150.

Török, Á. in print. Leithakalk-type limestones in Hungary: an overview of lithologies and weathering features. In R. Přikryl & P. Siegl (eds), Architectural and sculptural stone in man-cultivated landscape. Prague: The Karolinum Press.

Weishauptová, Z. & Medek, J. 1990 Methods of real and apparent densities determination (in Czech). In *Proc. of 23rd Coke Conference*, Reka: 88–96.

Bowing of natural stone cladding (TEAM session)

Dimension Stone 2004, Přikryl (ed)
© 2004 Taylor & Francis Group, London, ISBN 90 5809 675 0

Influence of rock and mineral properties on the durability of marble panels

L. Alnæs
SINTEF Civil and Environmental Engineering, Department of Rock and Soil Mechanics, Trondheim, Norway

A. Koch
Geoscience Center of the University of Göttingen, Department of Structural Geology and Geodynamick, Göttingen, Germany

B. Schouenborg & U. Åkesson
SP Swedish National Testing and Research Institute, Borås, Sweden

K. Moen
NTNU, Faculty of Engineering Science and Technology, Department of Geology and Mineral Resources Engineering, Trondheim, Norway

ABSTRACT: The use of thin marble and limestone for facade cladding has increased substantially during the last five decades. The durability of such thin slabs (often only 30 mm thick) has been assumed to be satisfactory based on centuries of successful use as a structural building stone. Nevertheless, all over the world, the long-term deformation and strength loss of some claddings have led to concerns about its safe and durable use. The detailed assessment of marble and limestone within TEAM (see also www.sp.se/building/team), is used to develop a hypothesis for the observed deterioration and to develop remedial actions. This paper considers the influence of rock and mineral properties on the behaviour of marble in claddings. A selection of results is presented, illustrating the influence of properties like mineralogy, grain size and boundaries, preferred orientation and pore structure. Of the intrinsic parameters, the interlocking of the grains and the lattice preferred orientation seem to be very important influencing factors. The final evaluation of the results will be presented in 2005.

1 INTRODUCTION

The long-term deformation and gradual loss of strength of thin marble claddings has resulted in safety problems and increased maintenance costs at many building sites. To date known damaged buildings in Europe are expected to account for repair costs of more than 240 Million EURO unless other solutions can be foreseen. This has also a large negative impact on the entire stone trade and the willingness of architects to use marble as a cladding material.

It has for long been known that some marble types from Italy show this behavior. Earlier research has often been concentrated on marble from the Carrara district. Not surprisingly though, since this marble is the most used and therefore also most well known worldwide. Even though a majority of buildings exhibiting bowing are clad with Carrara marble, it is very important to note that many buildings with marble from the

Carrara region have performed well as cladding! At the same time, there are also several buildings clad with other marbles that show the same deterioration pattern. An additional fact is that the bowing shows different extent and patterns on different buildings, as there may also be pronounced differences (e.g., strong/ weak and convex/concave bowing) on the same facade on the same building as shown in Figure 1. This implies that the deterioration expressed as the strength loss ($\Delta F\%$) is a function of several potential parameters (1):

$$\Delta F\% = (\text{Type}) \times (\text{Production}) \times (\text{Exposure}) \times (\text{Years}) \quad (1)$$

where (Type) indicates the characteristics for the marble type, (Production) is the character of extraction and processing, (Exposure) includes both climatic (ΔRH, ΔT, $\Delta micro$) and construction specific parameters (fixing, installation), (Years) is the time of exposure.

Figure 1. Convex and concave bowing in the same façade.

Table 1. Selected stone types from quarries (also represented in buildings). A coding system have been used to keep the exact quarry location confidential. Bowing figures from bow test are included (see section 5 and Grelk et al. 2004).

Name	Country of extraction	Rock Type[1]	Apparent grain size	Bowing (mm/m) after 40–50 cycles
Chq1	Switzerl.	cc m	medium-coarse	0.25–0.45
Chq2	Switzerl.	cc m	medium-coarse	–
Grq1	Greece	dol m	medium-fine	0.02
Itq1	Italy	cc m	fine	−0.15 to 0.77
Itq2	Italy	cc m	fine	2.5 to 4.5
Itq3	Italy	cc m	fine	−0.05
Itq4[2]	Italy	cc m	fine	1.2 to 2.4
Noq1[3]	Norway	cc m	fine	0.4
Plq1	Poland	lms	fine-medium	–
Poq1	Portugal	cc m	medium	1.32 to 1.95
Poq2	Portugal	cc m	coarse-medium	0.38 to 0.48
Poq3	Portugal	cc m	medium-fine	0.08
Poq4	Portugal	cc m	fine	0.27 to 0.70
Seq1	Sweden	dol m	fine	−0.03
Siq1	Slovenia	lms	fine-medium	−0.03
Siq2	Slovenia	dol m	fine-medium	−0.05

[1] cc m = Calcitic mar245ble, dol m = dolomittic marble, lms = limestone. [2] Arabescato type. [3] Contact metamorphic type with high content of silicates.

In order to determine which, how and how much rock and mineral properties influence the bowing, expansion and strength loss, a large selection of different marble and limestone varieties is needed. For the purpose of the TEAM project, a selection of 14 marble and 2 limestone varieties have been selected for a comprehensive study. This paper describes some parts of the investigation program that deals with the influence of rock and mineral properties on the durability of marble claddings.

2 SAMPLING AND SAMPLES

2.1 Sampling from buildings

Within the project about 150 European marble and limestone buildings have been registered. Of these, 7 of them have been investigated in detail (see Yates et al. (2004), including sampling of slabs for laboratory analyses.

For the purpose of the investigation, one major goal has been to select a broad variety of marble types; both those that show and do not show bowing on the buildings, and marbles of different composition and rock fabric. Of great importance has also been to select buildings where both exposed and unexposed (spare) slabs are available.

2.2 Sampling from quarries

Based on the above mentioned selection criteria, 14 marble and 2 limestone types have been sampled from European quarries (Table 1).

The sampling activities in the quarries have pointed out the great importance of detailed instructions for

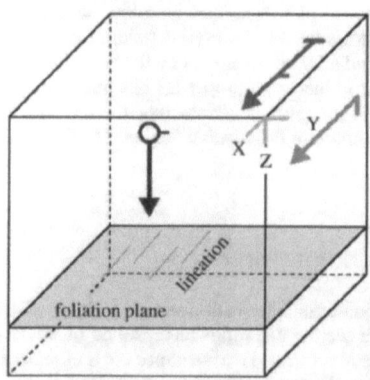

Figure 2. Reference coordinate system for block sampling. X, Y, Z are selected parallel to visible fabric elements.

the sampling location, providing a reference coordinate system for the orientation in relation to rock fabric and sample identification as shown in Figure 2. Directional cut specimens has also been of importance in order to study the influence of preferred orientations (foliation, principal rock stresses, crystal lattice). A complete concept for marking of test specimens in accordance to the orientation of blocks will be used as an important input to the European

246

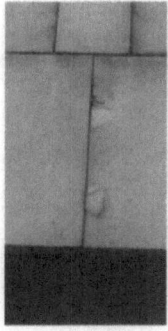

Figure 3. Severe bowing and total failure can be experienced on some marble claddings.

standardisation. It provides a possible solution for the traceability from the quarried blocks to the production units.

3 MATERIAL CHANGES IN CLADDINGS

The deterioration phenomenon under investigation in the TEAM project express itself as bowing, warping or dishing, volume and porosity increase, brittleness and loss of strength, and ultimately in detachment of the thin marble slabs from the fixings.

The most obvious and visual sign of change in the marble is the bowing (Fig. 3). Whether it is convex or concave, the bowing of some of the investigated marbles exceed 70 mm/m in buildings with an age of 6 to 36 years.

The flexural strength of the marble decreases (see Yates et al. (2004), as does the pull-out strength at fixing points (Fig. 3). In addition, all marbles exhibit a distinct reduction in ultrasonic velocity.

All petrographic investigations of thin sections from exposed marble slabs show that the grain boundaries are opening up, i.e. granular disintegration as shown in Figure 4. The granoblastic marble has both a higher initial grain boundary porosity and get a more pronounced intergranular decohesion during exposure.

4 ROCK FABRIC AND INVESTIGATIONS

4.1 Fabric nomenclature

An extensive terminology exists for the description of the geometry of grains and fabrics in metamorphic rocks. In this study, the terms micro-fabric and microstructure are used in accordance with Passchier and Trouw (1996) meaning the complete spatial and

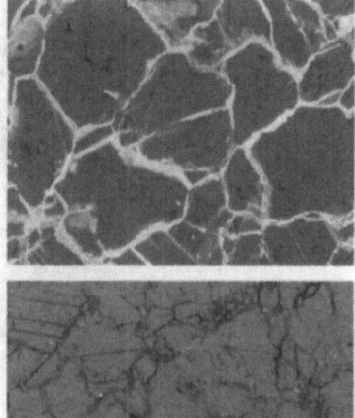

Figure 4. Thin sections impregnated with fluorescent epoxy, taken from exposed granoblastic (upper picture) and xenoblastic marble (lower picture). The light lines in the figure show where fluorescent epoxy has penetrated into the grain boundaries.

geometrical configuration of all components a rock consist of, including grain size distribution, grain aspect ratio, grain shape preferred orientation, grain boundary geometry, size and orientation of microcracks and lattice preferred orientation (texture).

With respect to shape of grain aggregates in marbles, one may say that there are two extremities; the equigranular-polygonal microstructure, often termed as *granoblastic* marble (homoblastic has also been used by many authors for these marbles) and at the other end the seriate or amoeboid structure – xenoblastic marble, see examples in Figure 5.

4.2 Analyses of microstructure

Investigations of the microstructure by polarizing microscope, SEM etc. have been done by using both 30 µm thin sections and polished thin sections with and without fluorescent epoxy and small pieces of rock, see EBSD below. By preparing sections or cores in directions according to visible fabric elements, it is possible to get an overview of both grain shape preferred orientation, size and orientation of microcracks and lattice preferred orientation.

For the quantification of the lattice preferred orientation (texture) of calcite, neutron diffraction has been applied (see Koch and Siegesmund, in press).

247

Figure 5. Calcitic marbles. Left: Equigranular-polygonal (granoblastic). Right: seriate-interlobate to amoeboid (xenoblastic). Size: 2.0 × 2.8 mm.

This method allows to measure relatively large sample volumes even of coarse-grained marbles and it is also possible to calculate the three-dimensional orientation distribution function which represents the complete statistical description of the texture, allowing the recalculation of the pole figure of any desired lattice direction.

To quantify the pore space of fresh and weathered marbles in terms of pore radii distribution and porosity, mercury (Hg) porosimetry has been applied.

Quantification of microstructure in marbles is challenging, especially due to the presence of twins disrupting or masking the grain boundaries. Several attempts with various image analyses techniques and various sample types (ultra-thin sections etc.) have been tried out.

By using adjacent grain analysis, AGA (Åkesson et al. in prep.) it has been found possible to numerically describe the different microstructures and to get a very good correlation between bowing observed on buildings and microstructure.

Electron BackScattered Diffraction (EBSD), which detects texture by means of crystallography, is also being used and gives promising results, both with respect to lattice preferred orientations and quantification of grain size distribution, grain shapes etc. The EBSD technique is based on the weak diffraction pattern that forms when a focused, stationary, primary electron-beam strikes a polished sample, backscatters and diffracts. The diffraction pattern is formed on a fluorescent screen and transferred by a camera to the computer (Hjelen 1990). The diffraction pattern is characteristic for the crystal structure and space orientation of the crystal (Moen et al. 2003). Quantitative data is obtained by mapping the specimen by a certain step size.

4.3 Testing of bowing potential

In addition to investigations of microstructure and changes in the marbles during natural exposures on claddings, the accelerated test of bowing potential described in Grelk et al. (2004) has been used. By performing bowing tests under the same controlled

conditions and by comparing with the results of microscopical features it is possible to identify the relative importance of individual rock and mineral parameters.

5 THE INFLUENCE OF VARIOUS FACTORS

In Table 1, results of the bowing test on fresh quarry material of the sampled marbles are given and Grelk et al. (2004) shows several examples of typical bowing curves obtained for various marbles.

5.1 Mineralogy

All true marbles are mainly composed of either calcite or dolomite or a combination of the two. Most of the research referred to in the literature on the topic of bowing has focused on calcitic marble.

As mentioned earlier, the TEAM investigations reveal that both pure dolomitic and calcitic marbles and also polymineral varieties may behave similarly on the buildings. This proves that marbles with different compositions may display bowing. It also indicates that bowing of marble is a relatively common phenomenon and that other factors have a large influence on the bowing and strength loss in addition to the rock composition.

5.2 Grain size

The grain size of marble has been regarded as a less important factor for marble degradation. Tschegg and others (1999) assume that this is attributed to the fact that larger grain sizes on the one hand leads to a stronger deterioration along grain boundaries and, on the other hand, the total grain boundary area per volume decreases. Zeisig and others (2002) demonstrated in a large collection of different marble types that marbles with a large grain size exhibit the same deterioration rate as marbles with smaller grain size.

In Figure 6, examples of the relationship between grain size distribution (GSD) measured during EBSD analyses and bowing figures are given for two marbles with the same mineralogy. The differences in grain size and GSD is much less pronounced than the bowing figures for these two marbles. However, TEAM investigations have revealed a very good correlation between grain size and the bowing actually observed on buildings (Åkesson et al., in prep.).

5.3 Grain boundaries – shape and interlocking

A micro structure dominated by straight boundaries and weak grain interlocking is by many authors believed to be more susceptible to bowing and other kinds of deterioration than marbles with sutured and

248

Grain Size (diameter)

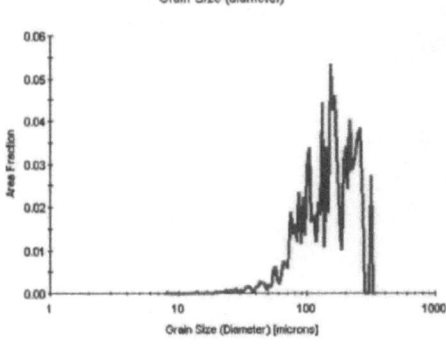

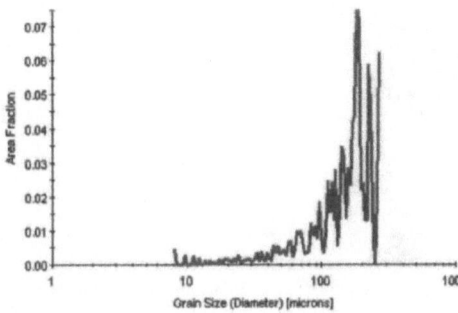

Figure 6. Grain size distribution from EBSD analyses. Upper graph: Itq2, with 3.38 mm/m bowing after 50 cycles in the bow test. Lower graph: Itq1 with 0. 86 mm/m bowing after the same number of cycles.

complex grain boundaries (Bain 1941). This is strongly confirmed by the work of TEAM. In Figure 7, examples of building slabs of the same marble type but with various microstructures are given, together with field measured bowing figures.

5.4 Grain shape and lattice preferred orientation

The directional dependence on bowing and expansion has been demonstrated by laboratory testing on several marbles investigated so far. Both lattice preferred orientation (texture) and the grain shape preferred orientation influence the anisotropy of marble degradation. Examples of this are given in Grelk et al. (2004).

Two marbles with distinct foliation and strong texture (Chq1 and Poq4) show a clear correlation between the c-axis preferred orientation and the direction of maximum deterioration observed in all experimental results (Koch and Siegesmund, in press), see Table 2. The direction of maximum deterioration is linked to the orientation of the c-axis maximum. However, a similar relationship has so far been difficult to demonstrate on samples from buildings.

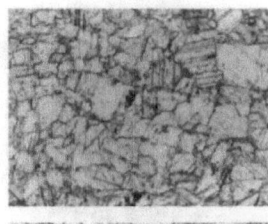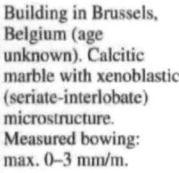

Building in Brussels, Belgium (age unknown). Calcitic marble with xenoblastic (seriate-interlobate) microstructure. Measured bowing: max. 0–3 mm/m.

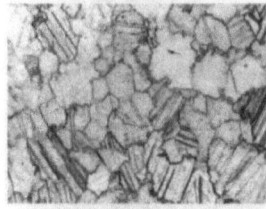

Building in MalmÖ Sweden (1983). Calcitic marble with xenoblastic (inequigranular - interlobate microstructure. Measured bowing max. 0–3 mm/m.

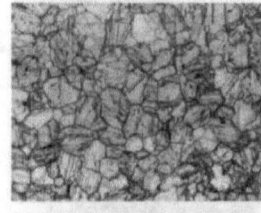

Building in NykÖping, Sweden (1969). Calcitic marble with granoblastic (equi- to slightly inequigranular polygonal) microstructure. Measured bowing ca. 30 mm/m.

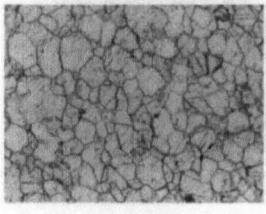

Building in MalmÖ, Sweden (1970). Calcitic marble with granoblastic (equi- to slightly inequigranular polygonal) microstructure. Measured bowing ca. 30–40 mm/m.

Figure 7. Microphotographs from exposed marbles on investigated buildings. All pictures are taken in ordinary light and represent an area of 2 mm × 2.8 mm.

By the time of writing, EBSD analyses on two marbles (Itq1 and Itq2) cut in various directions according to measured in situ rock stresses are also under investigation. There is a clear tendency that the most pronounced bowing occur in directions where the effective shear stresses of the marbles are highest, due to the internal friction of the rock. Whether the relation between bowing and orientation of the slabs is solely related to rock stresses, to micro structure/texture or to both/a combination is a question under present study.

EBSD analyses of the same marbles indicate that Itq1 shows a bigger amount of grains with preferred orientation while Itq2 shows more random orientation (Fig. 8). This trend seems to be general for any orientation (with respect to rock stresses) in these two marbles and confirms the statement of Barsottelli et al

249

Table 2. Expansion and bowing behaviour of two marbles parallel and perpendicular to a strong c-axes preferred orientation. After Koch & Siegesmund (2004) (in press).

Orientation of the specimens	Irreversible residual strain (mm/m) after heating cycles (20–90°C)		Bowing (mm/m)
	After 1 dry cycle	After 8 dry and 10 wet cycles	After 40 cycles
Chq1 c-axes	0.13	0.25	0.14 ± 0.05
Chq1 c-axes	0.30	0.59	0.27 ± 0.07
Poq4 c-axes	0.03	0.27	0.12 ± 0.04
Poq4 c-axes	0.25	0.76	0.36 ± 0.05

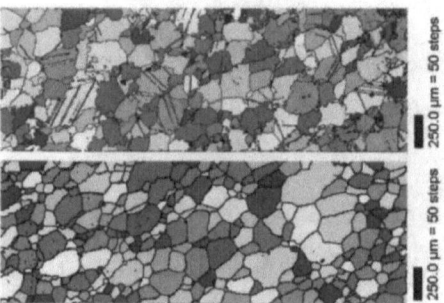

Figure 8. EBSD analyses of Itq1 (upper) and Itq2 (lower). The maps are colored according to the Euler angles. Pole figures indicate that Itq1 shows a stronger preferred orientation, while Itq2 shows a more random orientation.

(1998). The same trend is also found by neutron diffraction analyses (Fig. 9).

5.5 Porosity, pore size distribution

Garzonio et al. (1995, 2000) found a clear relationship between porosity and bowing. Barsotelli et.al. (1998) state that not all pores behave in the same way with regard to water. The interaction depends on their dimensions and shape.

In Figure 10, pore size distribution (PSD) and porosity for most of the investigated quarry marbles are given. As an example of porosity versus grain size versus bowing, one can mention that Chq2 and Itq2, which have similar porosity and PSD, differ in grain size and microstructure. Chq2 has large grains and show weak bowing, while Itq2 is fine grained and show a very strong bowing (see Table 1). A clear correlation between bowing and porosity/PSD is not found. The bowing versus porosity relationship reveals a coefficient of correlation of only R = 0.79.

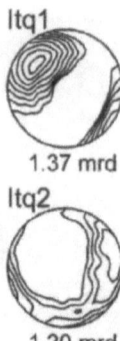

Figure 9. C-axis distribution of Itq1 and Itq2 gained by neutron diffraction method. Itq1 has a stronger preferred orientation (1.37 mrd – multiples of random distribution) Itq2 has a gridle distribution, its maximum is therefore less pronounced (1.20 mrd).

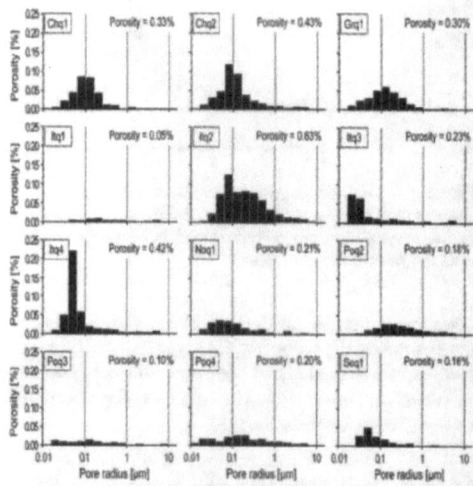

Figure 10. Pore size distribution (PSD) and porosity of investigated marbles.

6 CONCLUDING REMARKS

A variety of marble types have been investigated with respect to the influence of intrinsic rock and mineral properties on the deterioration phenomenon under study. Limestones have also been included. The selected stone types give a necessary and good spread in both physical and petrographical properties.

It is commonly assumed that the deformation of marble originates from a combined effect of intrinsic and extrinsic parameters. With reference to

Equation (1), one might say that giving the "right" marble type; i.e. a resistant microstructure, the bowing and strength loss will be much smaller regardless of the severity of extrinsic parameters. Even though deterioration will take place even in these, the weathering rate will be much slower and the service life time correspondingly longer.

Of the intrinsic parameters, the interlocking of the grains and the lattice preferred orientation seem to be the most important influencing factors.

The durability differences between the polygonal (granoblastic) and lobate to suture structured (xenoblastic) marbles are obvious, the first ones being those most prone to bowing and strength loss.

Completion of the comprehensive laboratory research within TEAM is necessary in order to give conclusive answers with respect to deterioration mechanism and influencing factors. The goal is thus to avoid the dilemma of including marbles with potential poor performance and excluding marbles that will perform well in new, modern buildings and to give guidance on how to choose, produce and use marble and limestone cladding panels without the risk of deterioration due to bowing or expansion.

ACKNOWLEDGEMENTS

The authors would like to thank the European Commission for financial support of the European research project "Testing and Assessment of Marble and Limestone" (Contract no. G5RD-CT-2000-00233). The co-operation with the project partners is gratefully acknowledged.

REFERENCES

Bain, G.W. 1941. Measuring grain boundaries in crystalline rocks. *Journal of Geology* 49(2): 199–206.

Barsotelli, M., Fratini, F. & Giorgetti, 1998. Microfabric and alteration in Carrara marble: a preliminary study. *Science and technology for Cultural Heritage* 7(2): 115–126.

Garzonio, C.A. 2000. Analyses of the physical parameters correlated to bending phenomena in marble slabs. *9th International Congress on Deterioration and Coinservation of Stone*, Venice 19–24 June.

Garzonio, C.A., Fratini, F., Manganelli, C., Giovannini, P. & Blasi, C. 1995. Analyses of geomechanical decay phenomena on marbles employed in historical monuments in Tuscany. In H.-P. Rossmanith (ed.), *Mechanics of Jointed and Faulted Rock*: 259–263. Rotterdam: A.A.Balkema.

Grelk, B., Golterman, P., Schouenborg, B., Koch, A. & Alnæs, L. 2004. The laboratory testing of potential bowing and expansion of marble. *In prep. Proc. Dimension Stone 2004.*

Hjelen, J. 1990. *Teksturutvikling i Aluminium, studert ved Elektronmikrodiffraksjon (EBSP) i Scanning lektronmikroskop*. Ph.D. Thesis. Trondheim, Norway: University of Trondheim, 1990. (In Norwegian).

Malaga, K., Schouenborg, B., Alnæs, L., Bellopede, R. & Brundin, J.-A. 2004. Field exposure sites and accelerated laboratory test of marble panels. In R. Přikryl (ed.), *Dimension Stone 2004*. Rotterdam: A.A.Balkema.

Moen, K., Malvik, T., Hjelen, J. & Leinum, R.L. 2003. Automatic Material Characterization by means of SEM-techniques. *Proceeding, 9th Euroseminar on Microscopy Applied to Building Materials*, Ed: M.A:T:M: Broekmans et al. and In press Material Characterisation 2004.

Koch, A. & Siegesmund, S. 2004. The combined effect of moisture and temperature on the anomalous expansion behavior of marble. In S. Siegesmund, H. Viles & T. Weiss (eds), *Stone Decay hazards; Environmental Geology*.

Passchier, C.W. & Trouw, R.A.J. 1996. *Microtectonics*. Berlin: Springer Verlag.

Yates, T., Brundin, J.-A., Goltermann, P. & Grelk, B. 2004. Observations from the inspection of marble cladding in Europe. In R. Přikryl (ed.), *Dimension Stone 2004*. Rotterdam: A.A.Balkema.

Zeisig, A., Siegesmund, S. & Weiss, T. 2002. Thermal expansion and its control on the durability of marbles. In Siegesmund, S., Weiss, T. & Vollbrecht, A. (eds) *Natural Stone, Weathering Phenomena, Conservation Strategies and Case Studies*: 65–80. London: Geological Society.

Åkesson, U., Grelk, B., Lindqvist, J.E. & Schouenborg, B. (in prep.): Relationship between microstructure and bowing tendencies of marbles from claddings.

Dimension Stone 2004, Přikryl (ed)
© 2004 Taylor & Francis Group, London, ISBN 90 5809 675 0

The laboratory testing of potential bowing and expansion of marble

B. Grelk & P. Goltermann
Rambøll, Copenhagen, Denmark

B. Schouenborg
SP Swedish National Testing and Research Institute, Building Technology and Mechanics, Borås, Sweden

A. Koch
Geoscience Centre of the University of Göttingen, Department of Structural Geology and Geodynamics, Göttingen, Germany

L. Alnæs
SINTEF Civil and Environmental Engineering, Department of Rock and Soil Mechanics, Trondheim, Norway

ABSTRACT: The use of thin marble and limestone panels for facade cladding has increased substantially during the last five decades. The durability of such thin slabs (often only 30 mm thick) has been assumed to be satisfactory based on centuries of successful use as a structural building stone. Nevertheless, all over the world, the long-term deformation and strength loss of some cladding panels have led to concerns about its safe and durable use. The detailed assessment of marble and limestone within the TEAM-project (see also www.sp.se/building/team), is used to develop a hypothesis for the observed deterioration and to develop remedial actions. This paper presents test methods for the bowing and expansion potential of marble. The method is discussed and the relevance of the test exposures is demonstrated by comparisons with on-site observations on marble cladding. A number of results are presented, illustrating the influence of the temperature cycles, the humidity and the stonetype on the bowing and expansion.

1 INTRODUCTION

Marble and other natural stone have been used for cladding in many structures and have usually performed very well. There are also cases throughout the world, where the panels have developed severe bowing (Fig 1.) or expansion (Fig. 2), leading to severe loss of strength, failure of joints, fixings (Fig. 3) or collapse of the panel itself. Such cases will usually lead to an expensive replacement of the cladding and the fixing system.

The research, inspection (see Yates et al. 2004) and testing show that a great many factors may influence the risk of bowing and expansion as e.g. stone type, panels thickness, joint design, fixing methods as well as the environment. The inspections showed, that an environment with temperature cycles and moisture as in Northern and Central Europe increased the risk of bowing and expansion.

The problem has become increasingly important as the international trade has lead to an increased use of

Figure 1. Bowing panels on a facade in Copenhagen.

new stone types in new environments, without the proper testing. The problem can be solved or reduced substantially by screening the selected stone types with e.g. a realistic, accelerated testing of the bowing and expansion potentials. The partners in the TEAM

Figure 2. Expansion of panels leads to contact stresses and compaction of joint material, which can be squeezed out.

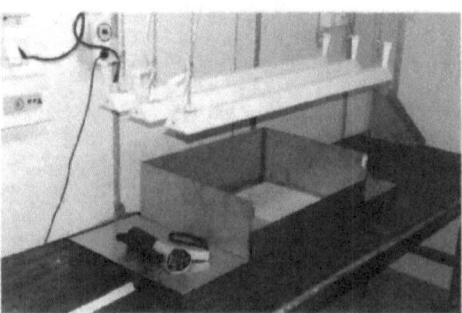

Figure 3. Illustration of test set-up, NT Build 499.

project decided therefore to develop test methods for bowing and expansion and to correlate these to conditions and performance in practise.

This paper presents the developed test methods for bowing and expansion potentials, a few typical results and a correlation to observations from practice.

2 TESTING OF BOWING POTENTIAL

The test method NT Build 499, developed by Schouenborg et al. (1997), uses a standard test specimen of 400 mm length, typically 100 mm width and a thickness, similar to the panel's thickness (or 30 mm in standard tests).

The specimen is conditioned by drying at 40°C until a stable weight is achieved (usually within 7 days), followed by cooling to 20°C and a partial submerging in water for 24 hours at this temperature.

The specimen is placed in an insulated container, where it is placed on a tray, filled with a layer of filter cloth or sand (Fig. 3). The tray is filled with distilled

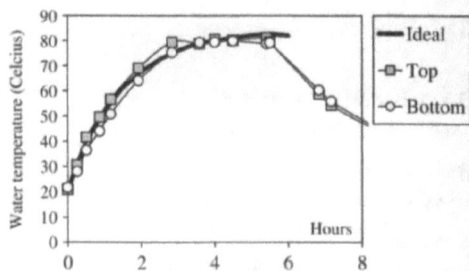

Figure 4. Ideal temperature variation in bowing test and monitored temperatures on the upper and lower side of a test specimen.

or demineralised water up to approx. 10 mm below the upper surface of the test specimen.

The specimens are exposed to a number of cycles. Each cycle begins with an exposure to infrared heating from above, which lead to an increase of the surface temperature from the ambient room temperature to 80°C over a period of 1–3 hours. The surface temperature is maintained at 80°C for 2–3 hours, after which the infrared heating is turned off and the specimen allowed to cool to ambient room temperature, (20°C) until at least 24 hours has passed from the start of the exposure cycle. The surface temperature on the upper side shall therefore follow the ideal curve shown in Figure 4 with a tolerance of ±5°C. The Figure 4 shows also the actual temperatures measured on the top and the bottom of a test specimen, which shows that the temperature is almost identically the same through the specimens at any given time.

Temperature measurements on panels on buildings by Perrier & Boineau (1997) have revealed that the daily magnitude of the temperature cycles can be up to 60°C and that most of this temperature increase happens in within 3 hours.

The bowing of the specimen is measured after a number of cycles. This is carried out by placing the specimen in the bow-test rig (Fig. 5), where the specimen is supported in fixed positions with a distance (L = 350 mm), which ensures that all measurements of the deformation (Δh) are carried out in the same position on the specimen. The bowing is then defined as the ratio $\Delta h/L$.

The test method has been applied on a large number of different stone types during the TEAM-project and has yielded very important results.

The bowing will usually grow unlimited in an environment with temperature cycles and moisture (Fig. 6). The same tests have been carried out on test specimens from the same sample, but without any water in the tray, corresponding to a dry exposure. The Figure 6 shows that the bowing after a few cycles reaches a stable level in a dry environment, after which the

254

Figure 5. Measuring bowing in the laboratory.

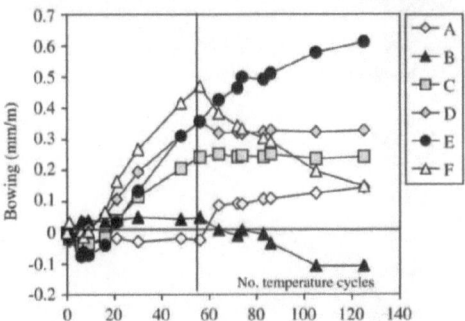

Figure 6. Bowing on Norwegian marble versus cycles; wet and dry exposure. A: Dry exposure until 56 cycles, then wet; B: As A, but with wetting of upper surface; C: Wet exposure for the first 56 cycles, then dry; D: Wet exposure for the first 56 cycles, then constant temperature in dry environment; E: Wet exposure; F: Wet exposure, but sample is turned after 56 cycles.

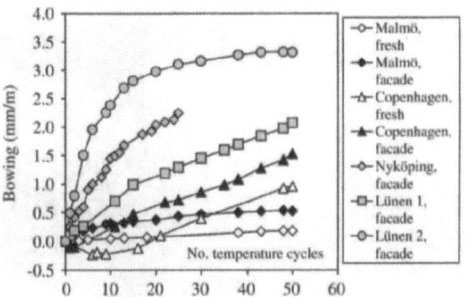

Figure 7. Bowing of marble versus cycles; wet exposure of samples from exposed panels from the buildings facades and unexposed (fresh) panels.

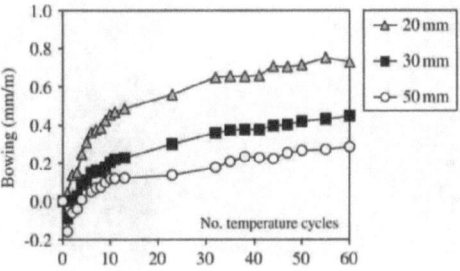

Figure 8. Bowing of samples of Norwegian marble with different thickness versus cycles; wet exposure.

bowing does not increase. This illustrates that bowing only becomes critical in environments, where moisture is also present.

The bowing potential differs significantly between different stone types, as it can be seen in Figure 7, where samples have been cut from exposed panels from the buildings facades and some samples from unexposed panels, stored inside the buildings. The unexposed panels came from the same delivery as the exposed panels, but have been kept in storage or at indoor positions, where they are not exposed to temperature cycles or moisture.

The thickness of the panel will also have an effect on the development of bowing as shown in Figure 8.

The testing has so far not been able to identify a "pessimum", defined as a critical thickness, where the bowing would be largest, but have shown that a larger thickness leads to a slower development of the bowing.

Parallel petrographic studies showed that the microstructure influences the bowing potential significantly. It has been shown from the above tests and

from the numerous other tests on other marble samples, that the calcitic marbles with interlobale grain shapes have low bowing potentials and that the marbles with granoblastic-polygonal grain shapes have a much higher bowing potential. The influence of the mineral properties on the bowing and expansion potentials is described in details by Alnæs et al. (2004)

The microstructure of a marble with typical interlobale grain shapes is shown in Figure 9, top (taken from the Malmö panels) and the microstructure of a marble with typical granoblastic-polygonal grain shapes is shown in Figure 9, bottom (taken from the Nyköping panels). The terms granoblastic-polygonal and interlobale grain shapes in this paper are used in accordance with Passchier et al. (1996) and Spry (1983).

Bowing and loss of strength have been reported by Yates et al. (2004) to correlate in samples from building claddings. Flexural strength have therefore also been measured on a number of bowing laboratory specimens. A correlation of loss of flexural strength and bowing has been observed on the laboratory specimens.

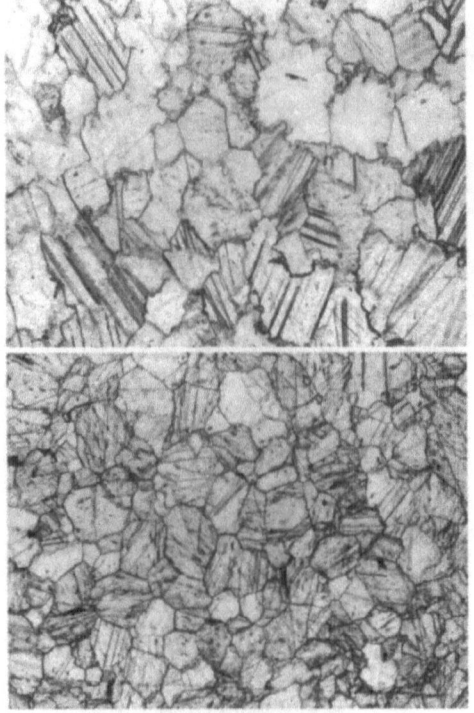

Figure 9. Microstructure of tested marbles; top is Malmö (interlobale) and bottom is Nyköping (granoblastic-polygonal). Each image is 2 × 2.8 mm.

Figure 10. Expansion test-setup (right) and measurement of the length of the specimen (left).

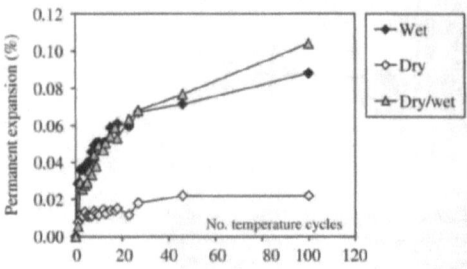

Figure 11. Expansion versus cycles; wet and dry exposure of Norwegian marble.

3 TESTING OF EXPANSION POTENTIAL

One of the first to report irreversible thermal expansion of different marble types was Kessler (1919), but many researchers have later shown, that repeated heating cycles lead to permanent expansion of marble. The TEAM-project has therefore developed a test method for permanent expansion of marble, exposed to temperature cycles.

This test method uses a standard test specimen of 30 × 30 × 300 mm. The specimen is conditioned as the bow test specimens. The specimens are then placed in a tank, filled with distilled or demineralised water with an ambient temperature of 20 ± 5°C and left until constant mass is reached, at which stage the length and mass is measured again.

The temperature of the water in the tank follows the same variation as the surface temperature in the bowing test (Fig. 4) with a tolerance of ±5°C.

The length of each specimen is measured 2 hours after 80 ± 5°C in the water has been reached (Fig. 10). The temperature is then decreased to the ambient temperature and the length is measured again at least 2 hours after 20 ± 5°C in the water has been reached.

The expansion can be mapped as a function of the number of exposure cycles as shown in Figure 11, where it can be seen that the expansion seems to grow continually with the number of exposure cycles. The similar exposure can also be carried out in a dry environment and will lead to an expansion, which after a few cycles reach a permanent level.

This difference between the wet and the dry exposure corresponds to the differences observed in bowing testing.

The expansion testing can also be carried out with every second cycle dry and every second one wet, simulating an environment, where moisture is only available in some periods. The Figure 11 shows that this exposure (wet/dry) will lead to approx. the same expansion as a constantly wet environment.

The permanent expansion may lead to large forces in the panels and joints and may lead to failure of the panels or the fixings (Fig. 12).

256

Figure 12. Failure over anchoring fixing (dowels), due to permanent expansion of the panels.

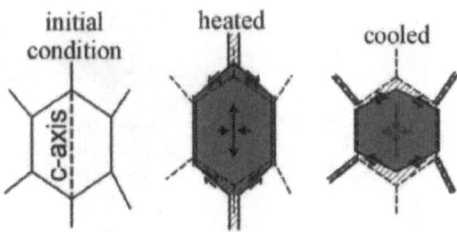

Figure 13. Anisotropic thermal behaviour of a single calcite crystal (after Ruedrich et al. 2001).

4 EFFECT OF DIRECTION

The samples normally tested in the expansion or bowing testing are cut in a direction, determined by the panels direction, which again were determined by the producer.

This may have an effect, since the extremely anisotropic coefficient of thermal dilatation of calcite (26×10^{-6}/°K parallel to the crystallographic c-axis and -6×10^{-6}/°K perpendicular to it according to Kleber (1990)) is well known from literature to be responsible for the development of thermal micro cracks in marble due to diurnal atmospheric heating-cooling-cycles (Fig. 13). If the bulk marble shows a preferred orientation of the c-axes its thermal dilatation anisotropy should approximate that of calcite.

Moreover, the orientation of newly formed microcracks is controlled by the lattice preferred orientation as well leading to a directional dependence of the permanent length change.

As a consequence, the cut direction may have an effect on the bowing and expansion potentials if the microfabric of a calcitic marble is anisotropic.

The TEAM project has therefore looked into the directional dependence on bowing and expansion as reported in Koch & Siegesmund (2004). It was found that some calcitic marbles with a strong lattice and grain shape preferred orientation display an anisotropic bowing and expansion behaviour. A calcitic marble with an average grain size of 0.2 mm and a broad grain size distribution from 0.1 mm up to 1.5 mm and mostly curved or irregular, rarely straight grain boundaries was used for the testing. Thin sections of different orthogonal orientations revealed a distinct grain shape preferred orientation, which is perpendicular to the c-axis maximum of the calcite crystals. To detect the anisotropy test specimens were cut in two different orientations (Fig. 14).

The detection of thermal expansion anisotropy was tested under modified conditions. Cylindrical specimens of 50 mm length and 15 mm diameter were

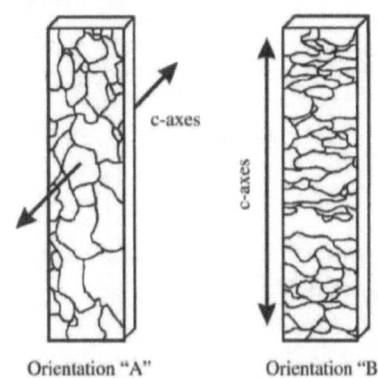

Figure 14. Scheme of bowing test specimens cut in two different orientations in relation to c-axes and grain shape preferred orientation of calcite.

measured in a pushrod dilatometer in two orthogonal directions coinciding with the directions of maximum and minimum thermal expansion. The specimens were exposed to eight dry and ten wet temperature cycles at the same time. Each cycle started at 20°C followed by a heating phase by 0.5°C/min, an equilibration phase of two hours under dry conditions or eight hours under wet conditions at 90°C, a cooling phase by 0.5°C/min and a final equilibration phase of two hours at 20°C. One hour before the heating-up of each wet cycle the climate chamber of the dilatometer was filled with demineralised water, which evaporated during the heating period.

The specimens cut in orientation "B" (Fig. 14) showed a three times higher bowing than those in the "A"-orientation after 40 cycles in the laboratory bowing test (Fig. 15). A similar observation was made in the modified expansion test (Fig. 16). As described above the expansion reached a constant level after a few cycles under dry conditions and increases continuously under wet exposure. The expansion parallel to the c-axes (orientation "B") was found to be three times as high as the expansion perpendicular to the c-axes (orientation "A").

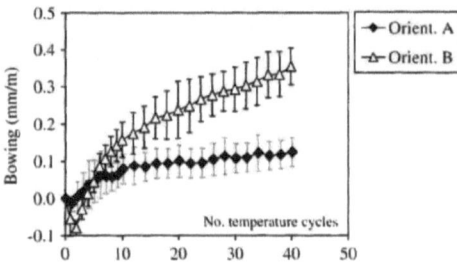

Figure 15. Bowing of marble slabs versus number of heating cycles in Koch & Siegesmund (2004). Each curve represents the mean bowing trend of three specimens.

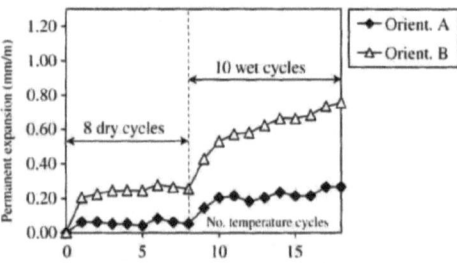

Figure 16. Irreversible expansion of marble versus number of heating cycles (20–90°C) under dry (cycle 1 to 8) and wet (cycle 8 to 18) conditions in Koch & Siegesmund (2004).

This shows that the preferred crystallographic direction may have a marked influence on the bowing and expansion.

5 LAB TESTING VERSUS FIELD CONDITIONS

The purpose of the laboratory testing is to provide a relevant and realistic prediction of what would happen to the panels under the field conditions in a structure. The results of the testing of the bowing potential and the expansion potential must therefore be correlated to the observations on structures, which have been carried out during the TEAM-project.

The TEAM has obtained both exposed and non-exposed panels from a number of the inspected buildings. Samples from both groups of panels have been tested.

A number of other buildings were inspected, but non-exposed panels were not available and it was therefore necessary to identify the quarry, producing the original panels. A number of test panel were obtained from these quarries and used for the laboratory testing.

A very good correlation was observed between the observed bowing problems and the laboratory bow tests; all stone types, which had been observed or reported to bow on the facade did also bow in the laboratory. Stone types, which did not bow in the laboratory, has not been observed or reported to bow on any facade.

The correlation between bowing and loss of strength, observed in exposed building panels has also been observed in laboratory test specimens.

The bowing test method therefore provides a good assessment of the risk of bowing potentials in real structures.

6 CONCLUSION

The two test methods have now been tested over more than 5 years and on 75 different stone types, which covers marble, sandstone, granite, chalkstone, etc. and originate from over 15 countries. The test methods have been found to work very well in the laboratories involved. The results of the test methods correspond to the observed behaviour in buildings with marble cladding. The results of the test methods provide also a necessary link to other material data and to the microstructure of the marble and will thus facilitate the understanding of the deterioration mechanism.

The test methods are able to distinguish between stone types with low, medium and high bowing and expansion potentials, thus providing a very much needed tool in the selection of suitable marbles. The methods are being discussed as potential CEN-test methods and could form a part of the later, mandatory product control.

ACKNOWLEDGEMENT

The authors would like to thank the European Commission for financial support of the European research project TEAM "Testing and Assessment of Marble and Limestone" (Contract no. G5RD-CT-2000-00233). The co-operation with the project partners is gratefully acknowledged.

REFERENCES

Alnæs, L., Koch, A., Schouenborg, B., Åkesson, U., Lindborg, U. & Moen, K. 2004. Influence of rock and mineral properties on the durability of marble panels. In R. Přikryl (ed.), *Dimension Stone 2004.* Rotterdam: A.A.Balkema.

Kessler, D.W. 1919. *Physical and chemical tests on the commercial marbles of the United States.* Technological papers of the Bureau of Standards No. 123.

Kleber, W. 1990. *Einführung in die Kristallographie.* Berlin: VEB Verlag Technik.

Koch, A. & Siegesmund, S. 2004. The combined effect of moisture and temperature on the anomalous expansion behaviour of marble. In S. Siegesmund, H. Viles & T. Weiss (eds), *Stone decay hazards*, Environmental Geology, in press.

Malaga, K., Scouenburg, B., Alnæs, L., Bellopede, R. & Brundin, J.-A. 2004. Field exposure sites and accelerated laboratory test of marble panels. In R. Přikryl (ed.), *Dimension Stone 2004*. Rotterdam: A.A.Balkema.

Passchier, C.W. & Trouw, R.A.J. 1996. *Microtectonics*. Berlin: Springer Verlag.

Perrier, R. & Bouineau, A. 1997. La descohesión Térmica en mármoles y calizas marmóreas, *Roc Maquina*.

NT Build 499 2002. Cladding panels. Test for Bowing.

Ruedrich, J., Weiss, J.T. & Siegesmund, S. 2001. Deterioration characteristics of marbles from the Marmorpalais Potsdam. *Zeitschrift der Deutschen Geologischen Gesellschaft* 152: 637–664.

Schouenburg, B., Grelk, B., Brundin, J.-A. & Alnæs, L. 2000. *Bow test for facade panels of marble*. Nordtest project 1443–99.

Spry, A. 1983. *Metamorphic textures*. Pergamon Press.

Yates, T., Brundin, J.-A., Goltermann, P. & Grelk, B. 2004. Observations from inspection of marble cladding in Europe. In R. Přikryl (ed.), *Dimension Stone 2004*. Rotterdam: A.A.Balkema.

Dimension Stone 2004, Přikryl (ed)
© 2004 Taylor & Francis Group, London, ISBN 90 5809 675 0

Field exposure sites and accelerated laboratory test of marble panels

K. Malaga & B. Schouenborg
SP Swedish National Testing and Research Institute, Borås, Sweden

L. Alnæs
SINTEF Civil and Environmental Engineering, Trondheim, Norway

R. Bellopede
Politecnico di Torino, Dipartimento di Georisorse e Territorio, Torino, Italy

J.-A. Brundin
JAC, Engelholm, Sweden

ABSTRACT: The use of thin marble and limestone for facade cladding has increased substantially during the last five decades. The durability of thin slabs (often only 30 mm thick) has been assumed to be satisfactory based on centuries of successful use as a structural building stone. Nevertheless, all over the world, the long-term deformation and strength loss of some claddings have led to concerns about its safe and durable use. The detailed assessment of marble and limestone within TEAM (see also www.sp.se/building/team) is used to develop a hypothesis for the observed deterioration and to develop remedial actions. This paper describes the field exposure sites, the accelerated test, and presents parts of the results. The results indicates that all marbles bow, but at different magnitudes. The marbles demonstrate diurnal and seasonal variation in bowing and dependence on the thickness and impregnation. The final evaluation of the results will be presented in 2005.

1 INTRODUCTION

Durability assessment of a building façade clad with natural stone panels describes its capability to maintain serviceability over a specified time. Besides several aesthetical aspects of a cladding, the main property deciding the suitability of the selected stone is its stability in a number of mechanical properties. The well know examples of mechanically weak performance of marble claddings on the Finlandia Hall in Helsinki (Royer Carfagni 1999) and the Amoco building in Chicago (Logan et al. 1993, Winkler 1994, 1996) brought about a detailed research about variation of mechanical properties and the mechanisms behind the bowing. The cladding panels on the mentioned buildings lost a lot of their strength and changed their original form due to expansion and started to bow either convex or concave. This behavior has been observed all over the world and current knowledge was insufficient in order to explain this feature.

In March 2000 started a European research project TEAM Testing and Assessment of Marble and Limestone. The main goal of this project is to answer the question; how to choose, produce and use marble and limestone cladding panels without the risk of deterioration due to bowing or expansion.

This paper describes the parts of the investigation program that deals with the exposure of marble panels under field and laboratory conditions.

The main objective with the field exposure is to establish a sound correlation with accelerated laboratory tests and to establish comparable outdoor testing places in a variety of European climates. The main purpose of the climatic chamber test is to study the effect of various impregnation chemicals on fresh and natural weathered/old marble building slabs.

2 MATERIALS

The properties of marble differ not only between the different quarries but also within the same one. The same commercial name might include several types of marble. Most of the marble blocks in this study were sampled under control of at least one geologist and member of the TEAM research group. All samples were quarried in Europe; see Alnæs et al. (2004). The marble types used in this project are both dolomitic

Table 1. Description of the investigated marbles. (See also Alnæs et al. (2004)).

Name	Color	Mineralogy
Itq1	Light gray	Calcitic
Itq2	White	Calcitic
Itq3	Light gray	Calcitic
Itq4	White/dark veins	Calcitic
Grq1	White	Dolomitic
Seq1	Grey/green	Dolomitic
Noq1	Dark grey	Calcitic
Chq1	Grey	Calcitic
Chq2	Grey	Calcitic
Poq1	Grey/green	Calcitic
Poq2	Grey	Calcitic
Poq3	Light pink	Calcitic
Poq4	Dark grey	Calcitic

Figure 1. Field exposure rack in Italy.

and calcitic, with either granoblastic or interlobate microstructures are homogeneous or with pronounced veins, and of different colours ranging from the very white to dark grey (Table 1). The names of all samples are confidential therefore an abbreviation system is used. The first two letters design country of the origin (It- Italy, Gr- Greece, Se- Sweden, No- Norway, Ch- Switzerland, Po- Portugal), q stands for quarry and numbers from 1 to 4 for the number of the quarry.

Two different impregnation agents: GypStop (GS) and Anti Graffiti System (AGS) were used in order to analyse their effect on bowing on the samples exposed to field conditions. GS called sols of colloidal silica is an inorganic, water-based product, which after gelation and drying, forms a chemically inert, highly stable protective silica gel product. Due to the small particle size (5–150 nm), GS easily penetrates pores of a stone material.

AGS is a water-based all-weather long-term impregnation agent, formulated for most types of surfaces that need protection against spray-paint or graffiti. It also provides strong protection against airborne pollution and moisture for a minimum of 5 years. AGS is permeable and allows the protected surface to breath. Correctly applied, life expectancy against graffiti is approximately 5 years.

AGS, GS and a mixture of them and one additional hydrophobic treatment acrylic-copolymer with various fluorinated, silane, non-ionic and ionized chains called Faceal Oleo HD were included in the test program for the accelerated climatic studies.

3 EXPOSURE SITES AND ACCELERATED AGEING TEST

3.1 Field exposure test site

The purpose of this site is to analyse changes in the mechanical properties of chosen types of marbles

when exposed to four corners of the compass. Figure 1 illustrates how the exposure site located in Sweden is constructed. The exposure racks are tilted at 45° in order to prevent accumulation of snow or rainwater and to receive the maximum amount of sun and rain. The set up is in accordance with ISO 8565 (1992). The dimensions of the samples are $400 \times 500 \times 30$ mm. Two samples of Itq2 with thickness 20 mm and two samples of 40 mm are also present. The exposure sites are located in Sweden, UK, Italy, Poland and Slovenia.

Five different types of marbles (Itq2, Seq1, Chq1, Noq1 and Grq1) have been chosen for exposure towards SE, SW, NE and NW at the Swedish site and towards S, N, W and E at the Italian site. The S and SW directions are expected, based on the several observations within the TEAM, to give highest changes in bowing therefore most of the samples are placed towards them. In UK, Poland and Slovenia panels are exposed toward south.

The observed variables that are anticipated to influence the mechanical properties of the marbles are: the corners of the compass, type of marble, thickness, impregnation, finishing and orientation/fabric. Air temperature, precipitation and humidity are monitored continuously. Several measurements of temperature on the front- and backside of the panels are also performed. The changes of the mechanical properties are controlled by repeated non-destructive measurements of bowing and ultrasonic velocity. At the end of the exposure additional destructive tests on some of the specimens will be performed. It is planned that the majority part of the marbles will continue to be exposed even after the end of the TEAM project.

The readings are taken about four to eight times a year *in situ* and additionally all specimens are taken indoors and dried twice a year for "reference" measurements. The daily variations on a façade clad can be quite large and the only way to eliminate this variation

Figure 2. Full-scale exposure site in Nyköping, Sweden. The original white marble panels have been exchanged by several marble types. Three of the panels have an on-line monitoring of strain, temperature and humidity.

Figure 3. Test wall site for anchoring system at fisherwerke.

is to measure the permanent changes after a specified conditioning indoor.

3.2 Full-scale in situ exposure site

A full-scale *in situ* exposure site with 15 different types of marble has been installed on the SW wall of the City Hall in Nyköping, Sweden. The dimensions of the samples are 927 × 910 × 30 mm. Itq2 is also present in two other thicknesses; 20 mm and 40 mm. In addition, several original panels are also included in the study. For further information of monitoring programme on this and other buildings, see also Yates et al. (2004).

The City Hall was clad with white marble panels in 1969. The old panels display both concave and convex bowing on all facades. The highest values of 23 mm/m were observed on the WSW façade. Severe decrease in joint width has been observed and the strength of some measured panels is between 25 and 50% of the original.

Besides periodic measurements of bowing and ultrasonic velocity on all panels, air temperature in the gap behind the panel, surface temperature, humidity and strain in two directions on the external and internal surface are continuously monitored on three panels (Fig. 2). The factors investigated here are: types of marbles, thickness, impregnation and comparison between new and old panels.

3.3 Anchoring system

The literature (Cavallucci et al. 1997, Bain 1940) deals with several cases of external cladding mechanically fixed and placed in mortar backing concerning with different types of bowing or expansion. An exposure on a wall where a marble concerned as bowing type is tested for different anchoring systems has been installed at fisherwerke, Waldachtal, Germany (Fig. 3). The

dimensions of the panels are: 1350 × 1000 × 30 mm. The types of anchoring system used here are: continuous kerf, mortised, back face fixings FZP, dowels at vertical edges and dowels at horizontal edges.

The purpose of this test-wall is to investigate if there is any difference in bowing pattern between the panels fixed with the different systems. Another purpose is to evaluate whether some fixing system are more favorable than others to utilize the remaining strength and therefore safer.

Bowing measurements at fisherwerke are performed once a month since mid 2003. Each panel is measured for the two diagonals, along three vertical lines and three horizontal lines – along edges and midfield. Several additional panels are exposed like the test wall panels and taken down for flexural strength testing at scheduled time intervals. Once the long-term test period is ended it is planned also to test the façade panels after dismounting.

3.4 Laboratory exposure to accelerated climatic strains

As a supplement to the field and building inspections described above, accelerated aging in a special climatic chamber at the Norwegian Building Research Institute (NBI) has been included in the test program of TEAM. The main purpose of the ageing test is to study the effect of various impregnation chemicals on fresh and naturally weathered/old marble building slabs. One important question is whether impregnation may stop or reduce the bowing and strength loss and hence prolong the service life of vulnerable marble types. Another important question is the potential of correlating accelerated test and the field exposures and thus being able to define the acceleration factor for laboratory testing.

263

Figure 4. Climatic chamber for the accelerated aging of marbles at the Norwegian Building Research Institute, NBI, in Trondheim, Norway.

Of special interest is also the possibility this accelerated test gives for assessing freeze-thaw durability.

The test is performed according to Nordtest-method NT BUILD 495 (2000) and is designed specially to study the degradation of vertically positioned materials and components in the building envelope (Fig. 4). The objective of the method is to concentrate the individual climatic factors so that they in total produce a cycle of strains giving degradation results similar to natural exposure but in a much shorter period of time. Under the accelerated ageing test the specimens are exposed to the following strains:

1. UV-radiation from artificial sunlight tubes perpendicular to and at a distance of 500 mm from the test specimen. The surface temperature raise from room temperature to 65°C within approximately 5 minutes by IR-radiation.
2. Wetting with a spray of de-ionized water: 15 ± 2 l/(m² h).
3. Cooling and freezing to an air temperature of −20 ± 5°C.
4. Thawing at ambient laboratory climate: 23 ± 5°C, 50 ± 10% RH. Under the thawing the specimens may be inspected, rearranged and/or changed.

The test specimens are exposed to the climatic strains 1, 2, 3 and 4 in turn and in this order and the duration in each position is one hour. The apparatus runs continuously day and night, and a total exposure period of 100 days (600 cycles) have been selected for the purpose of the TEAM project.

Twelve already deteriorated and bowed marble slabs from the City Hall in Nyköping were cut into appropriate size and included in the test. In addition,

twelve slabs of fresh Itq2 and two slabs of Itq1 were installed.

Before installation, the following properties were characterized: weight, ultrasonic velocity, color and bowing.

The bow measurements were performed the same way as described above for the slabs *in situ* and full-scale exposures. During exposure the temperature was measured on the front of each slab and on the back of three slabs. In addition, measurements of bow and color were performed. Strength measurements will be performed after completion of the test.

The results of the accelerated exposure will be compared with the long-term building, the field tests described above and laboratory bow-tests. The exposure started in January 2004 and the results will be available in June 2004.

4 MEASUREMENT TECHNIQUES

Bowing is measured by use of a bow-meter that follows Nordtest Method NT BUILD 500 (2002) Cladding Panels: Field method for measuring of bowing. The measurements of ultrasonic velocity on Nyköping City Hall and at the Swedish and Italian field exposure sites are performed with an indirect method. The transmitter and receiver transducers are positioned in contact with the front-side surface of the exposed panel and the velocity of propagation of ultrasonic wave in the stone is measured. The generated pulse frequencies are 60 and 50 kHz and the time resolution 0.1 µs.

5 RESULTS AND DISCUSSION

Only selected results from the field exposure sites located in Sweden and Italy and full-scale *in situ* exposure site in Nyköping, Sweden, will be discussed here. The reason is that these sites have been in operation for a much longer time than the others (ca. 1.5 year). However, it should be taken into account that these results are not conclusive yet, though, trends observed at these places are worth a discussion. We also need to bear in mind that the number of different samples exposed to all directions is small.

It is assumed that temperature variations in combination with humidity are the external factors required for bowing to occur. Screening tests using the bow test described in Grelk et al. (2004) for samples exposed solely to heat resulted in no bowing.

Results from measurements of temperature on the front- and backsides on white (Itq2) and dark (Noq1) marbles at the field exposure indicated significant differences between them. The differences in temperatures between front- and backside were only visible during the sunny hours. The difference was clear for

264

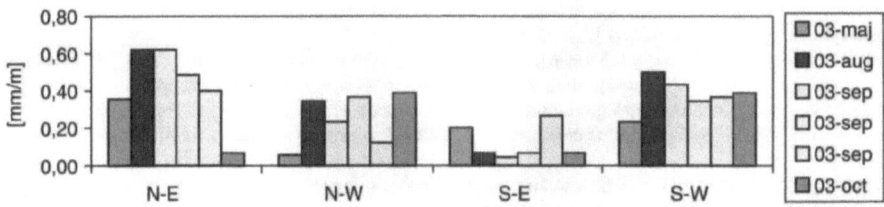

Figure 5. Absolute bowing of the Chq1 in relation to the exposure direction. Data from the field exposure test site in Sweden.

the white marble front- and backside but was much weaker for the dark marble. The highest temperature was recorded for the dark marble and was about 10°C higher on both sides compared to the white marble.

Until now, measurements of bowing indicate that all marble types bow, however of different magnitudes. The values vary for the fresh panels from 0.2–1.3 mm/m and for the old panels from 6–10 mm. The reproducibility in using the bow-meter has been determined to be less than 0.2 mm (NT Build 499 2002). Interpretation of very low values and small differences should take this uncertainty into consideration. The results for bowing of the Chq2 presented in Figure 5 show how the bowing varies depending on the exposure to the corners of the compass. The values are presented as the absolute values, which means that only deviations from the original flatness/bowing of a panel are presented and all minus values are converted to plus values.

The Anova (analysis of variance) tests have been applied in order to analyze if the results have a statistical significance. Results from bowing measurements indicated that the bowing of the Chq1 and Grq1 depends on exposure direction, however the bowing of the Itq2, Noq1 and Seq1 does not depend on it.

Ultrasonic velocity decreased for all exposed samples though; analysis of variance showed no dependence on exposure direction. Measurements of ultrasonic velocity are very sensitive to humidity changes. All panels from the field exposure sites are, therefore, twice a year dried indoors and measurements of permanent changes are performed. All panels showed a permanent decrease in ultrasonic velocity. The decrease in ultrasonic velocity is demonstrated for the Itq2's different samples in Figure 6.

Analyses of how thickness and impregnation influence bowing were performed only on the Itq2 panels. The Anova tests indicated that bowing depends on the thickness and impregnation. The highest bowing was observed for the 20 mm thick panels and lowest for the 30 mm impregnated thick panels. Both impregnation agents showed to have an inhibiting effect on bowing process, however, it should be born in mind that changes were small and in most cases within the uncertainty of measurement.

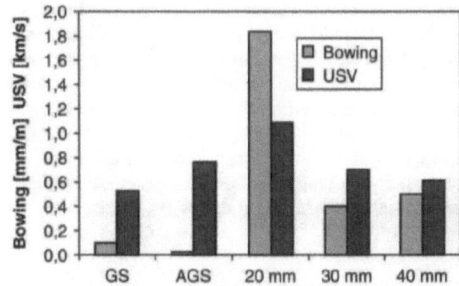

Figure 6. Absolute and permanent bowing and decrease in ultrasonic velocity (USV) for Itq2 after one-year exposure.

There is a considerable difference between bowing magnitude of the same samples measured at field conditions and after drying indoors. All results from *in situ* measurements gave higher bowing compared to measurements on dried samples. In some cases, marble panels normally showing bowing at the field exposure; for example highest measured bowing for the Noq1 was ca. 1.3 mm/m, after drying at indoors conditions showed no signs of bowing. This indicates that no permanent bowing could be observed for these panels. Seq1 and Chq1 showed no permanent bowing either. Only Itq2 and Grq1 demonstrated clear permanent bowing. Comparison of bowing amounts taken at different conditions illustrates how sensitive the marbles are to changes in humidity and insolation.

Another important observation was made for the panels when bowing was measured on three consecutive days, at the same time of the day. It was observed that bowing fluctuates slightly in most cases, but for some panels this fluctuation was quite substantial; up to 1.2 mm/m. The same pattern was observed at the full-scale field exposure site in Nyköping where the on-line monitoring equipment detected an ongoing movement of the panels.

This observation gives important information for interpretation of the results. Bowing or movement of a panel observed during short-term exposure is not a continuously increasing movement. It fluctuates noticeable even during one day. This means that single

265

measurements of bowing only give rough results. This hypothesis applies to fresh stone panels. Bowing measurements done on the old marble panels on the façade at Nyköping indicated week though stable increase in bowing. This means that fluctuation range of bowing decreases with the time of exposure of a panel.

The comparison between the different anchoring systems measured at the mid centre point and in the mid bottom point, indicated that the highest bowing values have been assessed at the mid bottom point. The dowel-system vertical joint is characterized by the highest increase of bowing respect the first measurements both for in the mid centre point and in mid bottom point. However, the bowing results are characterized by a high standard deviation, in some cases the same magnitude as the bowing: 0.2/0.3 mm/m. In addition, a high spread of values within the slabs with same anchoring system in different positions has been observed. The evaluation of the results will be for that reason possible after the end of the TEAM project.

6 CONCLUDING REMARKS

The 1.5-year monitoring at several field exposure sites indicates that all marble types show an internal movement that leads to bowing of different magnitudes. Most of the marbles show evidence of diurnal fluctuation in the bowing. In addition, it has been observed that some marbles demonstrating bowing in the field, showed no bowing after the drying indoors. It has been also observed that the rate of bowing is higher for the fresh panels than for the old panels. Diurnal fluctuations also seem to be larger for fresh panels compared to old ones. The thickness and impregnation of the marble panels have shown to have an influence on the bowing. However, it would be investigated in more details.

All panels demonstrated a permanent decrease in ultrasonic velocity. However, no correlation with the bowing results could be found.

We need to bear in mind that the number of different samples exposed to all directions and the observed changes in bowing are very small. In addition the exposure time is relatively short. It is therefore planned to continue monitoring of these sites after the end of the project in 2005.

ACKNOWLEDGEMENTS

The authors would like to thank the European Commission for financial support of the European research project "Testing and Assessment of Marble and Limestone" (Contract no. G5RD-CT-2000-00233). The co-operation with the project partners is gratefully acknowledged.

REFERENCES

Alnæs, L., Koch, A., Schouenborg, B., Åkesson, U., Lindborg, U. & Moen, K. 2004. Influence of rock and mineral properties on the durability of marble panels. In R. Přikryl (ed.), *Dimension Stone 2004*. Rotterdam: A.A.Balkema.

Bain, G.W. 1940. Geological, chemical and physical problems in the marble industry. *American institute of mining and Metallurgical Engineer*, Technical publications n.1261, p. 16.

Cavallucci, F., Garzonio, C.A., Giovannini, P., Fratini, F. & Manganelli Del Fa, C. 1997. Mechanical decay process in lapideous materials (Carrara marbles): Preliminary study of creep phenomena. In *Proc. 4th Int. Symp. On the Conservation of Monuments in the Mediterranean*.

Grelk, B., Goltermann, P., Schouenborg, B., Koch, A. & Alnaes, L. 2004. The laboratory testing of potential bowing and expansion of marble. In R. Přikryl (ed.), *Dimension Stone 2004*. Rotterdam: A.A.Balkema.

ISO 8565, 1992. *Metals and Alloys – Atmospheric Corrosion Testing. General Requirements for Field Test.*

Logan, J.M., Hasted, M., Lennert, D. & Denton, M. 1993. A case study of the properties of marbles as building veneer. *Int. J. Rock. Mech. Min. Sci. and Geomech. Abstr.* 30: 1531–1537.

Nordtest Method NT BUILD 495. 2000. *Building materials and components in vertical position: Exposure to accelerated climatic strains.*

Nordtest Method NT BUILD 499. 2002. *Cladding panels: Test for Bowing.*

Nordtest Method NT BUILD 500. 2002. *Cladding panels: Field method for measurement of bowing.*

Yates, T.J.S., Brundin, J.-A., Goltermann, P. & Grelk, B. 2004. Observations from the inspection of marble cladding in Europe. In R. Přikryl (ed.), *Dimension Stone 2004*. Rotterdam: A.A.Balkema.

Winkler, E.M. 1994. *Stone in Architecture Properties, Durability*. 3rd ed., Berlin: Springer Verlag.

Winkler, E.M. 1996. Properties of marble as building veneer. *Int. J. Rock. Mech. Min. Sci. and Geomech. Abstr.* 33: 215–218.

Dimension Stone 2004, Přikryl (ed)
© 2004 Taylor & Francis Group, London, ISBN 90 5809 675 0

Observations from the inspection of marble cladding in Europe

T.J.S. Yates
BRE Ltd, Watford, UK

J.-A. Brundin
JAC, Engelholm, Sweden

P. Goltermann & B. Grelk
Ramboll, Copenhagen, Denmark

ABSTRACT: The use of thin marble and limestone for facade cladding has increased substantially during the last few decades. The durability of such thin slabs (often only 3–4 cm thick) has been perceived to be satisfactory based on centuries of successful use as a structural building stone. Nevertheless, all over the world, the long-term deformation and strength loss of some claddings have led to concerns about its safe and durable use. The detailed assessment of facades, within TEAM, at several study sites, using monitoring systems, risk assessment and lifetime prediction, are used to develop a hypothesis for the observed deterioration. Repair techniques will be developed for existing buildings displaying damages. The TEAM project consortium represents nine countries and comprises sixteen partners (see also www.sp.se/building/team). The project has a budget of 4 M€ and is partly funded by the European Commission under the contract no. G5RD-CT-2000-00233. This paper presents the results of the parts of the concerned with the inspection and monitoring of marble clad buildings identified as being effected by deformation and loss of strength.

1 INTRODUCTION

Despite more than 100 years research and being a world wide problem, the solution to the problems with bowing marble has not yet been found.

The main objectives of the TEAM-project are therefore to find the mechanisms of bowing façade claddings of marble, the expansion of marble and limestone and the connected loss of strength. In addition, a field monitoring, evaluation and repair guide for facade cladding, which will include risk assessment and service life prediction shall be developed and various solutions to the problem will be tried. Drafts of laboratory test methods shall be delivered to the European Standardisation of Natural Stone: CEN TC 246.

The project has developed from its main activity; focussing on bowing, towards the causes and interaction of bowing, expansion and strength loss. However, reproducing the conditions of the bowing is a prerequisite to find both the driving mechanisms and the response of different marble types to these conditions! The problems are clearly of interdisciplinary character. The project therefore engages experts from all parties concerned, e.g. stone producers, trade associations, testing/research laboratories, and standardisation bodies, consultants, building owners and caretakers and producers of fixing and repair systems.

This paper describes three aspects of the overall TEAM project that relate to the inspection and monitoring of some selected buildings in Europe that are known to be affected by bowing and loss of strength. This work is underpinned by a comprehensive literature review of the 'state of the art' relating to the bowing and distortion of marble. The first section of this paper reviews the geographical distribution of buildings with marble cladding based partly on the findings of the literature review and partly on observations made during the visual inspection of marble cladding – both those with and without apparent problems – and the types of measurement that have been made on these buildings. The second section describes the development of long term monitoring systems in Copenhagen (Denmark), Göttingen (Germany) and Nyköping (Sweden). The final section shows some of the results from 15 months of monitoring of the deformation of panels at Nyköping City Hall.

267

This paper is one of four in this volume relating to the TEAM project. The other three cover:

- Field exposure sites and accelerated laboratory testing of marble panels (Malaga et al.).
- The laboratory testing of potential bowing and expansion in marble (Grelk et al.).
- Influence of the macro- and microstructure on the durability of facade material (Alnæs et al.).

2 'STATE-OF-THE-ART' LITERATURE REVIEW

The first result from the project was a state-of-the-art report on 'Deterioration mechanism hypotheses' (see TEAM, 2000). In this state-of-the-art report more than 190 papers from the last 100 years were identified, collected, and reviewed in an attempt to extract the key findings with particular emphasis on the causes and mechanisms responsible for the bowing of natural stone cladding.

Despite the existing copious numbers of publications about the behaviour and deterioration of limestone and marble – particularly the bowing phenomenon, it was clear that no conclusive explanations exist concerning the influencing causes or mechanisms. However, two key parameters are agreed on: temperature variations and moisture. Interestingly the latter has only recently been acknowledged as a key factor.

3 BUILDING INSPECTION

One of the first tasks of the project was to collect data examples of bowed marble cladding from Europe. This data collection has continued throughout the project and has been expanded to include examples from the US. To date around 150 building projects around Europe have been recorded and about 50 of these are reported to have bowing problems. It is likely that many more examples remain un-noticed.

There is, for natural reasons, a concentration of buildings affected by bowing reported from northern Europe – particularly northern Germany, Denmark, Finland and Sweden, but there are also examples from the rest of Europe including Slovenia, Switzerland, Austria, France, UK, Belgium, Italy and Poland. This is merely the reflection of the partner's initial knowledge, and ease of access to sites and information. The most well known example in Europe is the Finlandia Hall in Helsinki, Finland. There is no evidence that any particular climate is typical of the conditions that can result in long term bowing and expansion. Cases of bowing have been reported from the most different climates, from Libya in the south to northern Sweden/Norway in the north but a temperature range and a source of moisture are common to all locations.

A wide range of buildings were visited and the condition recorded along with details of the type of stone used, the panel dimensions, the fixing system and the local environment. The main findings from this stage of the work were:

3.1 Façade compass direction and height over ground

Bowing is observed to all façade directions as well as on all heights over ground. On same façade section there is a tendency that bowing amplitudes are more significant higher up on the façade. There is also a tendency that facades facing south and west have higher bowing amplitudes then is the case for facades facing north and east.

3.2 Colour of marble

Among reported projects different colour marble types with bowing problems are included such as light types from Italy, Switzerland, Russia, Macedonia and Sweden, green type from Portugal, dark grey types from Portugal and Norway. It shall be stressed that same types of marble on other reported projects does not show any bowing. Thus there is no clear indication as to the effect of colour range of the marble related to deformation by bowing or not bowing.

3.3 Panel face dimensions and thickness

Very large marble panels ($>2 \, m^2$) have been recorded with perfectly plane and unaffected surfaces while on other buildings small ($<0.1 \, m^2$) have deformed, deteriorated and fall from the façade. No correlation has been observed between bowing tendencies and stone panel thickness. It might however be noted that marble panels on Nyköping City Hall (Sweden) which are 30 mm thick have deformed in concave direction while adjacent smaller panels which are 60 mm thick on the same façade elevation have deformed in convex direction.

3.4 Convex or concave bowing

It has not been possible to relate the options of convex or concave to any other influencing factor. Panels on one side or one location of the same building might deform in concave direction while on an other side or location the bowing is convex, for example at Lünen Hospital (Germany), Nyköping City Hall.

It has also been noted that the original marble façade panels on Finlandia Hall in Helsinki (Finland) were mainly concave with extremely high amplitudes after 30 years of exposure. However, the panels installed some 3–4 years ago of same white Italian marble type (however not from same quarry) have started to deform in a convex direction.

Another striking example is Torsviks Torg in Lidingö (Sweden) – here three tower buildings which are covered on all sides with marble panels (600 × 900 × 30 mm) show convex and concave bowing on alternate rows of panels.

3.5 Anchoring system

Most of the inspected facades have 30 or 40 mm thick marble panels on various types of anchors and with a ventilated air cavity 20–40 mm wide between the stone panel back face and layers of insulation. Some facades, are however, installed on mortar beds with or without restraint anchors – this means that there is no ventilated cavity. Some projects have part of the installation done on freely standing railing systems allowing both the front and rear panel faces to be 'exposed' to the surroundings. To date no link has been observed between the type of anchoring system and the bowing of marble panels with the exception panels which are bedded on mortar which do not seem to be susceptible to bowing.

3.6 Building measurements

Some preliminary measurements were made at a number of these buildings using a modified version of the 'bow meter' that had been developed in the earlier NORDTEST project (see NT BUILD 500). The 'bow meter' is basically a 1200 mm straight edge with a digital dial gauge that allows the distance from the edge to the panel surface to be measured very accurately. Detailed site investigations, measurements, sampling etc, have, so far, been performed at 8 buildings (seven are recorded in Table 1, the eighth is the University Library at Göttingen, Germany). In addition, panels were removed from 10 buildings for further investigations in the laboratory, seven of these have been completed to date. Petrographic thin sections have been produced and described from all these samples and they have been subjected to a laboratory bowing test and the flexural strength determined. A summary of the results of the site measurements and laboratory bowing is given in Table 1.

Probably the most important physical change observed is the loss of flexural strength. The results of changes in strength recorded during the literature review and the testing of samples from buildings are summarised in Table 2.

In addition to the bowing and loss of strength there are also examples of spalling of stone around the fixing points (often associated with the movement of the panel against the fixing) and erosion of the surface as the marble becomes less durable.

Table 1. Summary of the results from site measurements and laboratory tests on seven buildings in Europe.

Building name	Marble type	Bowing site/laboratory
City Hall, Nyköping, Sweden	B. Carrara (I) Calcitic	20–30 mm/ strong
Hospital, Lünen, Germany	Trigaches E. (P) Calcitic	30–50 mm/very strong
Bank Building, Copenhagen, Denmark	Porsgrunn (NO) Calcitic	10–20 mm/ medium
Office Building, Copenhagen, Denmark	Porsgrunn (NO) Calcitic	10–20 mm/ medium
City Hall, Malmö, Sweden	B. Carrara (I) Calcitic	1–2 mm/none
Hotel Terazza, Ljungby, Sweden	Ekeberg (S) Dolomitic	5–15 mm/ medium
City Hall, Lyngby, Denmark	Calcitic (Greenland)	0–2 mm/weak

Table 2. Summary of recorded losses of flexural strength.

Building	Marble type (origin, country)	Age (Year)	Loss of flexural strength	References
Finlandia Hall Helsinki, FIN	B. Carrara (I)	21	~85%	Mustonen (1993)
Amoco Chicago, USA	B. Carrara (I)	15	~40%	Hook (1994)
Office building, Nyköping, S	B. Carrara (I)	31	~75%	Schouenborg (2000)
Hospital Lünen, D	Trigaches/Escamado (P)	14	~30%	Stocksiefen (1996)
		28	~75%	
Office building, CH	B. Carrara (I)	3	~40%	Jornet (2000)
Bank building, Copenhagen, DK	Porsgrunn (NO)	23	~45%	Lekso (2002)
Office building, Copenhagen, DK	Porsgrunn (NO)	41	~75%	Lekso (2002)
Office building, Lyngby, DK	Marmorilik (DK)	60	~45%	Lekso (2002)
Office building, Paris, F	B. Carrara (I)	11	~50%	Grelk (2002)
Office building, Malmö, S	B. Carrara (I)	20	~10%	Brundin & Grelk (2004)

269

Figure 1. Distribution of buildings where bowing or distortion of the marble cladding has been recorded by the TEAM partners.

4 LONG TERM MONITORING ON BUILDINGS

4.1 *Selection of monitoring parameters and sites*

Once a wide range of geographical locations, materials and buildings had been identified and recorded three sites were selected for continuous monitoring of deformation/movements for panels and monitoring of environmental conditions in order to determine if bowing is occurring and the changes induced with time. This necessitated the development of equipment for long term monitoring, installation of equipment at three locations and processing of the results. The data is being used to calibrate model of the risks of failure associated with the changes in the properties of the panels over long periods of time in order that these risks can be better understood and managed.

The selection of parameters to be measured, selection of equipment and selection of suitable buildings were all given careful consideration before being agreed. It was agreed that ideally at each location monitoring would include:

- surface temperature on the external surface of the stone;
- time-of-wetness/condensation on external surface of the stone;
- strain in two directions on the external surface of the stone;
- surface temperature on the internal surface of the stone;
- air temperature in the gap behind the panel;
- relative humidity in the gap behind the panel;
- time-of-wetness/condensation on internal surface of the stone;
- strain in two directions on the internal surface of the stone;
- shade air temperature;
- shade relative humidity.

There were problems with finding ideal sites and so there was a need to 'compromise'. The final list of sites is:

- Bank Building, Copenhagen, Denmark;
- University Library Göttingen, Germany;
- Nyköping City Hall, Sweden.

4.2 *Installation of equipment*

The first two sets of equipment were installed in Copenhagen at the beginning of September 2001, two sets of equipment were installed in Göttingen in November 2001, and three sets at Nyköping in October 2002. The 'sophistication' of the equipment has progressed as the team has gained further experience of the sites and have overcome the initial problems relating, for example, to the fixing of the strain gauges.

4.3 *Bank building in Copenhagen*

Amongst the most interesting results from a detailed inspection of a bank building in Copenhagen is the fact that there is very little difference between the temperatures on the front and back of the panels – only a few minutes 'lag'. On the eastern side the front can be 4°C warmer when the sun first shines on this face. The full temperature rise can take 6 hours but in more extreme cases the rate is around 0.3°C per minute and this information has been used in the development of the bowing test.

4.4 *Göttingen University Library*

The equipment has been in place since November 2001. The initial aim of the logging at this site was to evaluate a new type of temperature and humidity logger and to evaluate the first strain gauges for their thermal stability and the magnitude of movement.

Most of the equipment seems to be working well but there are concerns about the strain gauges as the data shows sudden changes in strain. One set of covers have been added but it does not seem to have solved the problem. Despite some problems with the downloading of data interesting results have been found with diurnal cycles clearly visible against a background of longer term changes with reliable stable data being obtained over short periods – for example 7–8 days. However, over longer periods the 'drift' and background strain readings still seem to show sudden changes.

4.5 Nyköping City Hall

The equipment was installed on the City Hall at Nyköping in October 2002 – this was timed to coincide with the setting up of the field exposure site. The aims at this site are to evaluate the re-designed strain gauges, in particular to look at the magnitude of movement, expansion vs. bowing and the effects of surface treatments.

The strain gauges are a new design using 'Invar' to reduce the thermal expansion of the gauges and provide more stability in the results. There are three monitoring locations – one location is on an existing Carrara panels (that has been demounted and replaced), the second is on a new panel of Carrara with similar dimensions to the existing panel. The third location is on a panel that has been treated with a microcrystalline wax from Trion Tensid. In all three cases, strain gauges have been fixed to the rear and front faces to allow determination of movement in three dimensions. Further analysis allows this to be converted to a 'bowing' figure which can be compared directly to the 'bow

meter' measurements being made on site. All of the data is collected by a single data logger. Only one set of temperature and humidity measurements is being made behind the panels and in the air as experience from Göttingen has shown that there is very little variation between adjacent panels.

The diurnal movement in the strain gauges can be clearly seen and it is also clear that this daily movement is likely to 'mask' any longer term residual movement in the panels from the early expansion and/or bowing of the marble. As a result a considerable effort has gone into the development of methods to 'process' the data. Three approaches have been used:

- selecting and examining strain gauge data recorded over a narrow range of temperatures;
- selecting strain gauge data recorded at the same time each day;
- correcting the entire dataset (14,000 points) for thermal movement of the strain gauges.

Figures 2–3 show the results from applying the first method to one of the panels. The residual bowing

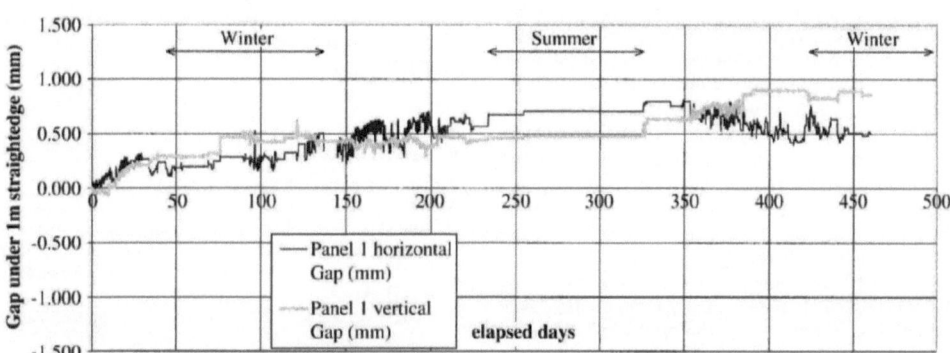

Figure 2. Bowing in Panel 1 at Nyköping, This was removed from the building in November 2002 and replaced on new fixings.

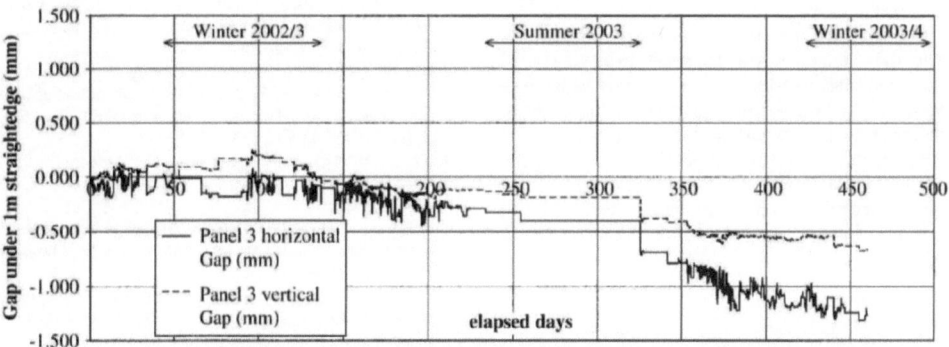

Figure 3. Panel 3 at Nyköping – a new panel of Carrara marble.

271

of the panels can be seen over the 15 months of the monitoring. Figure 2 shows Panel 1 which was removed from the building in November 2002 and replaced on new fixings. The pattern of movement seems to show a fairly rapid bowing immediately after the panel was replaced – possibly as it moved back against the fixings- followed by a long period with very little movement. Figure 3 shows Panel 3 which was a panel of new Carrara marble. Note that the panel is bowing in the opposite direction to Panel 1 and that the magnitude of the bowing is also much greater. Panel 2 at Nyköping was a panel of new Carrara marble that was treated with a hydrophobic coating. The treatment was intended to reduce the bowing. Initial the rate seemed to be reduced but it is not clear now quite what is happening except the panel seems to move far more on the horizontal axis than on the vertical axis.

4.6 Evaluation

The key finding since the initial installation requirements were determined is that there are no 'off the shelf' systems available for this type of monitoring. In particular, all the strain gauge systems required installation conditions which cannot be achieved on existing marble panels – for example a polished and clean surface or heating to 70°C to set the adhesive for the gauges. As a result considerable effort has been put into the development of a system which is suitable for measurement on an existing building. Evaluation of the initial observations from Copenhagen and Göttingen have been used to design more stable equipment for the measurements at Nyköping and this new equipment seems to be working successfully. The earlier work at the ban building in Copenhagen has shown that it is possible to reduce the data logging frequency with no loss of precision in the results and the analysis of the data from Nyköping has show that it is possible to determine the bowing of the panels by filtering the data to specific temperatures and so removing the need to correct all the data points for thermal movement of the gauges.

Comparison of the strain gauge monitor and the bow meter measurements shows the results to be very similar which indicates that the 'real time' strain gauge measurements can be used to supplement and strengthen the findings from the bow meter measurements.

The methods to 'filter' diurnal changes have provided important input to the interpretation of data from field exposure on Nyköping City Hall and at the various field sites established at the TEAM partners.

5 CONCLUSIONS

The early tasks in the project have shown that bowing and distortion of marble cladding can occur on a wide range of buildings and stone types in many different locations. There are examples of buildings that show both concave and convex bowing and also ones where panel of different thicknesses seem to behave in different ways. The monitoring carried out to date has show that it is possible to measure the movement of the panels *in situ* but it required the development of a new and purpose made system. The first results are now available and are being analysed to convert the movements and expansions into 'bowing' in vertical and horizontal directions. The exposure site at BRE, combined with the other sites, will provide long term exposure data for different marbles, different panel sizes and different orientations. It is planned to continue monitoring these sites after the end of the project in 2005.

ACKNOWLEDGEMENTS

The authors would like to thank the European Commission for financial support of the European research project 'Testing and Assessment of Marble and Limestone' (Contract no. G5RD-CT-2000-00233). The co-operation with the project partners is gratefully acknowledged.

REFERENCES

Alnæs, L., Koch, A., Schouenborg, B., Åkesson, U., Lindborg, U. & Moen, K. 2004 Influence of rock and mineral properties on the durability of marble panels. In R. Přikryl (ed.), *Dimension Stone 2004*. Rotterdam: A.A.Balkema.

Brundin, J.-A., Grelk, B. 2004. Besiktning av naturstensfacade på Malmö Stadshus. *Malmö Stad, Stadsfastigheter, Internal report, February 2004.*

Garzonio, C.A., Fratini, F., Manganelli, C., Giovannini, P. & Blasi, C. 1995. Analyses of geomechanical decay phenomena on marbles employed in historical monuments in Tuscany. In H.-P. Rossmanith (ed.), *Mechanics of Jointed and Faulted Rock*: 259–263. Rotterdam: A.A.Balkema.

Grelk, B. 2002. *Pers. comm.*

Grelk, B., Goltermann, P., Schouenborg, B., Koch, A. & Alnæs, L. 2004. The laboratory testing of potential bowing and expansion of marble. In R. Přikryl (ed.), *Dimension Stone 2004*. Rotterdam: A.A.Balkema.

Hook, G. 1994. Look out below – The Amoco Building Cladding Failure. *Progressive Architecture*, Feb., 75, pp. 58–62.

Jornet, A. & Rück, P. 2000. Bowing of Carrara marble slabs: a case study. *Quarry-Laboratory-Monument Int. Congress*, Pavia, Vol. I, pp. 355–360.

Leksø, H. 2002. Holdbarhed af marmor facadeplader. *Byg•DTU, Technical University of Copenhagen, Department of Civil Engineering.*

Malaga, K., Schouenborg, L., Alnæs, L., Bellopede, R. & Brundin, J.-A. 2004. Field exposure sites and accelerated laboratory test of marble panels. In R. Přikryl (ed.), *Dimension Stone 2004*. Rotterdam: A.A.Balkema.

Mustonen, J. 1993. Finlandia-Talon julkisivujen korjaus. *Rakennusinsinööri-päivät, RIL K160-1993, 1993*: 61–8.

Nordtest Method NT BUILD 500. 2002. Cladding panels: Field method for measurement of bowing.

Schouenburg, B., Grelk, B., Brundin, J.-A. & Alnæs, L. 2000. Bow test for facade panels of marble. *Nordtest project 1443–99*.

Stocksiefen, W. 1996. Marmorschäden: Analyse und therapie. *Naturstein* 12.

TEAM, 2000. Task 1.2: Deterioration mechanism hypothesis – State-of-the-art report, 2. draft, *TEAM report, WP1.2-RBL-TD001128-stateofthereport.doc, November 2000*.

Environmental aspects and technological research of dimension stone

Dimension Stone 2004, Přikryl (ed)
© 2004 Taylor & Francis Group, London, ISBN 90 5809 675 0

The qualification policy of VCO stone products

L. Antonazzo, E. Badiali, P. Laurenge & R. Bruno
Dept. of Chemical, Mining and Environmental Technologies Eng., University of Bologna, Italy

M. Proverbio
Provincia del Verbano Cusio Ossola, Verbania, Italy

ABSTRACT: The Italian Province of VCO promoted an original characterisation work all over the finished products of the provincial stone quarries, aimed to qualify and certify the origin and the aesthetical properties of the different materials. The work, committed to the University of Bologna, is based on the image analysis characterisation.

1 GENERALITIES

1.1 *The purpose of the study*

The project we are presenting have been promoted by the Verbano-Cusio-Ossola (VCO) Province and developed by the Department of Chemical, Mining and Environmental Technologies Engineering (DICMA) of University of Bologna. This study tends towards an aesthetical characterization of the ornamental stones extracted in that Province (Antonazzo et al. 2003). The aim is both, scientific and commercial. In fact, we tried to characterize objectively the aesthetical properties of stone polished tiles in order to differentiate each product type and to assign them a quality mark.

1.2 *The customer*

The territory of VCO Province, Piedmont Region, north of Italy, even though limited in size, is especially rich in stone resources, with some of them being unique and centuries old in prestige.

This area represents an important economic reality in terms of production and commerce for Piedmont and Italy. It is worth remembering that Italy accounts for about 15% of the world quarrying production and stone exchange market. The territory of VCO itself represents about 7% of the Italian quarrying production. Nowadays there are about 100 activities involved in quarrying and working ornamental stones in the VCO territory. They are distributed in 22 of the 38 VCO townships and directly employ about 700 people, while the induced activities employ about 600 people.

Annual average turnover reaches approximately 55 million Euros with capital investments in machinery and equipment of over 75 million Euros.

The importance of the stone sector in the local economy and the need of qualifying policy for stone products induced the VCO Province to support this research.

2 DESCRIPTION OF RESEARCH WORK

2.1 *Acquisition system*

The digital system includes a vertical linear colour-scan camera, a conveyer belt and software based on Morphological Image Analysis. The system allows evaluating and measuring the natural variability of some characteristics, such as colour, luminosity, veins, granulometry, etc.

Stone tiles are scanned and their digital image is sent to the computer. One of the first differences between each type of stone is the colour distribution. The chromatic composition of each pixel is measured with reference to the HLS or to the RGB system. Each pixel on each channel has a luminous intensity which varies between 0 and 255, 0 being the complete absence of light (or black) and 255 full light (or white).

2.2 *Processing techniques*

The percentage of pixels with a certain amount of luminosity can be synthesised in a distribution curve, obtained for each channel of colour. Thereby a first

Red Channel Histogram
for a Serizzo Antigorio tile

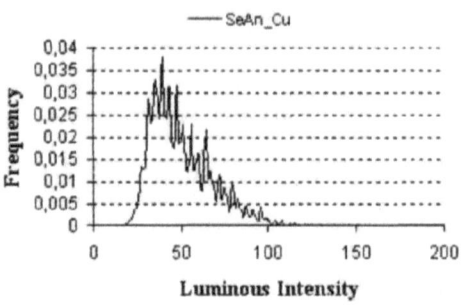

Figure 1.   Histogram of luminous intensity of a tile image.

Curva Cumulata canale rosso: confronto tra materiali diversi

Figure 3.   Ditribution functions of red colour intensity for White Baveno. Grey Beola and White Beola sample images. Existence area as envelope of min/max values of distribution functions.

Media e Varianza (canale Blu)

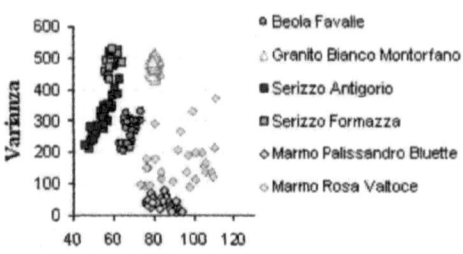

Figure 2.   Mean-Variance scatter of luminous intensity of tiles images for 6 different stone types.

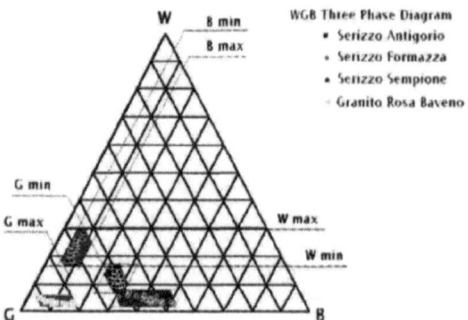

Figure 4.   WGB three-phase diagram of blue channel intentisty. Plot of representative points of Serizzo Antigorio/ Formazza/Sempione and of Pink Baveno samples.

indication of the visible characteristics of each tile is given by the histogram (Fig. 1).

The histograms examination of each different stone type already allows pointing out some differences among colour distributions. And particularly meaningful are the statistical parameters as mean and variance (Fig. 2).

But also the analysis of cumulative histogram of each stone type samples reveals that specific areas are typical of a given material. Therefore, it is possible to identify and numerically define an existence field for cumulative histograms, which allows differentiating the stone materials (Fig. 3).

In order to represent this information in a three-phase diagram WGB, it is necessary to transform the image by assigning a ternary code to each pixel, depending on the colour intensity. Between 0 and 85 the code is black, from 86 to 170 the code is grey and from 171 to 255 is white. The same valuations can be made, instead of considering the colour intensity, on

each colour channel. The percentage of the white, grey and black pixels on each transformed image generates a point on the three-phase diagram WGB. That point represents the tile (Fig. 4).

The introduction of the morphological functions, allows differentiating stone materials by other operators such as the so called "granulometry" (Soille 1999). Of course it is not physical granulometry, that needs the material be reduced in grains. The methodology applies to the image and uses a geometrical element, which can be isotropic or anisotropic: by increasing its dimension and doing what is called "a closure" step by step, curves can be found that resume the granulometric fraction of pixels having a certain dimension. Extending the process to the samples of each different stone type, differences and similarities can be noticed (Fig. 5).

In addition, a variografic analysis of each image completes this first level study. In Figure 6 a classical

278

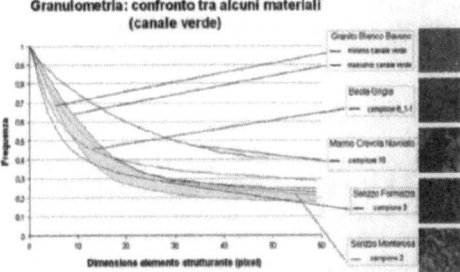

Figure 5. Granuletric diagram of green channel intensity for White Baveno, Grey Beola, Crevola Marble, Serizzo Formazza/Monterosa samples.

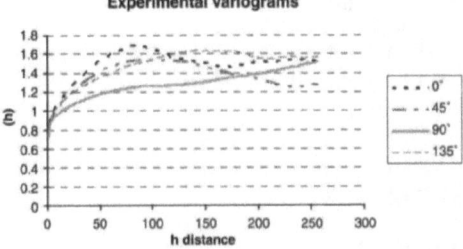

Figure 6. Experimental geometric covariogram of luminous intensity in the four directions of the image plane for one Crevola Marble sample.

geometrical covariogram is shown, calculated in the four directions, each 45 degrees. It shows relevant variability characteristics: an isotropic variability, correlated and uncorrelated, as well as a zoned anisotropy. This type of study is more indicated as far as carbonatic materials have to be characterised, typically a marble with veins.

2.3 Identification methodology

Image analysis produces parameters values or parametrical functions that can or cannot refer to different rock characteristics. But for each stone type a set of values is identified, so that for each studied parameter, minimum and maximum values are established. For example colour intensity mean and variance.

The systematic execution of this processing on the tiles of the same material, allows pointing out if there are zones with higher samples concentration of the same stone type. For example, on a WGB diagram, such an area can be defined by the intersection of grey, white and black minimum and maximum intensity values. Each type of stone can have its own characteristic area on the diagram as shown in Figure 4.

In case of parametric functions a variability area, or existence area, can be defined as the envelope of max and min values of experimental functions for each parameter value. This definition of upper and lower curve releases them to be true parametric functions, actual functions that must be totally included in the existence area.

Of course, whatever image of different materials can have the value of a given parameter, for instance the colour intensity mean, which falls inside the variability interval of another stone type. But even if you consider a 50% probability of incertitude, by taking into account just two parameters, for instance mean and variance, and the three-colour channel (RGB), the misclassification probability reduces to few percents. And it is sufficient to consider a couple of parametrical functions for cancelling any possibility of classification error, without resorting to any particular discriminant analysis.

The result of this research can be reassumed as the identification of an 'x' dimensional space for each material, which permits us to establish with certainty the origins of each tile and which type of particular stone it belongs to. This is possible through a simple intersection of the coordinates of a stone tile image with the identification domain of a stone type.

The petrographic definition and the chemical, physical and mechanical analysis still remain the starting point for such an image analysis, as they suggest the fundamental parameters to be referred for choosing the multi-dimensional space.

3 THE CASE STUDY

The object of our work has been the set of stone polished tiles coming from 60 VCO quarries and of 20 different commercial stone types. Each active quarry supplied 30 samples with minimum dimensions of $20 \times 20 \times 2$ cm. A total of about 1800 tiles have been studied.

In order to have a unique "support" for the characterisation study, that is the same sample image dimensions, all acquired images have been cut with a square template having a 510 pixels side. Given the experimental nature of the project, a 30 tiles sample was considered sufficient for characterising a stone type.

The results of the study confirm the good performance of the approach. On Figure 2, six different stone types are very well identified by simply the average-variance scatter of the intensity.

Remark that not only different geological formations as igneous and metamorphic stones are well distinguished; not only siliceous from carbonatic materials are well distinguished; but even the same petrographic material, marketed as different products, can be univocally discriminated. This is shown on Figure 4,

279

**Mean and Variance Red Channel
Stone vs Ceramic tiles
Bianco Montorfano Granite**

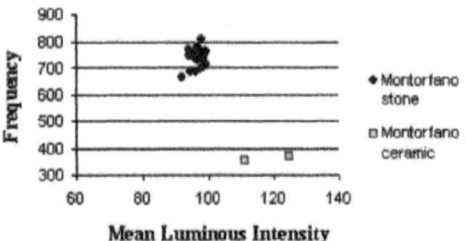

Figure 7. Mean-Variance scatter of red channel intensity of White Montorfano stone samples and ceramic imitations.

where a WGB diagram univocally distinguishes three commercial varieties (Formazza, Antigorio, Sempione) of the same petrographic material, the Serizzo stone.

By considering more and more characterization functions, the risk of possible misclassification reduces immediately. For example on Figure 3 three stones aesthetically quite similar as Grey Beola, White Beola and White Baveno have the intensity histograms incompatible. If more complicated functions, as the granulometry, are considered, the residual doubts disappear. On Figure 5 the existence area of the White Baveno is clearly incompatible with similar or different stone types as Beola orthogneiss, Crevola marble or Serizzo orthogneiss.s

The proposed processing is useful not only for certifying a stone type product and for distinguish it from natural competitors, but it is an important tool for facing the competition against the ceramic imitations.

On Figure 7 the simplest processing of a mean-variance scatter and of the distribution function existence area reveal the evident differences between the natural White Montorfano granite and the ceramic imitation marketed with the same name.

4 CONCLUSIONS

This paper presents an original application of a new approach and technique for characterizing the ornamental stones: the image analysis (Bruno et al. 1998, Bonduà 2002).

In the study performed, the aim was to characterise finished product images so to be able to distinguish them univocally. Remark that such discrimination overcomes just the petrographical identification. In fact it allows distinguishing different market products of the same petrographic material. There are experiences that image analysis utilisation solves also the problem of quality class identification in case of the same market product.

The VCO Provincial Administration will exploit the study results for the valorisation of the territory stone products.

REFERENCES

Antonazzo, L. et al. 2003. Esthetical characterisation of Provincia VCO ornamental stone production. In FIL (ed.) *"International Congress on Natural Stone"*, Lisbon.

Bonduà, S. 2002. *New Instruments for Aesthetical Valorisation of Ornamental Stones: Geostatistical Simulation and Morphological Characterisation*. PhD thesis, DICMA-University of Bologna.

Bruno, R. et al. 1998. *Characterisation of ornamental stone standards by image analysis of slab surface*. Project final report, SMT4 CT95 2028, European Commission, Standards Measurement and Testing Programme.

Soille, P. 1999. *Morphological Image Analysis, Principles and Applications*. Berlin: Springer Verlag.

Dimension Stone 2004, Přikryl (ed)
© *2004 Taylor & Francis Group, London, ISBN 90 5809 675 0*

Staining of natural stone: test methods and proposals for preventive and curative measures

V. Bams & F. de Barquin
Belgian Building Research Institute, Limelette, Belgium

ABSTRACT: Because stone is a natural building material, problems regarding aspect and maintenance may appear shortly after application. One of the most occurring problems on natural stone is staining. This article will report the preliminary results of a recent study at BBRI concerning surface treatments aiming at reducing the risk of staining. As a preventive solution three kinds of products were tested on different types of stone. For each sort of stain the best precautionary measure is sought, to exclude an alteration of the natural stone during his application in the floor, wall, kitchen tablet, etc. Because innumerable cases exist where the aspect of the stone is changed by stains, both preventive and curative measures are necessary. Therefore methods for removing the different sorts of staining on natural stone are developed. The most progressive achieved one is the destaining of Carrara-marble.

1 INTRODUCTION

1.1 *Problematic nature*

Nowadays people expect natural stone to have the same properties as ceramic tiles regarding aspect and maintenance.

Complaints concerning esthetical problems with natural stone originate from:

- a natural variation in aspect (colour, veins, texture, etc.) between the different elements of a furnishing. This very frequent issue is a question of contractual agreement between the customer and the supplier and will not be discussed here.
- changes in the aspect of the finished products after application. Sources of staining can be divided into three types. (see section 1.3)

1.2 *Selection of natural stone*

In this research, several different stone types were included to determine a general idea concerning staining. Here are only a few stones considered. These gave the most distinctive and mentionable stains.

1.3 *Type of stains*

Firstly there are stains, caused by oxidation of iron-bearing minerals present in the stone. These minerals are mainly ironsulfides (pyrite, marcassite, etc.),

Table 1. Sensitivity to stains of the tested stones.

Stone type		Stain sensitivity		
		I	II	III
LM1	German fossilerous limestone	x		x
LM2	French oolitic limestone		x	x
LM3	French oolitic-crinoidal limestone		x	x
LM4	French coquinoid limestone		x	x
LM5	Spanish fossilerous limestone	x		x
M1	Italian marble	x		x
GR1	French granodiorite	x		x
GR2	Indian granite	x		x

I stains due oxidation of ironbearing minerals (cfr. 1.3).
II stains due reaction of organic matter and alkalis (cfr. 1.3).
III external stains (cfr. 1.3).

ironcarbonates (siderite, etc.) and ferromagnesium-silicates (biotite, hornblende, etc.). They occur as accessory minerals in natural stone, mostly well scattered or in relation with present veins. The oxidized iron will be transported by water to the surface of the tile, where it is deposited as rustcoloured ironhydroxides (limonite). This kind of staining develops after the placing of the stone elements (after 8 to 12 months), except when the elements are placed close to heating- or hot water piping. (Dugniolle & Muzzin 1997) This kind of stains, not only disturb the aspect of the stone

Figure 1. Dark coloured migration front due reaction of the organic matter and alkalis originated of the mortar below.

Table 2. Amount of water, that can be absorbed by the stone, in function of mode of placing.

Mode of placing	Amount of water (L/m^2)
traditional placing (mortar bed)	5–7
placing in fresh screed	10–13
placing with adhesive mortar on hardened screed	0.7–1

but it can also bring about physical damage to natural stones. Fissures appear due to the volume expansion of oxidized and hydrated minerals (Winkler 1973).

Many of the tested stone types contain minerals which have the potential to oxidize. The LM1 contains ironsulfides visible to the naked eye. The ironsulfides consist mostly of well cubic crystallized pyrite. In M1 there are also pyrite minerals present, but as two different qualities. On the one hand, they appear as rounded to hexagonal crystals with a size varying of 8 to 16 μm dispersed in the calcite matrix. On the other hand, they appear as small rhomboidal crystals (<1 μm) dispersed in the matrix but mostly concentrated in the gray "veins" of the marble. As possible reactive minerals can sulfides and biotite be found in GR1.

Stains can also be caused by the reaction of the organic matter, present in the natural stone, and the alkalis, originated from the cement. The precipitation of organic matter occurs simultaneously with sedimentation of the stones, and is thereby trapped within the stone. The structure of the organic matter is comparable with this of humus. (Dugniolle & Muzzin 1997) Organic matter consists of substances, which are soluble in water and have their own colour (generally brown-yellow) or get a color by the reaction with alkalis. During drying of the floor, water, originated from the mortar or the screed beneath, migrates to the evaporation surface (Fig. 1). The amount of water that can migrate towards the stone is in function of the mode of placing (Table 2). The water transports the soluble fraction of the in cement present alkalis, who on their turn react with the organic substances within the stone. The coloured reaction products will be carried along to the stone surface, where they concentrate and form more or less marked and homogenous stains. These stains emerge quickly (few days or weeks) after positioning the stones.

The stones which are susceptible to this kind of staining are sedimentary limestones, like LM2, LM3 and LM4 (Table 1).

The third type of stain is the external stain. An external stain can be defined as a reaction with substances that appear at a bad moment and a bad place. Such stain as a functional system generally exists of two components. The first component attaches the stain on the surface. This can be gum or simply dust. The second and most important component is the susceptibility for stains of the floor covering is, meaning the substances which penetrate into the material. These results in discoloration, which originates in two different causes: on the one hand a chemical reaction is observed, like acids damaging marble surfaces. On the other hand a purely physical alteration of gloss is noticed, like for example the infiltration of oil in gneiss (Fahrenkrog 2002).

2 STAINTESTS

For each kind of stain, a test is developed to determine whether the natural stone is susceptible for that kind of staining. To produce stains due to oxidation of ironbearing minerals, two different tests are evolved. The accelerated oxidation tests each create stains on different types of stone, with each stone containing a different kind of sulfide. Stains due to reaction of organic matter will be made by an adapted Venuat test. The test of the external stains brings forth stains with oil, vinegar, ink, etc.

2.1 Accelerated oxidation test

2.1.1 Thermal shock test (prEN 14 066)
In this test, natural stone tiles ($20 \times 20 \times 1$–2 cm) undergo 20 cycles of 6 h emerged in a demineralized water at $20 \pm 5°C$ followed by 18 h in an oven at $105 \pm 5°C$.

Of all the stones that have been tested, oxidation stains were only obtained on the natural stones where the ironbearing minerals were visible and outcrop the surface of the tile, like LM1 (sulfides) and GR1 (sulfides

282

and biotite). Contrary, the M1 displays no stains comparable with the one observed in situ. In addition, the test proved that the finish of natural stone elements plays a role in the intensity of oxidation. For example, the GR1 stone with a polished finish shows less oxidation stains than the one with a flamed finish. The difference, regarding stains after thermal shock, between a polished and honed tile is minor.

2.1.2 Alkaline water test

As there were no stains obtained with the thermal shock test on M1, there had to be sought for another stain test, for it is known from experience that this marble can display dispersed beige-brown stains (see section 5).

The most common ironbearing mineral present in natural stone is pyrite (FeS2). It is known and proved that the rate of the oxidation of pyrite is higher in an alkaline environment than in a neutral or acid one (Brown & Jurinak 1989). Therefore another accelerated oxidation test is developed. The test consists of 20 cycles of 6 h in an alkaline solution (1 M), obtained by dissolve $NaHCO_3$ in water, followed by 18 h in an oven at $55 \pm 5°C$.

Although the test produces stains similar to the one observed in reality with M1, there are no stains on the other stone types, even not on the ones who gave good results with the first accelerated oxidation test. Consequently, this test seems to be relevant for marbles containing fine dispersed pyrite minerals.

2.2 Adapted Venuat test

The experiment that has been carried out at BBRI, is an adapted version of the Venuat test. The original test prescribes prisms of stone in a square of mortar placed in demineralized water after hardening. Through experience is known that more reliable results, regarding stains due to reaction of organic matter of the stone and alkalis originated from the mortar underneath, are obtained when whole stone tiles are embedded in mortar. The aim of this test is also to examine the influence of the cement to the oxidation of ironbearing minerals.

Each time four stone tiles (13 × 13 cm) were embedded in mortar, made of white sand (quartz) and white cement (minimal content of alkalis), with joints of about 1 cm. After 7 days of hardening, the unit was placed in a container with demineralized water (the height of the water level is equal to half the height of the unit) and in a climate test chamber at 20°C and 60% RH.

The tested stones are LM1, GR1, LM2 and M1. The first two mentioned did not produced stains due to oxidation. M1 displayed a few beige stains but not like one can observe in reality. LM2 showed a dark brown discoloration of the whole surface of the stone tile.

2.3 External staintest

In this test the resistance against external staining of the natural stone is examined, inspired from NBN EN ISO 10545-14. There were three types of stone tested, namely a compact stone with a polished finish (GR2 and LM5), a compact stone with a honed finish (LM5) and a porous stone with a honed finish (LM3).

On a natural stone tile (40 × 40 cm) stains are provoked with substances, known to leave stains behind after wiping off. The staining substances are: red wine, ink, tomato ketchup, coffee, ammonia, vinegar and peanut oil. After applying the stains, the substances were wiped off after 15 minutes and after 24 hours.

All the applied substances, except the ammonia solution, made stains on the four stone types. Due to microfissures in the granite, the solutions are able to migrate through the stone away from the applied substance. It creates large coloured stains. This is also applicable for the porous limestone LM5. A polished finish seemed to be more resistant to acid solutions than a honed finish. This is not applicable for the other solutions.

3   TREATMENT PRODUCTS

The aim of the research was to find a way to protect natural stone tiles for any kind of staining. Since each type of staining is related to the presence of water, this factor has to be excluded. To achieve this, the natural stone tiles are treated with anti-graffiti products. This group of product is subdivided into three types, based on their water-repellent action.

The porefilling products seal the surface of porous materials, which restrict the exchange of vapor and fluid between the material and the environment. As porefilling products, an acrylic solution with a base of solvent (A_{sol}) and one with a base of water (A_{aq}) were chosen. The latter is not available as pure solution. It contains also a small amount of siloxanes. Siloxanes consist of weak polymerized molecules, which react after application and form a (polymer) silicone resin.

A silane without a solvent (Sil) and a fluoridated solution (Gfl) were opted as water-repellent products. Silanes are monomeric "organic silicon molecules" with very small dimensions. After application they hydrolize and polymerize to a silicon polymer. Fluoridated solutions can have an exceptionally low surface tension and are not only water-repellent, but also oleophobic (grease-repellent).

Polyurethane with a base of solvent (Psol) and with a base of water (Paq) were used as filmforming products. These solutions form a thin layer on the surface which makes every vapour and fluid transport through the surface impossible (Technical Note BBRI, 2002).

283

3.1 Performances of the treatments

Before investigating their potential efficiency towards staining, they were tested on their performances in several domains, namely waterabsorption at the surface, aspect (colour and gloss), resistance to skid and to abrasion. These characteristics were in fact considered to be the sine qua non conditions for any surface treatment.

3.1.1 Waterabsorption at the surface

The waterabsorption was measured under low pressure (pipe method). This method is described in the international prescriptions of RILEM-25 PEM, II-4. It involves measuring the amount of water, depending on time that can penetrate into the stone surface. Because of the low porosity of the stone, the measuring time was brought to 5 and 60 minutes. This enlarged the accuracy of the measures.

The test was carried out before and after the treatment of the tiles to examine the effectiveness of the product. In practice, the efficiency, in terms of %, will be determined with aid of the next formula:

$$Efficiency\,[\%] = \left(1 - \frac{\Delta_{(60-5)after}}{\Delta_{(60-5)before}}\right) \times 100 \qquad (1)$$

where $\Delta_{(60-5)after}$ = waterabsorption (mm) after treatment between the sixtieth and fifth minute of the test; $\Delta_{(60-5)before}$ = waterabsorption (mm) before treatment between the sixtieth and fifth minute of the test.

The products were applied on eight different stones, among which two types had two different finishes. The results, given in Table 3, represent the average of the results of the samples of one stone type:

3.1.2 Aspect: colour and gloss

The colour and gloss are two important optical characteristics to demonstrate a change in the aspect of natural stone. This is of great interest because the used product has to change the aspect of the natural stone tile as little as possible.

The measurements of colour were performed by a Minolta Chroma meter. The device works with an intern light source, therefore measurements will not be influenced by external light. The Minolta Chroma converts all colours, lying in the visual field, to a numeric code. "L", "a" and "b" are used. "L" is the term to determine the lightness of a colour. "a" refers to the red and green saturation of a colour and "b" to the yellow and blue saturation. The absolute difference in colour between two measurements, the ΔE-value, is the distance between the two measurement results, namely:

$$\Delta E_{ab} = \sqrt{(\Delta L)^2 + (\Delta a)^2 + (\Delta b)^2} \qquad (2)$$

The gloss measurements were carried out with a Novo-Gloss RHOPOINT 60°. Gloss is the optical property of a material that is determined as the amount of light reflected by a surface. The gloss meter measures the percentage light that under an angle of 60° is reflected by the measured surface (ISO 2813).

After application of the products, there were no large differences in colour to observe, in contrast to the gloss. The acryl solution with a base of solvent gave clearly a shinier surface, except on the polished tiles. With the same solution but with a base of water, there was a decrease of gloss present on the polished tiles. The same is observed with the fluoridated solution. The solution of silane changed nothing referring to the gloss (or colour). After applying the solution with a base of solvent, the tiles became more mat. On the other hand the same solution with a base of water had a favourable influence on the gloss of the natural stone.

3.1.3 Resistance to slip

This physical property is measured with a pendulum tester according to prEN 14 231. At the end of the pendulum, a small sole made of standardized rubber, is attached. By the swing of the pendulum, the frictional force between the sole and the surface of the sample will be measured by the reduction in length of the swing pointing out a value on a calibrated scale.

After treatment the values of all the different stones were similar, because it reflects the slip property of the applied product. Moreover it is important to compare the measurements before and after the application. A treatment should not change the slip enormously. Applying A_{aq}, Sil and P_{aq} will make the surface of the stone more slippery and P_{sol} rougher. Sometimes the behaviour in relation to the untreated samples is dependent of the finish of the natural stone.

3.1.4 Resistance to abrasion

To determine the resistance to abrasion of the treated samples, a PEI-apparatus was used. The PEI-apparatus makes a rotating movement. This specific movement

Table 3. Waterabsorption at the surface by low pressure.

| Stone type | Efficiency | | | | | |
	A_{aq} %	A_{sol} %	Sil %	Gfl %	P_{aq} %	P_{sol} %
Marble	100	100	75	83	100	100
Compact limestone	100	50	67	100	100	100
Porous limestone	33	60	50	47	48	100
Granite	100	57	92	90	94	100

284

Table 4. Selection of products.

		Polished		Honed	
		Glue	Mortar	Glue	Mortar
Compact	Impassable	P_{aq}/A_{aq}	A_{aq}/Sil	P_{aq}/A_{aq}	A_{aq}/Sil
	Low use	–	A_{aq}/Sil	–	A_{aq}/Sil
	Moderate Use	–	Sil	–	Sil
	Intensive Use	–	Sil	–	Sil
Porous	Impassable	–	–	P_{aq}	Gfl/Sil
	Low use	–	–	–	Gfl/Sil
	Moderate Use	–	–	–	Gfl/Sil
	Intensive Use	–	–	–	Gfl/Sil

– not relevant.

Table 5. Effect of the stain removing product on GR1 and LM1.

Stain removing product	GR1	LM1
CA20	Positive	Positive[1]
CA10	Positive	Positive[1]
PH50	Positive[2]	–[3]
PH25	Positive[2]	–[3]
ESP	Negative[2]	Negative[2]
S2-5	–[4]	Positive
HS10	Positive	Satisfying[5]
En196	Positive	Positive[5]
OXM		Positive

[1] Damage to the surface of the natural stone.
[2] Growing dark of the stone on the place of the application of the phosphoric acid.
[3] Not applicable on limestone by risking calcium phosphate crystallization.
[4] Not applicable on granite by damaging silicates by fluoride.
[5] Surface turns white.

provides sanding with a standardized abrasive of the surface. In the norm NBN EN 154 the sample after abrasion is visually compared to a reference sample and normalized light. For this research a new visual comparing method and class subdivision was created: the class subdivision of the norm only count for ceramic, enameled tiles.

Four samples (10 × 10 cm) a stone type were submitted to the abrasion test. The first measurements consisted of an abrasion test of a 100 revolutions. According to that result the number of revolutions of the next abrasion test was determined to get an idea of the abrasion behaviour of the treated stone in relation to the number of revolutions.

The polyurethane products are the most and the A_{aq} treatment the less wear-resistant. Another conclusion is that the finish of the natural stone has an influence of the wear-resistance, namely the treatments on a honed surface show more resistance than on a polished finish.

3.2 Selection of treatment products

To examine the effectiveness of the treatment products, a selection of one or two products of the already tested treatments will be made in function of four parameters:

• stone type (compact or porous);
• type of finish (polished or honed);
• utilization/placing (with mortar or glue);
• application (heavy passable place, kitchen tablet, etc.).

For each of these specific cases, the following criteria regarding the performance of the products will be considered:

• performance concerning abrasion (change in appearance);
• difference in gloss before and after treatment;
• slip resistance before and after treatment;
• efficiency in relation to the waterabsorption at the surface.

With the above-mentioned criteria took into account, the following selection could be made:

3.3 Efficiency of treatment products

The selected products (see Table 4) will undergo the same staintests as the not-treated natural stone elements (see section 2), in order to determine the efficiency against the staining.

This phase of the research is still in progress, therefore there are not yet any results available.

4 DESTAINING

The curative treatment aims to remove already existing stains without a considerably change of the aspect or the physical properties of the stone.

Several products were tested on their efficiency to remove oxidation stains on LM1 with honed finish and on GR1 with flamed finish. These two stone types were chosen, because their stains due to oxidation were the most distinct and therefore two different kinds of natural stone (limestone and granodiorite) could also be examined, as several products are not compatible with each kind of stone.

All the products were first applied in a liquid state to determine the effect of removing the stains. Subsequently the products with a base of organic solvent were used as a paste to prevent an early evaporation.

285

In the table below, a summary of the different products and their effect on the two stone types is given. The tested products are citric acid 20% (CA20), citric acid 10% (CA10), phosphoric acid 50% (PH50), phosphoric acid 25% (PH25), cataplasma of "blanc d'Espagne" (composed of glycerin, sodium citrate and "blanc d'Espagne") (ESP), solution 2–5 (containing the succession of the products: ammonia bifluoride (100 g/L), ammonia dihydrofluoride (100 g/L) + 10% hydrogenfluoride, oxalic acid (saturated)) (S2-5), hyposulfite 10% (HS10), solution En196 (mixture of sodium sulfite and sodium metabisulfite) (En196) and oxalic acid with methanol (OXM).

For the removal of stains due to reaction of organic matter and alkalis, often cleaning with water is satisfying. But for difficult to clean stains, a specific enzyme gel can be used. A thick layer of the gel will be applied on the clean surface and covered up with a piece of plastic to prevent early evaporation. The longer the exposing time, the better the brown stains will be removed.

5 CASE STUDY: MARBLE OF CARRARA

Numerous staining problems regarding Carrara- marble (M1) are notified. The aspect of the appearing stains is always similar, namely dispersed beige-brown that appear across the whole surface of the natural stone or concentrated along the "veins". They show up when the elements are placed with a traditional mortar, with adhesive mortar on screed and even when placed with white mortar (low amount of alkalis).

The marble of Carrara consists of nearly 100% of calcite ($CaCO3$), but also a small amount of opaque minerals is present. These minerals are identified by ore microscopy and SEM-EDAX (scanning electron microscopy) as ironsulfides, more specific crystals of pyrite (FeS_2). Pyrite appears in two different qualities in the marble (see section 1.2). On the one hand the minerals can be found dispersed in the calcite matrix. On the other hand they are rather concentrated in the gray "veins", although this is not confirmed by SEM-analysis. The "veins" consist of calcite crystals, which are smaller than the crystals found in the rest of the stone.

The existing sulfideminerals can oxidize and cause the macroscopic visual rust coloured stains. Especially the very fine minerals are very reactive because of their large reaction surface. The oxidation of pyrite is influenced by the following parameters:

- moisture and/or alkalinity: the rate of pyrite oxidation rises within an alkaline environment. (Brown & Jurinak 1989);
- mode of placing: utilization of alkali-riche cement or to much water, to short drying time, etc.;
- the maintenance of the natural stone: utilization of alkaline cleansing agents (see further);
- environment.

The proposition of the accelerated oxidation of pyrite in an alkaline environment, is proved by the two performed staintests, namely the alkaline water test (see section 2.1.2) and the Venuat test (see section 2.2). Given that pyrite is present in Carrara-marble, a humid and basic environment is the only verifiable factor. An alkaline environment can be caused by contact with a fresh binder (mortar or screed). Other sources of humidity can be: not fully dried concrete, leaking piping, etc. An improper maintenance of the natural stone can be a cause of an alkaline environment. Mainly the cleansing agents for cleaning greasy soiling are alkaline.

Because of the frequency of the problem, a product to remove these stains has been researched. On the market there is no such product available. The tested products were an enzyme gel, a paste with base of dithionite and the same paste with afterwards a surface treatment of siloxane, ammonia bifluoride, ammonia bifluoride with a crystallization product and corrosion inhibitors. Of all these products only, the paste with base of dithionite and the surface treatment of ammonia bifluoride and crystallization product gave a satisfying result. With the other products appeared the stains after a couple of days or weeks.

6 CONCLUSION

The aim of the research is to find a preventive treatment against all sorts of stains. A selection based on several properties concerning the performance of the products has been done (see section 3.2). But before concluding which product is the most qualified as a preventive treatment, one have to wait for the final results of the tests concerning the efficiency of the chosen products (see section 3.3).

By recommending the use of the tested treatment products and the application of the products in a stone processing company, this research intends to improve the surface properties of stone and to upgrade the image of natural stone used in the building industry.

ACKNOWLEDGEMENTS

This article is related to research funded partly by the three regional Governments in Belgium through the following institutions: IWT, DGTRE, Region Brussels. Thanks also to C. Callandt, D. Nicaise, A. Pien and K. Callebaut for their support.

REFERENCES

Brown, A.D. & Jurinak, J.J. 1989. Mechanism of Pyrite Oxidation. In Aqueous Mixtures. J. Environ. Qual. (18): 545–550.

Dugniolle, E. & Muzzin, G. 1997. Pose des pierres et marbres au mortier de ciment: comment parer au tachage? BBRI-magazine, spring 1997: 19–20.

Fahrenkrog, H. 2002. Mittel zum Fleck. Fliesen und Platten (9): 64–67.

Winkler, E.M. 1973. Stone: Properties, Durability in Man's Environment. Applied Mineralogy (4): 164–167. New York: Springer-Verlag.

Technical Note BBRI june 2002. Waterwerende oppervlaktebehandeling: 9–10.

prEN 14066 oct 2002. Natural stone test methods – Determination of resistance to ageing by thermal shock.

prEN 14231 nov 2001. Natural stone test methods – Determination of the slip resistance by means the pendulum tester.

NBN EN ISO 10545-14 sept 1997. Ceramic tiles – Part 14: Determination of resistance to stains.

NBN EN 154 june 1992. Ceramic tiles – Determination of resistance to surface abrasion – glazed tiles.

Dimension Stone 2004, Přikryl (ed)
© *2004 Taylor & Francis Group, London, ISBN 90 5809 675 0*

The exploitation of syenite in the Piedmont Alps (Italy): present relevance of the stone and future technological prospects for its sustainable exploitation

M. Cardu
IGAG-CNR, Torino, Italy "Geo-resources and Land" Department, Politecnico di Torino, Italy

E. Lovera & E. Michelotti
"Geo-resources and Land" Department, Politecnico di Torino, Italy

M. Fornaro
"Earth Sciences" Department, Università degli Studi di Torino, Italy

ABSTRACT: The syenite exploited in Cervo Valley is very appreciated in Italy and abroad, for both its good litho-applicative characteristics and its use in important architectural works. Many quarries were opened in the past, but only two are in activity today. Quarrying techniques changed from a systematic use of explosive to a mixed use of diamond wire sawing and dynamic splitting; diamond wire technology is more and more used, thanks to the improving performances of diamond beads. The design of specific diamond wires and the definition of theoretical-practical service parameters are ongoing within a EU research, which involves the Politecnico di Torino and an important producer of diamond wires. Furthermore, an underground exploitation of the rock could be experimented, using the diamond technology: if underground quarrying proves to be viable some old quarries, today closed because of the very thick overburden, could be re-opened.

1 INTRODUCTION

The syenitic rocks of Cervo Valley (Biella Province, Piedmont Region, Italy), commercially known as "Balma Syenite" or "Valle Cervo Syenite" or "Balma Granite," have been exploited since the Roman Age, but a full organized quarrying activity dates to the early 1800s, when large quarries, still visible, were opened in an area locally called "Balma di Quittengo". Today only two quarries are in activity, both located in the San Paolo Cervo municipality (Fig. 1).

Thanks to its decorative aspect and good mechanical characteristics, Balma Syenite has been widely used over centuries to build or clad monuments, churches and palaces, as well to create columns, staircases and other architectural elements; moreover its wear resistance makes it particularly suitable for roads paving.

Presently, the stone quarried in Cervo Valley is surely among the most prestigious Italian ornamental stones, especially due to its exclusive aspect: the chromatic tone is grey/violet, enhanced by the good attitude to polishing, with a dark speckled texture against a grey/white background.

2 GEOLOGICAL SETTING AND CHARACTERIZATION OF THE STONE

Val Cervo quarries exploit the largest Oligocene pluton (35 km^2) of the western Italian Alps, emplaced in the eclogitic micaschists of the Sesia-Lanzo Zone (Fig. 1).

The pluton consists of monzo-granitic rocks (Granitic Complex) in the core, surrounded by a discontinuous rim of syenitic rocks (Syenitic Complex) and, finally, by a wide rim of monzonitic rocks (Monzonitic Complex).

The subdivision in Complexes, rather than in lithological units, is preferred because of the heterogeneity of rock types: for example, locally syenites grade in composition into monzonites (Bigioggero et al. 1994). The monzo-granites, characterized by a reddish colour, medium to coarse grain size and pseudo-porphyritic texture, were exploited too in the past as ornamental stone, but the Balma Syenite has always been the most important quarried stone of the Cervo Valley pluton (Sandrone et al. 2004).

Balma Syenite shows a typical grey-violet colour, a medium grain size and a well developed magmatic

289

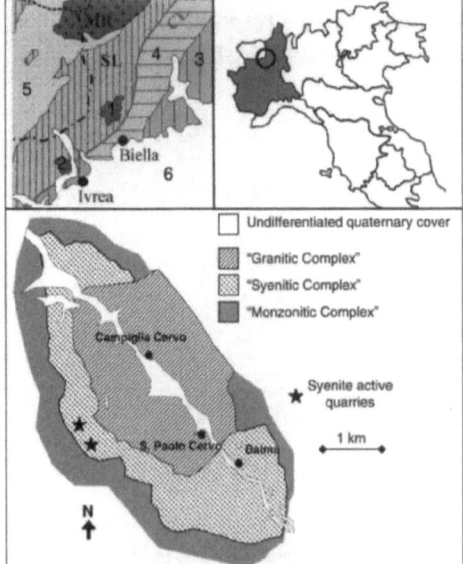

Undifferentiated quaternary cover
"Granitic Complex"
"Syenitic Complex"
"Monzonitic Complex"
Syenite active quarries
1 km

Figure 1. Above, localisation of the syenitic pluton in Piedmont and structural sketch map of the area. Legend – 1: Cervo Valley Syenite; 2: Vico and Traversella Diorite; 3: Serie dei Laghi Zone; 4: Ivrea-Verbano Zone (CL = Canavese Line); 6: undifferentiated basement and cover; SL: Sesia-Lanzo Zone; MR: Internal crystalline massifs of Monte Rosa; 5: Piemontese Zone. Below, simplified geological-petrographical sketch of the "Balma Syenite" quarrying area.

flow foliation. Its mineralogical composition is: K-feldspar (55% – orthoclase), plagioclase (15% – andesine), Mg-hornblende (18%), biotite (5%), diopside (3%) and scarce quartz (3%). Sphene is the distinctive accessory mineral (1%); the others are apatite, zircon and ores (Fiora et al. 1999).

K-feldspar, whose pink/violet colour mainly contributes to the colour of the rock, typically occurs in megacrysts with Carlsbad twinning and perthitic exsolutions. Such violet colour, which proves a durable fastness, makes this syenite a nearly unique material in the world scene, often preferred to other "granites" with similar mechanical properties.

Natural defects of the stone, which can be observed in the quarry face and in the processed material, are essentially due to clear lineation of aplite and to dark spots of femic minerals (horneblende and biotite).

These are mainly aesthetical defects but they point out plane of weakness which could tend to fracture during processing operations.

Within the commercial group of "granites", Balma Syenites has got high values of apparent density, good compressive strength, low water absorption coefficient, high Knoop hardness and fair flexural

Table 1. Technical characteristics of Balma Syenite.

Traditional denomination	Balma Syenite
Petrographic denomination	Syenite
Aspect	Grey-violet
Place of origin	S.Paolo Cervo
Apparent density	$2725\,kg/m^3$
Absorption coefficient	0.35%
Compressive strength	172 MPa
Flexural strength under concentrated load	14.7 MPa
Impact strength	5.1 J
Knoop hardness	5334

strength under concentrated load and impact strength (Tab. 1).

3 HISTORIC AND PRESENT USE OF BALMA SYENITE

Thanks to its physical properties the Syenite is well workable, it can be sculptured, and it is resistant to mechanical stress; generally it can be considered unalterable, even if used in external applications.

As said before, the colour and texture of the stone make it quite unique, at least in Europe, and very appreciated for ornamental purposes.

For all these reasons, Balma Syenite has been widely used since the 1800s in several applications. Of course, it was before used mainly in Biella province, but subsequently it spread extensively outside this area and in the 1900s it was well known in Italy and abroad. For example, among the most remarkable applications of the syenite there are the realization of pillars for the church of Notre Dame de Fourvières in Lyon (France), the columns of a palace in the Duomo Square in Milan, the columns of the "Stock Exchange" palace in Naples and the base of monuments in Milan and Rome (Fiora et al. 2000).

Especially in Turin, the first capital of the former Italian Kingdom, many notable uses of the syenite can be found: the votive column supporting the statue of the Madonna, in front of the "Consolata's Sanctuary", is made by a big monolith of syenite (Fig. 2), as well as the colonnades of important streets and many architectural elements, cladding and paving of historical public and private buildings are made with this stone.

Road paving and, more recently, urban furnishing are another important field of application for syenite.

Its wear resistance makes it suitable to create paving tiles or "Belgian blocks", which have been widely used in many roads and squares all around Piedmont Region.

A significant recent work has been the paving and furnishing of the large area in front of the Basilica of Superga, near Turin, where flame treated slabs of Syenite have been placed, together with benches and small decorative columns (Fig. 2).

290

Figure 2. On the left, the votive column of "Consolata" in Turin, built up with a monolith of syenite. On the right, the pavement made of Balma Syenite slabs in front of the Basilica of Superga, near Turin.

Figure 3. An application of Balma Syenite in a recent work of urban renewal in the centre of Turin.

Currently, in the context of the general urban renewal carried out by the Turin Municipality in view of the Winter Olympic Games of 2006, Balma Syenite features among the quality stone materials more used: the main squares of the city and some streets of the historical centre see the application, in different shapes and sizes, of syenite elements, often researching innovative application of a traditional material (Fig. 3).

Table 2. Yearly production data of the Balma Syenite quarrying basin (October 2003).

Excavated material	$5000\,m^3$
Dimensional blocks	$1000\,m^3$
Little blocks and kerb stones	$1000\,m^3$
Natural split crafting	$100\,m^3$
Armour stones and rip rap	$2500\,m^3$
Debris for crushing	$400\,m^3$
Waste	$0\,m^3$

4 PRESENT QUARRYING SITUATION AND PROSPECTS

The exploitation of Balma Syenite at present involves two quarries: the "Vej della Balma" quarry (reactivated in the year 2003) and the "Colombari" quarry, both located on the right side of the Cervo stream, in the San Paolo Cervo Municipality.

Actually the quarries are operated by the same enterprise, with a crew of 4–5 workers and the following machineries: 1 diamond wire cutter, 1 fix diamond mono-wire saw, 2 tracked excavators, 1 tracked loader, 1 wheeled loader, 3 derrick cranes, compressors and drillers.

Crew and machinery work alternatively in the two yards.

The production data of the quarrying basin are reported in the Table 2. It is not excluded by the quarry owner to increase the number of workers and machineries in a next future, in order to parally operate in the two quarries: in this case the production of the basin could virtually be doubled.

The estimated yearly production of the two quarries is about $2000\,m^3$ of "workable stone".

All the rest of the excavated material, about $3000\,m^3$, is used as rip rap or armour stones when the volume of the blocks is big enough ($>1\,m^3$), otherwise it is destined to crushing plants.

The "Vej della Balma" quarry (Fig. 4) is located at 1200 m above the see level, while the "Colombari" quarry (Fig. 5) is at 1350 m; the sites are reached through private ramps which depart from a steep and narrow public road.

Both quarries are open cast operated and exploit the syenitic pluton with a typical alpine hillside configuration: high vertical fronts (up to 100 m height in the "Colombari" quarry) and quite restricted horizontal spaces.

The deposit is exploited through a sequence of "vertical slices", which are progressively excavated by "descending" large banks ($1000–1500\,m^3$), usually with height of about 10 m, length of 20–25 m and width of 5–10 m. There is only one productive level, hence creating a single, very high step.

Bank detachment is performed by a mixed technology: diamond wire and explosive. Two vertical side

291

Figure 4. The "Vej della Balma" quarry. In the foreground it is visible part of the heap of big shapeless blocks selected to be used as armour stone.

Figure 6. The "old" dump below the "Colombari" quarry. Some trees naturally grew, but the big size of the stone wastes does not allow the creation of a stable coverage.

Figure 5. The "Colombari" quarry. The wedge-shaped configuration of the quarry face can be observed. Further open cast exploitation will be difficult because of the present morphology, given by a quarrying method, not sustainable in a long-term.

cuts are first made by diamond wire and then the bench is definitively split by explosive (about 150–180 kg of black powder, in different charges).

Sometimes detaching operations take advantage of natural preferential splitting surfaces, allowing a reduced use of explosive (<100 kg in a single blasting round).

The blasted volume is overturned on a suitable debris "bed" and lastly divided into commercial blocks using splitting wedges inserted into closely drilled holes. A derrick crane is installed in each quarry and it is used to handle the quarried materials and the machineries.

The selection of non-workable stone is directly made in the quarry floor: according to volume and shape, one wheeled loader heaps the material in a lateral part of the floor, ready to be loaded on trucks and taken away.

Such preliminary selection is essential in order to reach a rational management of stone wastes; if wastes were destined to dumps, it would be nearly impossible, for both economical and technical reasons, to recover them for commercial uses.

The overburden of the exploitable rock body is not relevant, and the altered and fractured stone excavated in the cortical part of the deposit is used to create ramps. Today no wastes are produced and this is really a great result because waste disposal was one of the main issues for these quarries.

The large "old" dumps (400,000–500,000 m³ of stone debris is estimated), created just tipping down the material along the slopes below the quarry floors, caused hydrological problems compromising the free-flow of the streams below (Fig. 6). The environmental rehabilitation of the old dumps is now needed, firstly restoring the natural flow of the streams. In this occasion, larger blocks can be recovered and used for the hydraulic work, while other stone material may be removed and exploited, before the final replanting of trees.

As far as technological development is concerned, the introduction of diamond wire cutters, still combined to explosive, allowed an increase of dimensional block yield and an improvement of health and safety condition, with less dust and noises and a better control of the final result of the bank splitting.

On the other hand, diamond wire technology involves higher operative costs, essentially due to the cost of the wire itself. It is therefore important to optimise the productive performance and the service life

292

of the wire in order to extend its applications (Mancini et al. 2003). Anyway, in the next future, diamond wire will not likely exclude the use of explosive in syenite quarries, especially where fractured portions of rock body are concerned. The combined and flexible use of the two technologies, in relation to the local characteristics of the stone to be quarried, will probably be best choice.

As regards quarrying methods, the creation of higher and higher vertical faces, without horizontal interruptions, probably will be no more sustainable, for both stability and environmental reasons. Hence, a further lowering of the present quarry floor level is not practicable. If the prospection is positive, a new quarry front could be opened starting the exploitation from the upper part of the authorized area and then descending again toward the lower level through the excavation of regular benches. Adopting as much as possible the diamond wire technology, a stepped configuration should be leaved.

A more prospective option is given by underground exploitation. So far, underground quarrying has been positively adopted only for "weak rocks", such as marbles, limestones and slates, while for "hard rocks" some technological improvements are still needed.

On the one hand, underground option has to face different onerous aspects of the creation and control of the structures: rock mechanics studies and verifications, monitoring and intervention in course of works are obviously demanded.

Higher costs in the early stages of the opening should be evaluated, considering also that a part of potentially exploitable material has to be leaved in pillars.

On the other hand, if the general condition of the rock mass are suitable, a selective underground exploitation allows to quarry mostly the soundest materials, with an higher overall blocks yield. Of course, less impacts on the landscape and local environment are other important positive aspects.

In particular, a further surface activity in the "Colombari" quarry will have to face many difficulties, due to the morphologic situation and to the occurrence of thick non-exploitable cap rocks covering the workable portion.

Given the uniqueness of this litho-type, an experimental underground exploitation is probably worthy.

The experimentation, as already done in other cases (e.g. for the green quartzite commercially known as "Verde Spluga" – Sondrio, Italy), has to rely on suitable cutting technologies, mainly the diamond wire.

First, a blind drift has to be driven, with a height of 6 m, a width of 5 m and a length of 5 m; then the excavation may advance with heading, lowering and widening of the first tunnel (Fig. 7). The technique so

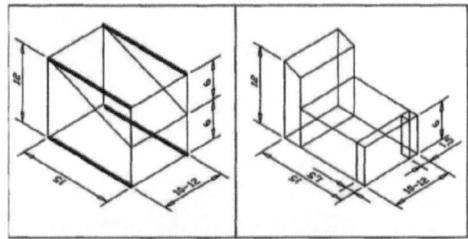

Figure 7. Sketch of the possible phases of the experimental underground exploitation (the 4 horizontal drills with big diameter are represented by thicker lines). Being impossible to make the back cut with the wire, it is necessary to extract a first wedge of rock, splitting it by explosives (on the left). Then the lower prism is quarried like in an open cast step, detaching a sequence of slices: side cuts are made by diamond wire and back and horizontal cuts by splitting.

far adoptable involves the drilling of 4 holes (diameter 255 mm), at the edges of the rock volume to be extracted: large diameter holes are necessary to hold the "down-the-hole pulleys" needed to realize blind cuts by diamond wire in "reverse catenary" arrangement (Fornaro & Lovera 2004).

Apart the hardness of the stone, the diamond wire has to work in hard conditions (stresses and wear) and therefore it is necessary to get wires made "to measure" on syenite and to adopt machines guided by suitable operative parameters (wire speed, wire pressure, water flow).

Diamond wire characteristics and operative parameters condition both safety and productivity (to say cost) of the system: only an optimal solution will make underground exploitation competitive with traditional open cast quarrying (see Chapter 5).

5 THE EXPERIMENTAL RESEARCH ON DIAMOND WIRES

An experimental research aimed to the improvement of the block cutting technique through the scientific analysis of the variables involved is the subject of an EC sponsored research program (EI 2280 "Innovative Stone Process Ecological Cutting").

Parameters of a small scale cutting operation, on a rock sample, are provided by an instrumented, laboratory size, machine, shown in Figure 8.

The machine can perform cuts at different, and measured in real time, values of the wire speed, wire pressure, water flow, and provide a continuous record of the electric power consumed, of the torque of the driving wheel, of the cut progression, of the tensile stress in the wire and of the cutting path; moreover, cuttings produced can be collected for further analysis.

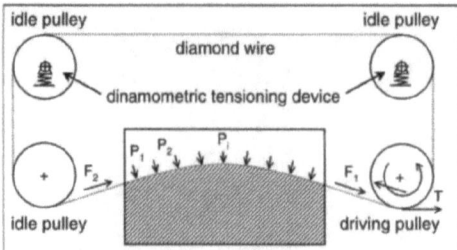

Figure 8.   Experimental wire saw machine.

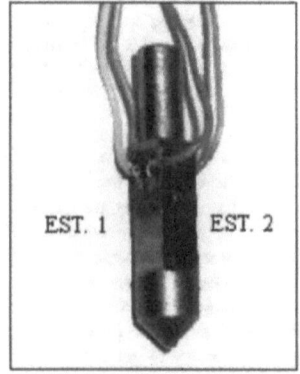

Figure 9.   Instrumented diamond holder.

Thanks to this machine, different wires and operating parameters can be reliably compared on samples of rock of decimetric size.

At a smaller scale, tests on centimetric size of the same rock are performed with an instrumented, single diamond tool, installed on a Leitz micro-durimeter, which provides also the conventional Knoop micro-hardness characterization of the material. The instrumented tool (Fig. 9) is a synthetic diamond of the 40/50 size (the same used in the wire beads) mounted on a special holder, to which strain-gauges are applied.

By this device, scratches can be made on the surface of the rock samples under selected average pressure values, recording continuously the tangential force and the geometry of the scratch obtained.

The aim of the tests is to develop a laboratory testing methodology providing rock characterization data that can be used to predict machine performance data, reproducing (within certain limits) the actual cutting mechanism and, at a more advanced stage, to support a model of the action of the bead.

Up to now the tangential and normal forces on a number of rock types have been measured; the correlation is linear, but not directly proportional (Montaldo 2003).

Specific tests on Balma Syenite are going on, and first results are promising for the foresee industrial application.

6   CONCLUSIONS

Balma Syenite, exploited from the pluton of Biella, is an appreciated ornamental stone, both for its unique decorative characteristics and its good mechanical properties.

It has been widely used in Italy in different applications: to clad monuments, churches and palaces, to build columns, staircases and other architectural elements, to pave roads and to make urban furnishing. Today, in the general urban renewal project of Turin city, seat of the 2006 Winter Olympic Games, syenite features among the quality stone materials more used.

In general, natural stone is a part of our historic-artistic heritage, as its use as a construction material still characterizes and adorns most of the public and private places throughout European cities.

It is therefore important to support the sustainable exploitation of traditional stones, whose application may represent a material and cultural link between past and future generations.

The exploitation of Balma Syenite at present involves two quarries, both located in the San Paolo Cervo Municipality, at about 1300 m above the sea level.

The total production of workable stone is approximately 2000 m³ per year, considering squared blocks to be cut in slabs, kerb stones and natural split crafting. The rest of the excavated material (about 3000 m³ per year) is no more destined to dumps, but it is selected on the quarry floor and then destined to other uses. Such management of quarry waste is essential to achieve an effective sustainability of quarrying activities.

On the other hand, technical and technological development have to supply the possibility of working safely and with high productive rates.

Balma Syenite is quarried using diamond wire cutters and explosive. In order to get higher block yield, the use of diamond wire should be extended, and so it is necessary to rely on wires which are able to cut this stone with excellent performances (service life and cutting speed).

A EU research program is ongoing with the aim of designing specific diamond wires and defining theoretical-practical service parameters: first results on syenite samples are promising for a future industrial application.

Diamond wires made "to measure" on syenite are furthermore needed to experiment an underground exploitation.

The present morphological situation of quarry faces is not favourable for a rational and safe long-term production and so the underground option should be tested, given the value and uniqueness of the stone.

Underground exploitation, if the experimental stage succeeds, will probably open the way to a sustainable future for the ancient quarrying basin of Balma Syenite.

REFERENCES

Bigioggero, B., Colombo, A., Del Moro, A., Gregnanin, A., Macera, P. & Tunesi, A. 1994. The Oligocene Valle del Cervo pluton: an example of shoshonitic magmatism in the Western Italian Alps. *Mem. Sci. Geol.* 46: 409–421.

Fiora, L., Fornaro, M., Manfredotti, L. & Marini, P. 1999. La sienite nel bacino del Cervo. *L'Informatore del Marmista*, 455: 27–34 and 456: 47–56.

Fiora, L., Fornaro, M. & Manfredotti, L. 2000. Impiego della sienite piemontese nell'arredo urbano: tradizione, attualità, prospettive. In G. Calvi & U. Zezza (eds), *Quarry-Laboratory-Monument* 1: 299–308.

Fornaro, M. & Lovera, E. 2004. Geological-technical and geo-engineering aspects of dimensional stone underground quarrying. In *Proc. Int. Conf. EurEnGeo 2004* (in press).

Mancini, R., Cardu, M., Lovera, E. & Michelotti, E. 2003. Diamond wire saws for different applications: require-ments and performances. In E. Yuzer, H. Ergin & A. Tugrul (eds), *Proc. Int. Symposium "Industrial Minerals and Building Stones" (IMBS 2003), 15–18 Sep. 2003, Istanbul, Turkey*: 167–174.

Montaldo, G. 2003. *Determinazione sperimentale dell' energia specifica di incisione e delle relazioni carico/ passata di utensili diamantati in diverse rocce*. Degree Thesis, DIGET, Politecnico di Torino.

Sandrone, R., Colombo, A., Fiora, L., Fornaro, M., Lovera, E., Tunesi, A. & Cavallo, A. 2004. Contemporary natural stones from the Italian Western Alps (Piedmont and Aosta Valley Regions). In B. Messiga, M. Franzini & G.M. Bargossi (eds), *Geomaterials: from science to applications* (in press).

Dimension Stone 2004, Přikryl (ed)
© 2004 Taylor & Francis Group, London, ISBN 90 5809 675 0

Application of schist waste for ceramics tiles

L. Catarino & M.T. Vieira
ICEMS, Mechanical Department, FCTUC, Polo, COIMBRA, Portugal

ABSTRACT: The samples of schistose rock were collected in three different geological units in north and central Portugal (Canelas, Marão and Piodão). The main mineralogical compositions of all schist are chlorite, white mica (muscovite and sericite) and quartz evaluated by optical microscopy of thin sections. A comparative study of the thermal evolution of the mineral components in the different schist and the physical and mechanical behaviour of these complex materials, are the aim of this study. It could contribute to establish the limits of mineralogical composition which led to efficient recovery schist wastes. Thermal cycle appropriated for tiles manufacture is also defined is close to that used for clays. The present study reveals that schist whose grade of metamorphism corresponds to chlorite can be used as raw material by the ceramics industry. The different accessory minerals have a minor role in the definition of thermal cycle used in ceramics production.

1 INTRODUCTION

The manufacture of flooring and roofing tiles from schistose rocks has created thousands of tons of waste material. This gives rise to numerous problems concerning economic aspects of production (sometimes lose of over 70% of raw material) and the environmental aspects, i.e. thousands of tons of waste usually end up in rivers or become part of the landscape as mountains of slurry.

The stone waste, although composed of the same material substances as the solid rock from which is derived, is not in the same physical condition because it has been disaggregated, mixed and moved to a different place. The surface area will be increased by the reduction of particle size and consequent increase of volume.

These wastes can be regarded as inert or non-hazardous and remain within the confines of the extraction site. The essential difference between them is the particle size. The range varies from blocks (derived from the extraction itself), small blocks, (produced by the grinding and finishing processes), to particles of small sizes, which can be just few microns, in the form of mud (resulting from the cutting and polishing in the finishing operations).

Using these wastes as raw material to manufacture ceramic tiles is a practical alternative that can consume local materials and create new jobs for interior villages where clay is not often an available raw material.

Schist with high argillaceous content, like shale, is often used as raw material in ceramics as it is similar

and more stiffness than clays. For this application where wet moulding is necessary the addition of water and their homogenisation yield lead to a good mixture (Loyola & Siedlecki 1993). The use of shale to obtain lightweight aggregates, because it expands when heated, is also well known (Burnett 1965, Bush 1973).

Brazil tried the use of bituminous schist wastes as a raw material for the ceramic industry, but some problems were found in wet moulding, due to the fragility of green bodies. Semi-dry pressing improved the results but, to obtain products without flaws, sintering must be performed under a very low heating rate to allow the escape of gases formed by the burning of organic matter (Fonseca 1981). The use of bituminous schist in bricks production was also tested. In this case the high plasticity, the easy extrusion and mechanical strength of green bodies was attributed to the content of organic matter (14%) still present in schist. The mechanical properties and the shrinkage of these products were within the limits imposed by Brazilian norms (Souza & Santos 1981). Other uses of this schist include ceramic-glass with good chemical and mechanical properties. This is produced by means of thermal treatment after melting, with the addition of a silica–carbonate rock at 1450°C (Fonseca & Santos 1991)

The use of waste slate powder as raw material for ceramic tiles application (dry process) has already been tested (Vieira et al. 1999, Catarino et al. 2003, Campos et al. 2004). The authors concluded that ceramic tiles manufactured from slate wastes give rise to products with better properties (water absorption,

297

resistance to abrasion and erosion wear and flexural strength) than traditional ceramic tiles.

However, other varieties of schist have not yet been studied for use in ceramics. This work describes the thermal evolution of the mineral components of schists from the different geological units and characterises the physical and mechanical behaviour of these complex materials, not particularly because of their high content in waste materials, but in order to extend the method applied to slate.

2 EXPERIMENTAL

The chemical composition of the samples was analysed using atomic absorption spectroscopy equipment, a UV/vis spectrophotometer and electron microprobe Cameca SX50 (EDAX). The critical temperatures of phase transformation or chemical evolution into schist powders were determined using differential scanning calorimetry (DSC) simultaneously with thermogravimetric analysis (TG) in Stanton Redcroft PL-STA 1500 equipment. Optical microscopy and X-ray diffraction were the techniques utilised to evaluate mineralogical composition. A Philips diffractometer Xpert with a cobalt anticathode were used. The ICDC patterns let to identify the phases present in the schist.

The samples were milled and sieved in order to obtain particles with particle size less than 100 μm. Real density of rock powders was measured through a helium picnometer. After sieving the samples powders were spray dried, at 8% humidity, uniaxially pressed in a metallic mould with dimensions 100 $\times$ 50 $\times$ 5 mm and sintered, as summarised in Figure 1.

The materials for flooring and roofing tiles must, after processing, exhibit certain properties, like low percentage of porous, low water absorption, flexural strength higher than 20 MPa and controlled shrinkage, lower than 10%. Apparent density was measured using the Archimedes method. Water absorption was evaluated by using the standard ISO 10545-3: 1995.

The final products were mechanically tested to determine the modulus of transverse rupture and flexural strength using a three-point bending test, according to ISO 10545-4: 1994. The value of rupture strength, and the Weibull modulus was calculated through the method described in standard ASTM C 1239-00.

3 RESULTS AND DISCUSSION

3.1 *Characterization of raw material*

The samples of schistose rock were collected in three different places in north and centre Portugal but they belong to different geological units. The schists from Canelas and Marão are from the Ordovician period (Llandeilian) and fit in equivalent lithostratigrafic formations. The Piodão sample was a Pre-Ordovician rock. The selection allowed enhancing the role of similar overall composition but different ages and accessory minerals, in the phase transformations or chemical reactions, during the heat treatment.

The main mineralogical compositions of all the schist samples were determined by optical microscopy of thin sections, and are: chlorite, white mica (muscovite and sericite) and quartz. However, as referred the accessory minerals are not the same (Table 1). The quantity of some accessory minerals observed under optical microscopy is less than 5% and are not discernable by X-ray diffraction. Concerning the mineral chlorite some differences between the schists studied has been detected: the chlorite present in Canelas and Piodão samples are a Mg-rich and the Marão sample is Fe-rich.

The chemical composition of schist samples determined by atomic absorption spectroscopy is mainly SiO_2, Al_2O_3 and FeO with small amounts of other oxides (Table 2).

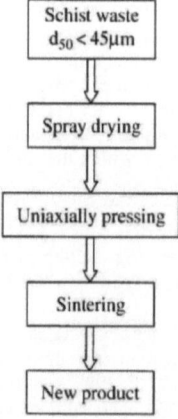

Figure 1.   Scheme for processing used.

Schist waste
$d_{50} < 45\mu$m

Spray drying

Uniaxially pressing

Sintering

New product

Table 1.   Accessory minerals identified in schist samples by optical microscopy.

Canelas	Marão	Piodão
Chloritoid	--	--
Tourmaline	Tourmaline	Tourmaline
Pyrite	--	Pyrite
--	Ilmenite	--
--	--	Albite
--	Magnetite	--
--	--	Andalusite

298

The median (d_{50}) particle size of the powder of all samples is from 7 to 10 μm, evaluated by laser diffraction particle size analysis (Table 3).

The real density for all samples is very similar, with the values of 2.87, 2.82 and 2.80 for Canelas, Marão and Piodão, respectively.

3.2 Evolution with temperature

The mineralogical composition of schist powders means that their evolution with temperature leads to phase transformations that can control the performance of ceramic products. Figure 2 represents the DSC/TG curves of the Piodão sample resulting from

a heating rate of $10°C.min^{-1}$. In order to evaluate intermediate transformations, powders of all samples were heated up to 600, 800 and 950°C and analysed by X-ray diffraction (Fig. 3).

The main minerals present that contribute to the phase and chemical evolution, on heating, are quartz, muscovite and chlorite.

Quartz α transforms to quartz β at a temperature of 570°C, but the reversibility of this transformation after cooling, means that it is not perceptible.

Muscovite dehydroxylates at temperatures higher than 800°C but is generally present up to 1000°C

Table 2. Chemical composition of schist samples.

wt%	Canelas	Marão	Piodão
SiO$_2$	53.27	53.89	59.01
Al$_2$O$_3$	24.45	23.78	20.01
FeO	7.21	7.82	5.78
Fe$_2$O$_3$	2.25	1.58	1.83
Others*	7.69	8.41	9.36
Loss of ignition	5.13	4.58	4.00
Carbon**	1.2	2.1	1.4

* K$_2$O, Na$_2$O, CaO, MgO, MnO, TiO$_2$, P$_2$O$_5$.
** Determined by EPMA.

Table 3. Particle size parameters exhibited by powder samples.

	d_{10}(μm)	d_{50}(μm)	d_{90}(μm)	Mode
Canelas	1.71	7.88	27.21	9.93
Marão	1.77	8.20	24.49	10.88
Piodão	1.78	7.32	20.57	9.93

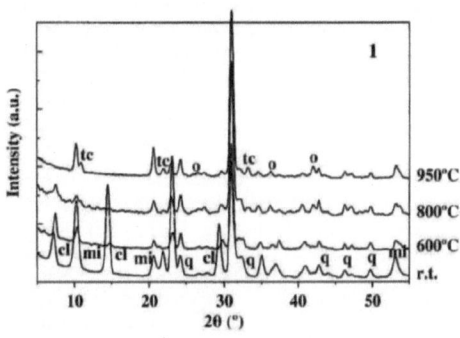

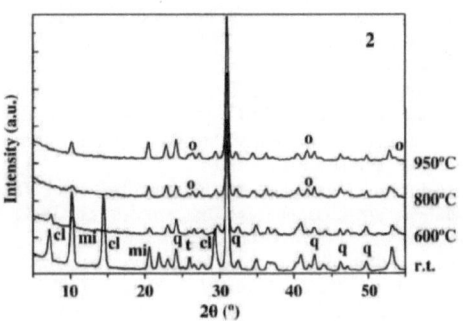

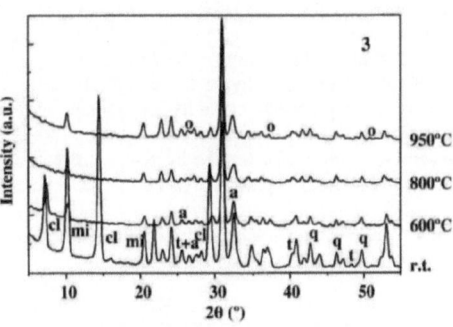

Figure 3. X-ray diffractograms of powder samples after heat treatment (1-Canelas, 2-Marão, 3-Piodão; cl-chlorite, mi-mica, q-quartz, t-tourmaline, tc-talc, o-olivine, a-albite).

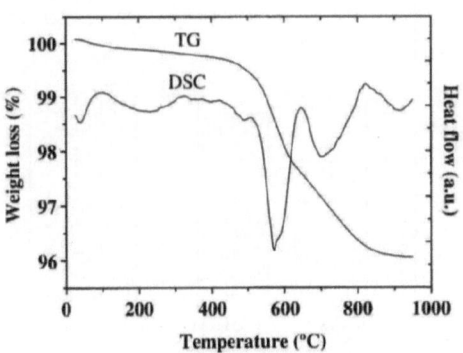

Figure 2. DSC/TG of Piodão powder sample.

299

(Yoder & Eugster 1955, Guggenheim et al. 1987, Mazzucato et al. 1999) and for higher temperatures it evolves in mullite and glassy phase.

Chlorite dehydroxylate on heating and also suffers a subsequent transformations that evolves to intermediate phases, before stabilize. This behaviour causes troubleshoots in the thermal cycle of sintering for temperatures below 1000°C. According to many researchers (Brindley & Ali 1950, Phillips 1963, Shirozu 1980, Bai et al. 1993, Villieras et al. 1994, Zhan & Guggenheim 1995, Barlow et al. 1997) at a temperature of 600°C dehydroxylation begins with the loss of OH groups, with the disappearance of most peaks in X-ray diffraction. In some cases the increase of the peak at d = 0.14 nm indicates dehydroxylate chlorite.

In the present study the first stage dehydroxylation temperature of chlorite is complete at 600°C. The decrease in the final temperature of dehydroxylation must be due to the presence of accessory minerals and the quantity of organic matter.

The second stage of dehydroxylation occurs at temperatures close to 800°C, removing the OH residual groups, while a talc phase ($Mg_3 Si_4O_{10}$) can also be present simultaneously with the last evidence of chlorite. After dehydroxylation, a new phase is formed, identified as olivine ($(Mg,Fe)SiO_4$ (Fig. 3). In Marão sample the olivine present is a Fe-rich olivine and in other samples is an Mg-rich olivine due to original composition of chlorite.

The presence of talc was only identified in the Canelas samples nevertheless olivine is always present at temperatures up to 950°C. Albite is another minor phase in the Piodão sample, but it is stable for temperatures up to 1000°C.

Chloritoid, tourmaline, pyrite, ilmenite, magnetite andalusite are minor phases in the samples, remaining unaltered up to 600°C, but at higher temperatures they were in such small percentage that the products of their transformation were not possible to be identified.

In Table 4 are summarized the start and finish temperatures of phase transformations that occur in different types of schist up to 1000°C. It must be enhanced that in Marão sample the 2nd dehydroxylation of chlorite and the olivine formation takes place simultaneously probably due to its iron content.

After uniaxial pressing and sintering at 1150°C, the samples were characterised by X-ray diffraction in order to evaluate the phases present. Figure 4 shows

Table 4. Temperatures of start and finish of phase transformations in different types of schist. (°C).

Phase transformation	Canelas	Marão	Piodão
1st dehydrox. chlorite	490–580	425–600	507–630
2nd dehydrox. chlorite	580–810	710–915	640–805
Olivine formation	810–850	–	805–860

that quartz remains from the original minerals, in all samples. All the phases evolves for high temperatures stable phases like mullite ($3(Al,Fe)_2.2SiO_2$), hercynite (($(Fe,Mg)Al_2O_4$) and hematite (Fe_2O_3).

A detailed analysis of the diffractograms show a large peak between 20 and 40°, that corresponds to a glassy phase, resulting from the phase evolution with temperature.

The main accessory minerals do not have much influence on the final product because the final phases after sintering are similar, but in different amounts. The presence of hematite is greater in the Canelas and Marão samples than in Piodão, as a result of a high FeO content, and Piodão exhibits a higher content of hercynite.

3.3 Characterisation of products

The values obtain for shrinkage in the three axes directions are presented in Figure 5. The values obtain

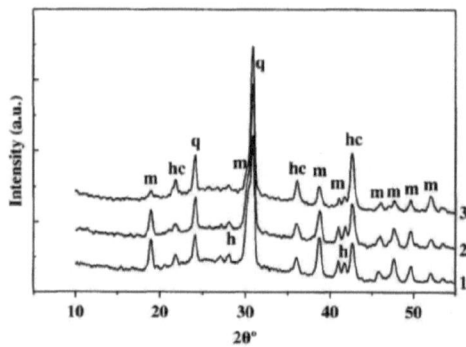

Figure 4. X-ray diffractograms of powder samples after sintering (1-Canelas, 2-Marão, 3-Piodão; m-mullite, h-hematite, hc-hercynite, q-quartz).

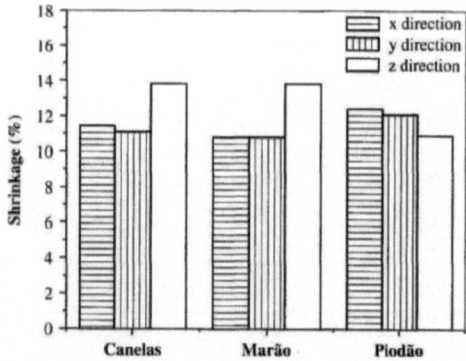

Figure 5. Shrinkage of samples after sintering in three principal directions.

300

in x and y directions are very similar corresponding to linear contraction in values between 10.8 and 12.3%. The retraction corresponding to z direction is higher in Canelas and Marão than in Piodão. This can be attributed to particle shape; Piodão exhibits a low tendency for planar schistosity, against the others places where a tabular shape is evident. Although the values obtained are light higher than those recommended for ceramics, this does not exclude their use in this industry, if the other properties present suitable values.

The apparent density determined for all shiest types is similar, with mean values between 2.54 and 2.61 (Fig. 6). These values are lower than real density of powders, but the schist evolves with temperature like slate (Vieira et al. 1999) and these values are similar to the real density of sintered powders.

The water absorption of sintered samples exhibit mean values below 0.8%, particularly in the Canelas and Piodão products, where the result is 0.3% and the

dispersion values is small (Fig. 7). This is due to the presence of the glassy phase. It must be considered that this is not due to occlusion in exterior pores but corresponds to the whole sample, because the values obtained intentionally include those for broken pieces.

The sintered schist powders present high values for flexural strength, similar to the slates treated at similar

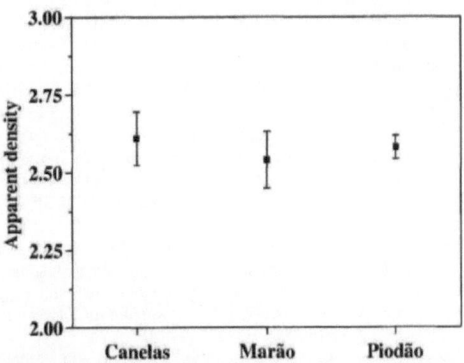

Figure 6. Apparent density of samples after sintering (mean and standard deviation).

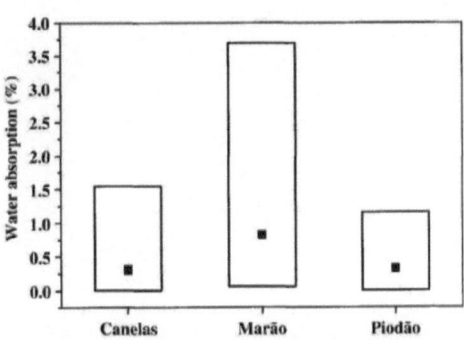

Figure 7. Water absorption of samples after sintering (minimum, mean and maximum values).

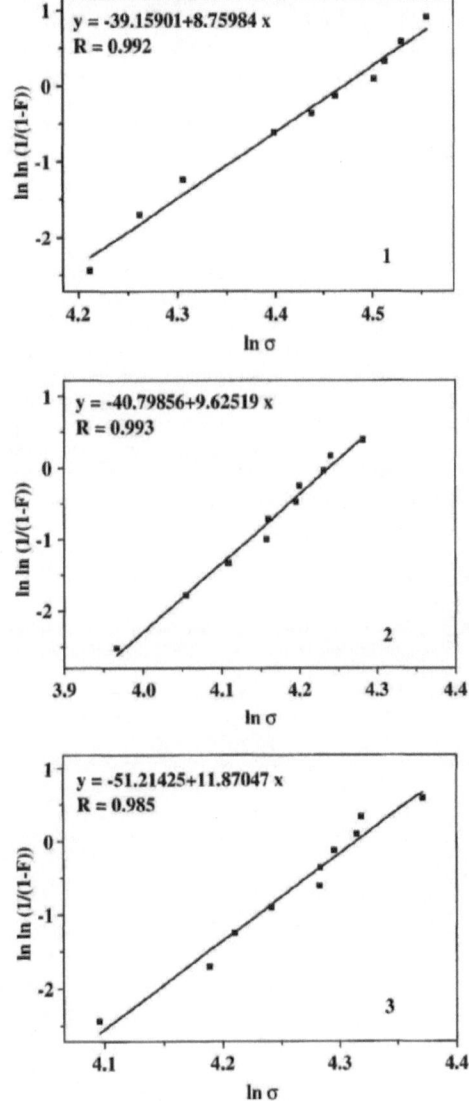

Figure 8. Estimation of Wiebull modulus (1-Canelas, 2-Marão, 3-Piodão).

301

Table 5. Flexural strength and Weibull modulus of sintered samples (T = 1150°C).

	σ_R(MPa)	m
Canelas	83	9
Marão	70	10
Piodão	75	10

temperatures. This behaviour is due to the presence of important quantity of mullite as in slates, than the present in products based on clays. The values of Weibull module are also higher than those resulting from traditional tiles.

The results of flexural strength are present in Figure 8 and Table 5.

4 CONCLUSION

This work shows that all the schist with a metamorphism grade corresponding to chlorite, muscovite and quartz, whatever the accessory minerals present can be used as a raw material to produce ceramic tiles. The thermal cycle that is recommended for firing is close to those used when clays/slates are the raw material. Small deviations of thermal cycle in according with differences between start and finish temperatures and values of the heat involve in phase transformations during the sintering must be considered.

REFERENCES

ASTM C-1239-00. Standard practice for reporting uniaxial strength data and estimating Weibull distribution parameters for advanced ceramics.

Burnet, J.L. 1965. Expansible shale resources of the San Jose-Gilroy area, California. California Division of Mines and Geology. Special Report 87, 32 pp.

Bush, A.L. 1973. Lightweight aggregates. U. S. Geological Survey Professional Paper 820: 333–355.

Campos, M., Velasco, F., Martínez, M. A. & Torralba, J.M. 2004. Recovered slate waste as raw material for manufacturing sintered structural tiles. Journal of the European Ceramic Society 24(5): 811–819.

Catarino, L., Sousa, J., Martins, I.M., Vieira, M.T. & Oliveira, M.M. 2003. Ceramic products obtained from rock wastes. Journal of Materials Processing Technology 143–144: 843–845.

Fonseca, M.C. 1981. Utilização do resíduo de pirólise do xisto do Irati; aplicações cerâmicas e recuperação de enxofre. Anais do Simpósio sobre Aproveitamento do Xisto, Academia de Ciências do Estado de São Paulo, Brasil, Publicação ACIESP 29: 17–27.

Fonseca, M.V.A. & Santos, P.S. 1991. Caracterização e desempenho de vidro e vitro-cerâmica obtidos a partir da reciclagem de rejeitos sólidos da industrialização do xisto. Anais do 35° Congresso Brasileiro de Cerâmica, Belo Horizonte, Brasil, 748–757.

ISO 10545-3: 1994 Ceramic Tiles – Part 3: Determination of water absorption, apparent porosity, apparent relative density and bulk density.

ISO 10545-4: 1994 Ceramic Tiles – Part 4: Determination of modulus of rupture and breaking strength.

Loyola, L.C. & Siedlecki, K.N. 1993. The shale of Campo do Tenente formation and its use in ceramic industry. Anais do 37ª Congresso Brasileiro de Ceramica, Curitiba, Brasil, 432–439.

Souza, J.V. & Santos, P.S. 1981. Pesquisas sobre a possibilidade de uso do xisto oleífero do município de Pindamonhangaba, Estado de São Paulo, na fabricação de materiais cerâmicos de construção. Anais do Simpósio sobre Aproveitamento do Xisto, Academia de Ciências do Estado de São Paulo, Brasil, Publicação ACIESP 29, 112–126.

Vieira, M.T., Catarino, L., Oliveira, M., Sousa, J., Torralba, J.M., Cambronero, L.E.G., González-Mesones, F.L., & Victoria, A. 1999. Optimization of sintering process of raw material wastes. Journal of Materials Processing Technology 92–93: 97–101.

Yoder, H.S. & Eugster, H.P. 1955. Synthetic and natural muscovites. Geochimica et Cosmochimica Acta 8: 225–280.

Zhan, W. & Guggenheim, S. 1995. The dehydroxylation of chlorite and the formation of topotactic product phases. Clays and Clay Minerals 43(5): 622–629.

Dimension Stone 2004, Přikryl (ed)
© 2004 Taylor & Francis Group, London, ISBN 90 5809 675 0

Water jets in dimension stone cutting and surface treatment

J. Foldyna, P. Martinec & L. Sitek
Institute of Geonics, Academy of Sciences of the Czech Republic, Ostrava-Poruba, Czech Republic

ABSTRACT: Traditional methods of dimension stone quarrying and working are becoming inadequate due to their proneness to cause negative effects regarding both the stone recovery and the quality of the material produced. Water jet technology proposes certain advantages over traditional methods of dimension stone quarrying and working. However, the performance of water jet techniques is not always competitive with existing conventional mechanical systems yet. Therefore, extending the use of water jets to the area of dimension stone quarrying and working requires significant improvement in their performance. One of possible approaches to increase the performance of water jets is pulsing the jet. A particular method of generating pulsed jets represents modulating a continuous stream of water. In this paper, possibility of utilization of water jetting technology in dimension stone cutting and surface treatment is discussed as well as progress in the development of novel technique based on the modulation of water jets.

1 INTRODUCTION

The world production of trade in dimension stones is experiencing an expansive trend as the result of a broad change in the building criteria of modern industrialized architecture. The demand for natural stone materials is also growing for diversified uses in urban and interior decoration and in production design. Consequently, dimension stone quarrying and working activity is undergoing a considerable technological progress during the last decades aimed at meeting both the demand of larger supplies and the requirement of better quality stone materials. Traditional methods of quarrying and working are becoming inadequate, since they are likely to cause negative effects regarding both the stone recovery and the quality of the material produced. In fact, major damages may be induced to the rock by the destructive action of the rude technologies of nowadays, resulting in a higher proportion of waste as well as in a possible weakening of the mechanical properties of quarried blocks and finished products.

In the field of soft materials like marbles, ophicalcites and travertine, a technological breakthrough took place with the availability of sawing machines using diamond tools like wires, chains or belts. Continuous production through the mechanization of the various quarrying operations allowed high productivity levels to be achieved, exceeding 1000 t/man-year in 1989

(Cipriani & Patrucco 1991). At the same time stone recovery substantially increased and a larger production of commercial blocks could be obtained from a given quarried volume, thus improving profitability (Fornaro et al. 1992, Pinzari 1989, Ciccu 1992).

In the case of granite, development was much slower due to some unfavorable characteristics of the rock, like hardness and abrasivity. Despite the widespread application of diamond wire, the use of continuous cutting technologies is not always justified on economical grounds, especially for those materials having a low market value. Thus, the granite sector is still characterized by the extensive use of explosive splitting, flame slotting and wedge shearing, which give rise to a considerable inaccuracy of block faces and a detriment of stone integrity to a significant depth. Therefore the advent of novel substitute technologies of commercial availability in the stone quarries of the future will be welcomed, provided that the technical advances are accompanied by a proven economical profitability.

For that reason, interest has recently been addressed to water jets, both among quarries and machine manufacturers. Experiments carried out in Canada and Italy have proven the advantages of this technology, promoting new efforts for the development of advanced systems capable of meeting the needs of an industrialized activity (productivity, flexibility, reliability, safety).

303

2 BACKGROUND

High-speed water jet technology achieved significant progress during last decades in applications such as cutting of wide range of materials, surface cleaning and removal of surface layers, and repair of concrete structures. Nowadays, number of commercial high pressure systems is available on the market, some of them generating pressures up to 420 MPa, other delivering up to hundred liters of water per minute.

Basic types of water jets are illustrated in Figure 1. Plain water jets (Fig. 1a) are generated by the change of pressure energy of water to kinetic energy by the acceleration of the water flow within the nozzle. Maximum pressures currently used to generate plain water jets are up to 700 MPa.

Cutting performance of plain water jets can be increased significantly by adding abrasive particles into the jet. There are two basic types of jets with abrasive: (i) abrasive water jets (AWJ) and (ii) abrasive suspension jets (ASJ). In case of AWJ (Fig. 1b), abrasive particles are added into the jet downstream the water nozzle orifice in mixing chamber and focused and accelerated in focusing tube. Pressures up to 420 MPa are commonly used to generate AWJ. In the case of ASJ (Fig. 1c), abrasive particles are mixed with water in pressure vessel and the abrasive jet is generated by the discharge of the suspension in the specially designed nozzle. ASJ is generated by pressures up to 250 MPa.

Almandine garnet almandine or olivine concentrates with grain size of about 0.2 mm are commonly used as abrasive materials in abrasive water jets (Martinec et al. 2002).

3 POSSIBLE UTILIZATION OF WATER JETS IN DIMENSION STONE INDUSTRY

3.1 Quarrying

Dimension stone quarrying requires production of very deep cuts in the material. Therefore, special cutting heads allowing generation of rotating and/or oscillating jets have to be used to produce such cuts. Plain oscillating jets, for example, were used efficiently in Rothbach (France) to quarry soft sandstone. The use of water jets resulted in stone recovery increase well over 80%. However, the use of the same system for quarrying of hard materials, such as granite, has not provided satisfactory performance and profitability.

Utilization of abrasive water jets in dimension stone quarrying is limited by the amount of abrasive material necessary to create cuts that has to be also disposed of.

Thus, the performance of water jet techniques is still not competitive with existing conventional mechanical systems. Therefore, extending the use of water jets to the area of dimension stone quarrying requires significant improvement in their performance.

3.2 Contour cutting and machining

Abrasive jets have a very good potential both for creating precise contour cuts in slabs of natural stones and machining of ornamental stones.

Carrino et al. (2000, 2001, 2002) studied marble production, starting from slabs, by means of two alternative technologies: diamond milling and AWJ. They have found that AWJ technology allows overcoming the

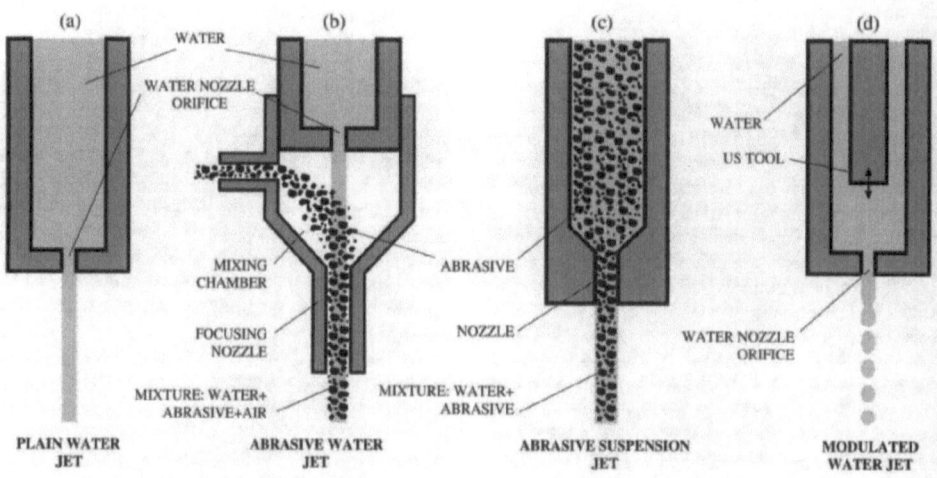

Figure 1.   Schematic drawing of various types of water jets.

main limits of traditional mechanical process, e.g. machining of small areas, such as the product edges, small thickness slabs, moving along curvilinear paths characterized by very small both radius and width, and thus reproducing the designed profile more accurately than traditional technology does. Moreover, they observed decrease of machining costs by more than 50% when using AWJ.

AWJ allows also 3D-machining of stone material. Figure 2 illustrates process of AWJ machining; Figure 3 shows final products of AWJ machining.

3.3 *Surface treatment*

Water jets can also represent suitable tool for non-traditional surface treatment and decoration. Rotating water jets can be used, for example, for gentle removal of weathered layers from the surface of historical stone constructions and statues.

Bortolussi et al. (2002) investigated the use of AWJ for reduction of slippery conditions of polished surfaces. This problem is normally solved by adopting

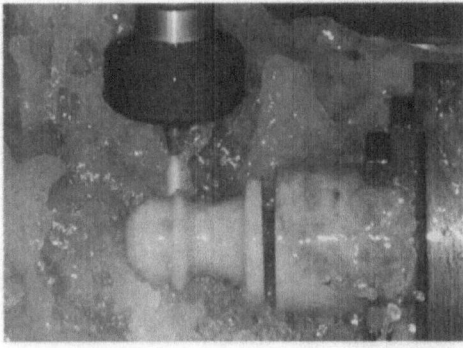

Figure 2.   Machining of marble by AWJ.

Figure 3.   Pieces machined by AWJ from marble and sandstone.

rough finishing methods (sand blasting, bush hammering) that however are often incompatible with the desired aesthetic features. Their results show that it is possible to use AWJ technology as a substitute to sand blasting for fulfilling the desired tasks, such as surface treatment, carving, and drawing.

3.4   *Projects of art*

There were reported several attempts to use water jets and AWJ as a tool for creation of artistic pieces.

Summers et al. (2001) described water jet cutting of large as well as smaller granite sculptures to make positive and negative images. In 1984, a half-scale version of the British Megalith known as Stonehenge cut by water jets from Georgia granite was erected at the campus of the University of Missouri-Rolla. The site was completed by Millennium Arch sculpture carved from red South-East Missouri granite using both abrasive and plain water jets in 2000.

Bach et al. (2002) demonstrated a special project of art – the making of a desk resting on three statues of 1.2 m height and 120 kg weight each created by high precision abrasive water jet. The project proved that abrasive water jets can serve as a tool of high potential for conversion of projects of art.

4   ULTRASONIC MODULATION OF WATER JETS

As was mentioned above, extending the use of water jets in the area of dimension stone quarrying is conditioned by significant improvement in their performance. One of possible methods to achieve this goal is represented by pulsing the water jet.

This follows from the fact that the impact pressure on a target generated by a slug of water is considerably higher than the stagnation pressure of a corresponding continuous jet. Pulsing the jet leads to an amplification of the impact pressure (see Foldyna et al. 2001).

$$z = \frac{p_i}{p_s} = \frac{2954}{v_0}, \tag{1}$$

where p_i = impact pressure, p_s = stagnation pressure, and v_0 = velocity of the jet.

Since velocities of continuous jets currently used in quarrying do not exceed $700 \, \mathrm{m \, s^{-1}}$, the impact pressure of pulsed jet will be at least 4 times higher at the same velocity and therefore significant improvement in cutting performance can be expected. Also additional effects such as fatigue failure of target material due to cyclic loading can contribute to the higher performance of pulsed jets.

Ultrasonically forced modulation of a continuous stream of water (Fig. 1d) represents the most promising

305

method of pulsed jets generation because of its simplicity, viability, and practicality. Modulated jet is generated by modulating a continuous stream of water by ultrasonic waves. An ultrasonic tool connected to a piezoelectric transducer is located axially inside a nozzle to induce longitudinal pulsations in the water. Ultrasonically modulated jet escapes from the nozzle as a continuous stream of liquid having unsteady velocity due to the modulation. The modulated jet breaks-up into slugs of water at certain distance from the nozzle, and starts to act as a pulsed jet.

At present, an extensive research program is in progress in the laboratory at the Institute of Geonics in Ostrava to understand the basic principles of the process and to optimize the nozzle design for applications of dimension stone quarrying and working. The program is oriented at the understanding of fundamental processes occurring both within and outside the nozzle during ultrasonic modulation of the jet as well as at the evaluation of potential of modulated jets in cutting of rock materials. Selected results obtained during laboratory tests of cutting of rock samples by modulated water jet are presented in following sections.

4.1 Ultrasonic nozzle device

Some of possible concepts and configurations of the ultrasonic nozzle can be found e.g. in Puchala & Vijay (1984), Vijay (1992), Vijay & Foldyna (1994), and Foldyna et al. (2001). Basically, ultrasonic modulation of the jet is produced by vibrations of ultrasonic tool located inside a nozzle. The vibration is generated by an ultrasonic transducer connected to the ultrasonic tool.

In our experiments, ultrasonic nozzle was driven by 20 kHz piezoelectric transducer connected to ultrasonic generator with maximum output power of 630 W. The ultrasonic nozzle was equipped with exponential ultrasonic tool with vibrating tip 10.0 mm in diameter.

4.2 Experimental facility

The experimental facility consisted essentially of a high-pressure water supply system, an ultrasonic nozzle device (see above), and X-Y table for traversing of the jet over testing samples. High-pressure water was supplied to the nozzle by a plunger pump capable to deliver up to 43 l min^{-1} of water at pressure up to 120 MPa. The traverse speed of X-Y table could be varied from 0.01 to 8.00 m min^{-1}.

Rock samples were cut by diamond saw to blocks with approximate dimensions of $100 \times 150 \times 60$ mm.

4.3 Results and discussion

Medium grained sandstone (locality Ráztoka), medium grained granodiorite (locality Žulová) and Carrara

marble were selected as testing materials for tests of cutting of rock samples by modulated jet. Basic physical and mechanical properties of rock samples can be found in Table 1.

Tests in rock samples were performed at pressure of 50 MPa, nozzle diameter was 1.19 mm, traversing speed 1 m min^{-1}, and ultrasonic power was set to maximum (630 W). The standoff distance was set to 60 mm for sandstone and granodiorite, 40 mm for marble. These standoff distances were found to be an optimum for individual rock materials and operating parameters specified above in previous tests.

In order to compare the performance of modulated jet with the continuous jet, tests with the latter were performed using standard nozzle for continuous jet at the same values of operating parameters. Average depth of cut calculated from five arbitrary measurements taken along the path of the jet was used as a measure of performance of the jet.

Results are presented in Figure 4 for sandstone, in Figure 5 for granodiorite, and in Figure 6 for marble samples. It can be seen that while modulated jet formed quite regular slots both in sandstone and granodiorite samples, slots produced by continuous jet are irregular both in shape and depth, being formed mostly by breaking out of rock chips. Average depths of cut produced by modulated jet are 2.2 to 3.1 times higher in sandstone and 2.8 to 3 times higher in granodiorite compared to that produced by continuous jet at the same operating parameters.

Slots produced by both types of the jet in marble are similar, as can be seen in Figure 6. Again, slots created by continuous jet are more irregular; however, the difference between slots created by modulated and continuous jets is not very distinctive. Typical are smaller irregularities both in shape and depth compared to that obtained in other tested materials and lower degree of chip breakage. Average depths of cut produced by modulated jet in marble are only 1.2 times higher compared to that produced by continuous jet.

Laboratory tests of cutting of rock samples performed so far proved superior cutting performance of

Table 1. Basic physical and mechanical properties of rock samples.

	Sandstone	Granodiorite	Marble
Locality	Řeka	Žulová	Carrara
Compressive strength	115 MPa	144.5 MPa	97 MPa
Tensile strength	5.8 MPa	11.3 MPa	7.8 MPa
Young's modulus	19.7 GPa	45 GPa	39.2 GPa
Density	2652 kg m^{-3}	2648 kg m^{-3}	2721 kg m^{-3}
Unit weight	2495 kg m^{-3}	2610 kg m^{-3}	2710 kg m^{-3}
Porosity	5.8%	1.41%	0.42%

306

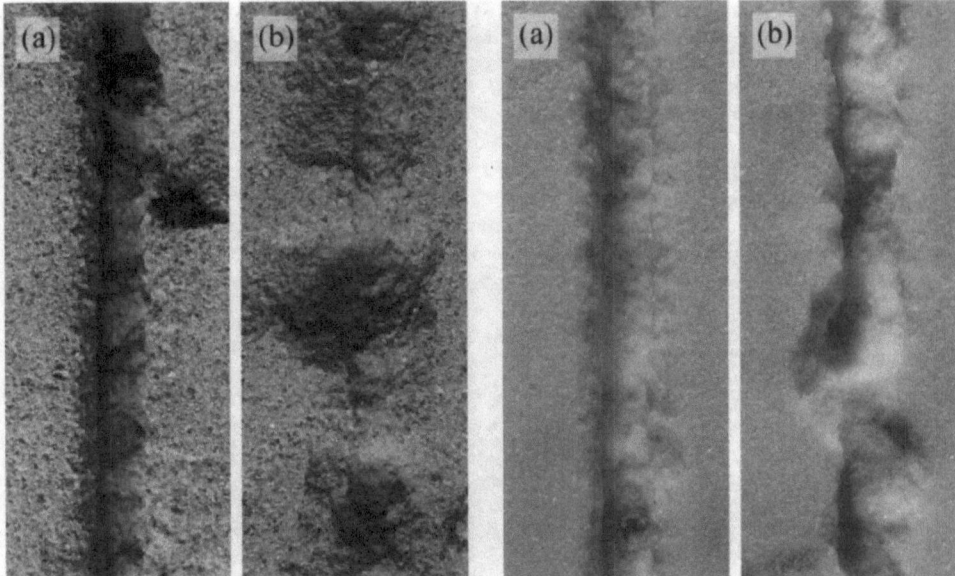

Figure 4. Sandstone exposed to modulated (a) and continuous (b) jets. Average depth of cut: 6.8 mm (a) and 3.2 mm (b).

Figure 6. Marble exposed to modulated (a) and continuous (b) jets. Average depth of cut: 4.8 mm (a) and 4.1 mm (b).

ultrasonically modulated jets over plain water jets. The tests show that the mechanism of disintegration of rocks by modulated jets is different than that by continuous jets. This fact can be also used in treatment of rock surface to obtain uncommon texture for various purposes.

Authors believe that the performance of modulated jet in rock cutting can be further improved by proper matching of operating parameters, such as operating pressure and traversing velocity. Investigations in these areas are still in progress in the laboratory.

5 CONCLUSION

Reasearch performed in the area of water jetting technology so far indicates that the technology has a good potential in applications of dimension stone cutting, machining, surface treatment, and as a tool for creation of artistic pieces. It requires, however, further development to improve its cutting performance to be successfully adopted also in dimension stone quarrying.

Laboratory tests of cutting various rock materials using ultrasonically modulated water jets proved superior performance of modulated jets over continuous ones. Therefore, enhanced performance of ultrasonically modulated jets can be expected in applications of dimension stone quarrying and working. The utilization of modulated jets will allow achieving higher

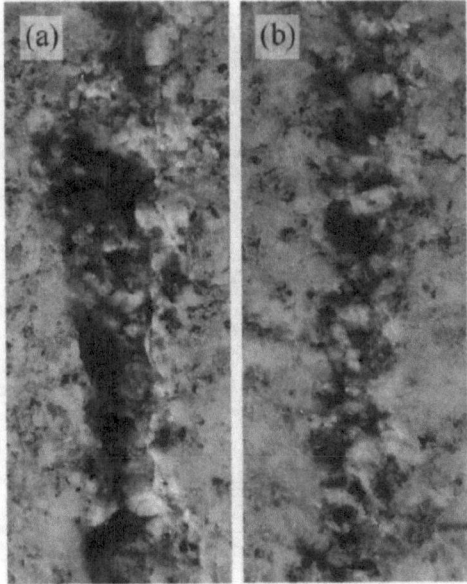

Figure 5. Granodiorite exposed to modulated (a) and continuous (b) jets. Average depth of cut: 5.1 mm (a) and 1.6 mm (b).

307

cutting performance at lower pressures due to impact pressure generated by impact of modulated jet on the rock surface.

Accordingly, it is authors' view that an enormous market potential awaits the commercialization of this development.

ACKNOWLEDGEMENTS

Presented work was performed within the framework of research projects of the Grant Agency of the Czech Republic, project No. 105/03/0183. Research of abrasive water jets and plain water jets is supported by the Czech Academy of Sciences, Project No. K3012103. Authors are thankful for the support.

REFERENCES

Bach, Fr.-W., Louis, H., Rad, Ch. von, Schuster, B. & Eggers, E. 2002. Application of abrasive waterjets for the creation of a desk of modern art. In: Lake (ed.), *Proceedings of the 16th International Conference on Water Jetting*: 51–57. Cranfield: BHR Group.

Bortolussi, A., Manca, M.G., Careddu, N., Ciccu, R. & Olla, S. 2002. Surface finishing of marble with abrasive waterjet. In: Lake (ed.), *Proceedings of the 16th International Conference on Water Jetting*: 425–435. Cranfield: BHR Group.

Ciccu, R. 1992. Technologie avanzate per la coltivazione e la lavorazione dei lapidei. *Atti Conv. Int. ANIM Eurocave '92, St. Vincent.*

Cipriani, L. & Patrucco, M. 1991. Problemi technologici e di sicurezza connessi con lo sviluppo di coltivazioni in sotterraneo di pietre ornamentali, *Atti II Convegno AMS Geoingegneria, Torino.*

Carrino, L., Polini, W., Turchetta, S., & Monno, M. 2000. Study of cutting quality and efficiency in stone. In: R. Ciccu (ed.), *Proceedings of the 15th International Conference on Jetting Technology*: 133–146. Cranfield: BHR Group.

Carrino, L., Polini, W., Turchetta, S. & Monno, M. 2001. AWJ to machine free form profiles in natural stone. In: M. Hashish (ed.), *Proceedings of the 2001 WJTA American Waterjet Conference*: 305–323. St. Louis: WJTA.

Carrino, L., Polini, W., Turchetta, S. & Monno, M. 2002. Surface processing of natural stones through A.W.J. In: Lake (ed.), *Proceedings of the 16th International Conference on Water Jetting*: 437–450. Cranfield: BHR Group.

Foldyna, J., Jekl, P. & Sitek, L. 2001. Possibilities of utilization of modulated jets in rock cutting. In: Filipowicz, Feliks (eds.), *Proceedings of the 1st International Conference Mining Techniques*: 85–96. Cracow: AGH.

Foldyna, J., Sitek, L., Jekl, P. & Nováková, D. 2001. Measurement of force effects of modulated jet. In: Jírová, Jiroušek, Kult (eds.), *Proceedings of the 39th International Conference Experimental Stress Analysis – EAN 2001 Tábor*: 63–68. Prague: CTU.

Fornaro, M., Mancini, R., Pelizza, S. & Stragiotti, L. 1992. Underground production of marble in Italy. Technology, economy and environmental problems. *Proc. XVth World Mining Congress, Madrid*

Martinec, P., Foldyna, J., Sitek, L., Ščučka, J. & Vašek, J. 2002. *Abrasives for AWJ cutting.* INCO-COPERNICUS No. IC 15-CT98-0821. Ostrava: Institute of Geonics.

Pinzari, M. 1989. La coltivazione dei lapidei carbonatici. Atti Conv. *Int. ANIM Situazione e prospettive dell'Industria Lapidea, Cagliari*

Puchala, R.J. & Vijay, M.M. 1984. Study of an ultrasonically generated cavitating or interrupted jet: Aspects of design. *Proceedings of the 7th International Symposium on Jet Cutting Technology,* Paper B2: 69–82, Ottawa.

Vijay, M.M. & Foldyna, J. 1994. Ultrasonically Modulated Pulsed Jets: Basic Study. In: N.G. Allen (ed.), *12th International Conference on Jet Cutting Technology*: 15–35. London: BHR Group.

Vijay, M.M. 1992. Ultrasonically generated cavitating or interrupted jet. *U. S. Patent No. 5, 154, 347.*

308

Dimension Stone 2004, Přikryl (ed)
© 2004 Taylor & Francis Group, London, ISBN 90 5809 675 0

Assessment of energy saving potentials in marble quarries

P. Konstantopoulou & M. Founti
National Technical University of Athens, School of Mechanical Engineering, Laboratory of Heterogeneous Mixtures and Combustion Systems, Athens, Greece

K. Laskaridis
Institute of Geology and Mineral Exploration, Dept. of Economic Geology, LITHOS laboratory, Athens, Greece

ABSTRACT: The work investigates energy saving potentials in typical medium sized operational marble quarries in the North of Greece (island of Thassos) producing white dolomitic marble. On the basis of electrical and thermal energy balances and water flows, potentials for energy savings are identified and expected cost savings are calculated. The techno-economic analysis focuses on the implementation of a compact Combined Heat and Power plant using Internal Combustion Engines and compares its performance to wind-turbine and photovoltaic applications.

1 INTRODUCTION

Marble constitutes an important rock of the mineral wealth of Greece not only due to its inexhaustible supply, but also thanks to the unique quality and variety of colours, which identify it. Greek marble and particularly the marble of Thassos, is known worldwide and forms the primary type of exported marble. Greece is an important producer, occupying the eighth position in the global production of primary rock (fourth in Europe). The Greek marble exports are constantly increasing since 1991, reaching almost 2.6 million tons during 1995–1997. In 2002, production made a new record.

The modern and dynamic Greek marble industry is classified among the top world producers of decorative natural stones, as it concerns both sizes of production and exports. The number of the companies engaged in the marble sector is estimated to be about 4,000 and includes small, medium-sized but also, several big units that they have realized important investments and rank among the best industrial units in Europe. The Greek marble companies are mainly engaged in one or more of the following fields: quarrying, cutting or/and processing, manufacture of art works, ecclesiastical elements and memorials, trade of marble blocks and products in home and foreign market, installation and applications.

The selected test case examines a typical, as far as equipment, quarrying methods and economic size are concerned, middle-sized marble quarry in Greece. For this reason, the systematic record of energy flows at the selected quarry comprises a prototype guideline for the energy overview of other middle-sized quarries (ACCI 2002, NSSG 2003).

2 DESCRIPTION OF SELECTED TEST CASE – QUARRY OVERVIEW

The quarry is located at Tsipoptsi-Limenas (Fig. 1) on the island of Thassos, at an amphitheatric location of 440–530 m altitude, with abrupt topographic variations and covering a total area of about 47,800 m². It

Figure 1. Quarry overview.

operates since 1997, as an "open pit" quarry with several fronts and seven beds of 6 m height each. It specializes in the quarrying of the white dolomitic marble which is available in the market with the patented commercial name "PRINOS". The annual productivity is up to 2,900 m³, while if the wastes (irregular small sized blocks) are added productivity reaches 3,960 m³ with a total value of 280–600 €/m³. The quarry has 20 employees and can operate about 10 months per year.

The chemical composition and the mechanical characteristics of the exploitable marble are shown in Tables 1 and 2, respectively (Laskaridis et al. 2000).

Quarrying operations basically involve isolating blocks from the parent ledge by cutting it free on all sides perpendicular to each other. The isolated stone block has dimensions suitable for sale and processing or it may be much larger so that further subdivision into smaller blocks may be made. The basic quarrying sequence includes the following working steps:

- pre-production operations;
- primary cuts;
- secondary cuts and finishing of blocks;
- removal and haulage of blocks.

The production efficiency of the operation increases if these four working phases may be done simultaneously (Milazzo & Blasi, 2004).

The mechanical equipment of the quarry includes eight BENETTI diamond wire saws, seven of which are model Alpha 840 and one is model VIP 910, two drill machines, three pneumatic top-hammers, two power generators; one DORMAN Diesel of nominal

power 450 kVA and one PERKINS Engines 2000 series of nominal power 250 kVA and two air-compressors; one diesel-powered Atlas COPCO XA146 and one electric INGERSSOL Rand SSR ML-110. There exist auxiliary drilling equipment such as short plugs, hydraulic jacks, air pillows and others. In addition, the quarry uses various vehicles such as excavators and loaders, truck/dumpers, cars and jeeps. All the equipment is property of the quarry.

The advantage of using diamond wire saws for stone cutting is demonstrated by reduced manpower (3–4 workers only), increased cutting speed (up to 10–12 m²/hour) and easy installation, regarding to Milazzo & Blasi (2004).

3 ECONOMIC OVERVIEW OF THE QUARRY

According to collected data for the year 2002 (Fig. 2), the energy costs for the quarry are 13.7% of the total operational costs.

According to Figure 3, the annual mean volume of sales is about 200 m³ per month. Considerable fluctuations are observed for the months of December and February, due to halt of quarry operations associated with weather conditions. During these months productivity is very low or zero. The recorded sales (Fig. 3)

Table 1. Chemical composition of PRINOS marble (Laskaridis et al. 2000).

Sample	SiO$_2$	Al$_2$O$_3$	Fe$_2$O$_3$	CaO	MgO
	0.6	0.08	0.07	30.4	21.9

Sample	K$_2$O	Na$_2$O	MnO	LOI
	<0.01	<0.01	0.01	46.93

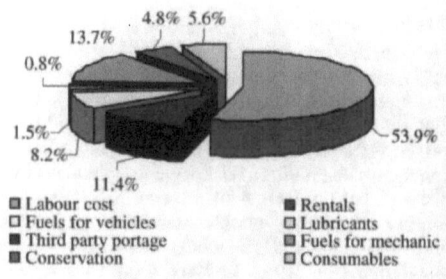

Figure 2. Distribution of typical quarry expenses.

Table 2. Physical characteristics of PRINOS marble (Laskaridis et al. 2000).

Apparent density [kg/m³]	2.846
Coefficient of open porosity [% vol.]	0.28
Absorption coefficient [% wt.]	0.10
Dynamic modulus of elasticity [GPa]	21
Compressive strength (dry condition) [N/mm²]	80
Compressive strength (wet condition) [N/mm²]	93
Flexural strength (dry condition) [MPa]	10
Flexural strength (wet condition) [MPa]	13
Resistance in frost (cycles of freezing and defrosting) [N/mm²]	103
Abrasion resistance [mm]	2.49
Impact strength [cm]	40

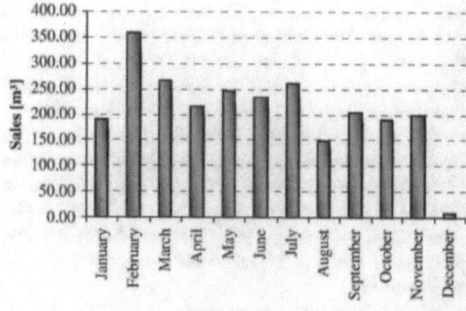

Figure 3. Annual mean value of monthly sales.

result from selling during these months, marble blocks already in stock.

4 OVERVIEW OF ENERGY FLOWS

Two diesel-powered generators cover the electric power requirements of the quarry. The local network of electric energy comprises the two generators, eight diamond wire saws, an electric air compressor and a water drill-pump. In particular, the DORMAN generator supplies the eight diamond wire saws and the water drill-pump. On the other hand, the PERKINS generator supplies only the electric air-compressor.

Compressed air is used for the operation of the drilling equipment and it is produced with the help of two air-compressors. In specific, the INGERSOLL Rand compressor supplies air to the two drilling machines, while the Atlas COPCO compressor supplies the three pneumatic top-hammers. In case only one of the two drills is in operation, the diesel-powered compressor stops and the electric one (INGERSOLL Rand) supplies the one drill and the three pneumatic top-hammers.

Considerable amounts of water, about 80 m^3, are consumed daily at the quarry in order to clean the diamond wire saws and for the drilling procedure. The water comes either from dams located on the mountain at a higher altitude than the quarry, or from the water-drill, especially during the summer months. The water is stored in three overflow-tanks with dimensions $6 \times 3 \times 2$ m, where from water is supplied to the quarry under free-fall flow.

5 ENERGY BALANCES – EVALUATION OF ENERGY CONSUMPTION

The energy balance was calculated taking into account the technical characteristics of the available equipment and the energy flows. The results are summarized in Table 3. 35.8% of energy produced remains unexploited, since both generators are over-dimensioned. The Dorman and Perkins generator produce 41.8% and 25.0% more energy than the consumption requirements, respectively. This is the result of non-optimal organisation of the local electric power network as well as the absence of energy management.

Further analysis of available data demonstrated that the diesel oil consumption is ca. 33.4% higher than the theoretically expected one. This fact is in agreement and supports the above initial finding of non-optimal energy management at the quarry.

Table 4 tabulates the operational conditions for the two generators, assuming that they could be operated at an optimal point. Energy saving potential (Table 4)

Table 3. Energy balances.

Dorman Electric Power Generator		
Production	Power [kW$_e$]	Energy [kWh]
Dorman Diesel Generators	360	576,000
Consumption		
7 Alpha 840 Wire saws	235	281,875
1 VIP 910 Wire saw	41	49,216
1 water-drill	10	4,000
Total consumption	286	335,091
Non utilized power [kW$_e$]	74	
Annual amount of non utilized energy [kWh]	240,909	
Percentage of non utilized energy [%]	41.8	
Perkins Electric Power Generator		
Production	Power [kW$_e$]	Energy [kWh]
Perkins Engines Generator	200	320,000
Consumption		
1 Ingersoll Rand Air compressor	150	240,000
Non utilized power [kW$_e$]	50	
Annual amount of non utilized energy [kWh]	80,000	
Percentage of non utilized energy [%]	25.0	
Total		
Power production [kW$_e$]	560	
Power consumption [kW$_e$]	436	
Non utilized power [kW$_e$]	124	
Energy production [kWh]	896,000	
Energy consumption [kWh]	575,091	
Annual amount of non utilized energy [kWh]	32,909	
Percentage of non utilized energy [%]	35.8	

is limited due to the fact that the two generators produce the amount of power that it is approximately required for the current quarry operation.

However, the operation of the quarry at its optimum energy consumption point, requires the reduction of power production by 6.6% and the maximum energy saving will be only 4.5%. Consequently, the expected saving in expenditure is 5,975 €, that it is only 6.1% of operational costs for the year 2002.

The above analysis indicates that for the examined case, any medium or high capital cost energy saving measures will not be viable.

6 ENERGY SAVING MEASURES

On the basis of the results of the energy audit, zero, medium and high capital cost measures can be proposed to improve the energy consumption at the quarry, but also to provide long term solutions for energy

311

Table 4. Estimation of optimum operational point.

Dorman Electric Power Generator	
Energy production [kWh]	549,120
Non utilized energy [kWh]	214,029
Percentage of non utilized energy [%]	39.0
Power production [kW$_e$]	343
Non utilized power [kW$_e$]	57
Percentage of non utilized power [%]	16.7
Produced power reduction coefficient [%]	4.7
Perkins Electric Power Generator	
Energy production [kWh]	288,000
Non utilized energy [kWh]	48,000
Percentage of non utilized energy [%]	16.7
Power production [kW$_e$]	180
Non utilized power [kW$_e$]	30
Percentage of non utilized power [%]	16.7
Produced power reduction coefficient [%]	10
Total	
Energy production [kWh]	837,120
Non utilized energy [kWh]	262,029
Energy saving [kWh]	58,880
Percentage of non utilized energy [%]	31.3
Power production [kW$_e$]	523
Non utilized power [kW$_e$]	87
Percentage of non utilized power [%]	16.7
Produced power reduction coefficient [%]	6.6

production. Such measures could permit operation of the quarry over a period of 12 months (instead of the current 10 months) with apparent financial benefits.

Zero capital cost measures involve only the existing personnel and have to do with the optimum use of the available equipment as far as energy concerned. Such measures are:

- Pause of machine operation when not in use.
- Examination of energy distribution system (insulation, leakages etc).
- Rearrangement of electric circuit to match the exact needs.
- Temporal re-adjustment of the various quarry procedures to reduce of electric energy consumption.
- Personnel training.
- Development action plan to include regular energy audits and servicing of the quarry machinery and electric network (generators, compressors etc).

Middle – low capital cost measures, which involve technological solutions of middle capital cost but are still highly dependent on personnel participation, are:

- Installation of additional hot water boiler for the winter months.
- Substitution of the currently used diesel fuel for bio-diesel.

Finally, the **High capital cost measures** focus on implementation of a technology solution with significant energy savings, but that also demands investment of considerable capital. Such measures are:

- readjustment/improvement of the generators' power-coefficient;
- replacement of the two generators for three smaller ones, to meet the exact electric energy supply needs;
- implementation of renewable energy technologies.

7 OVERVIEW OF PROPOSED INVESTMENTS

Three business plans are analyzed, concerning the installation of a biomass cogeneration plant, of a sequence of photovoltaic arrays and of a wind park. The proposed types of equipment have resulted from extensive literature and market surveys, in order to establish the most appropriate type for the given energy requirements and operational characteristics. The previous analysis established that the proposed investments should produce a minimum amount of electric power equal to 523 kW$_e$, when at the same time the energy demands are up to 837.120 kWh per year. In addition, in case of 12 months quarry operation, the energy demands rise up to 1.004.544 kWh. The use of a boiler is absolutely necessary as long as the quarry operates for the whole year. In case of wind park and photovoltaic arrays, approximately 80 m^3 of water must be heated daily to supply the quarry. The boiler must produce heat equivalent to 55,773 kWh/y, and therefore the electric energy consumption of the boiler rises to 65,569 kW$_e$.

Currently, the quarry is not connected to the national electricity grid. Recent legislation gives the opportunity to investors to become independent energy producers. The quarry, as an energy producer will be able to produce thermal and electric power to cover its needs. The excess of electric energy produced may be sold to the national Public Power Cooperation (PPC). In such a case, it is necessary to expand the current power distribution network for about 15 km in order to ensure full operation of the energy generation plants. In addition, the creation of a middle-voltage substation will be necessary.

Cogeneration plant: The proposed cogeneration plant (Table 5) comprises the installation of a Wärtsilä Internal Combustion Engine of nominal electric power of 1,000 kW$_e$ and nominal thermal power of 5MW$_{th}$ (Wärtsilä 2003).

The fuel proposed is olive-oil tree biomass (wood and oil-residuals), selected due to local availability. The requirements are 3,866 t/y of olive-oil tree "residual-wood" and 2,186 t/y olive-oil "residual-mass". The low calorific value of the fuel is 17,500 MJ/dry ton.

Photovoltaic arrays: The installation of photovoltaic arrays, type Shell SM110/24, of total electric

Table 5. Operational characteristics of proposed co-generation plant.

Conversion factor [%]	35
Thermal power demands [kW$_{th}$]	174
Electric power demands [kW$_e$]	523
Coefficient of availability [%]	90
Network losses [%]	3
Produced electric power [kW$_e$]	1,000
Produced thermal power [kW$_{th}$]	5,000
Electric energy excess [kWh]	
10-month quarry operation	6,810,360
12-month quarry operation	6,642,936
Thermal energy excess [kWh]	
10-month quarry operation	40,000,000
12-month quarry operation	39,944,267
Capacity [t/y]	6,052
Daily capacity [kg/d]	18,155

Table 6. Operational characteristics of proposed photovoltaic arrays.

Nominal power [W$_e$]	3,487
Conversion factor [%]	15.0
Coefficient of availability [%]	95.0
Network losses [%]	3.0
Energy production [kWh]	28,145,559
Electric power excess [kWh]	
10-month quarry operation	27,308,439
12-month quarry operation	27,075,446

Table 7. Operational characteristics of proposed wind park.

Number of wind turbines	3
Nominal power of wind turbine [kW$_e$]	600
Nominal power of wind park [kW$_e$]	1,800
Coefficient of exploitation	30.0
Coefficient of availability [%]	95.0
Network losses [%]	3.0
Energy production [kWh]	4,359,064
Electric power excess [kWh]	
10-month quarry operation	3,521,944
12-month quarry operation	3,288,951

Table 8. Cost characteristics of proposed plants.

Capital cost (€)		
Cogeneration plant	1,755,000	
Photovoltaic arrays	30,837,667	
Wind park	1,775,000	
Operational cost (€)		
Cogeneration plant	279,765	
Photovoltaic arrays	306,827	
Wind park	35,500	
Annual earnings (€)		
Quarry operation	10-month	12-month
Cogeneration plant	361,971	353,072
Photovoltaic arrays	1,693,396	1,678,948
Wind park	218,396	203,948

power of 3,487 kW$_e$ and of inverter of nominal power of 550 kVA, was evaluated (Shell, 2004). The operational characteristics of such a plant are shown in Table 6.

Wind park: The proposed wind park (Table 7) consists of three wind turbines of nominal electric power of 600 kW$_e$ each. The proposed wind turbines are manufactured by ENERCON-model E-40 (Enercon 2003).

8 ECONOMIC ASSESSMENT OF PROPOSED INVESTMENTS

The capital and operational costs of the three proposed plants have been based on manufacturers' data, available in open literature (Jurado et al. 2003, Thermie B, 2000, EU, IPD, 2002). The annual earnings have been calculated using the quarry operational data. The results are shown in Table 8.

9 RESULTS

Three energy production installations, a cogeneration plant, a sequence of photovoltaic arrays and a park with three wind turbines have been analysed.

The examined investments refer to the application of renewable energy technologies and are of major cost. Techno-economic assessment (Table 9), through calculation of the dynamic indices Net Present Value (NPV) and Internal Rate of Return (IRR) indicated that the three investigated business plans are of major capital cost, with low profit margins and of high risk since the level of uncertainties is significant, due to the assumptions made with regard to the operational and economic characteristics.

On the whole, the most profitable investment, on the basis of the techno-economic assessment, seems to be the installation of a wind park, consisting of three wind turbines. Second ranks the cogeneration plant using as fuel olive-oil tree residues and last the installation of p/v arrays. As far as technical feasibility is concerned, the cogeneration plant seems to be most viable under the assumption that the amounts of needed biomass can be secured. The wind turbines and the p/v arrays might fail to provide the quarry with the demanded amounts of energy due to the fact that it is not possible to always guarantee favourable weather conditions (sun or wind).

313

Table 9. Economic indices for proposed investments.

	10-month quarry operation	12-month quarry operation
Biomass cogeneration		
NPV 10-years	−563,293	−620.003
IRR 10-years	−12.2%	–
C/B	0.58	0.55
NPV 20-years	−397.217	−471.904
IRR 20-years	0.83%	−0.66%
C/B	0.65	0.61
Pay back period	–	–
Photovoltaic arrays		
NPV 10-years	−9,308,562	−9,400,636
IRR 10-years	−11.23%	−11.48%
C/B	0.70	0.70
NPV 20-years	−6,507,354	−6,628,617
IRR 20-years	1.09%	0.95%
C/B	0.79	0.79
Pay back period	–	–
Wind park		
NPV 10-years	−301,203	−364.218
IRR 10-years	−0.95%	−3.04%
C/B	0.83	0.79
NPV 20-years	68,291	−23.912
IRR 20-years	8.99%	7.65%
C/B	1.04	0.99
Pay back period	17.5	–

10 CONCLUSIONS

The paper investigated the energy saving potentials in typical medium sized operational marble quarries in the North of Greece (island of Thassos) producing white dolomitic marble. It has been demonstrated that considerable amounts, up to 35.8%, of energy produced in the quarry remains unutilized. The engines are designed for higher energy consumption than the actual quarry demands. However, the potentials of energy saving are limited due to the fact that the power the engines produce corresponds approximately to the power demands of the quarry. This indicates the importance of rearranging the local electric network distributing power to the various equipment of the quarry and optimizing the energy distribution network taking into account specific requirements of the quarrying process.

Consequently, the best practice for energy saving in the marble quarry is to create an action plan that consists of zero capital-cost saving measures. These actions focus on the optimization of the performance of the quarry by implementing training programs for the personnel, organizing audits, regular servicing and repairs of existing equipment.

Furthermore, the comparative assessment of three long-term business plans for independent energy production involving renewable-energy solutions indicated that all investigated technologies (cogeneration, wind parks, photovoltaics) require high capital investment and are of high risk due to technical uncertainties. Such high capital cost solutions are of investment interest rather than energy saving in quarrying operations.

REFERENCES

Athens Chamber of Commerce & Industry (ACCI), 2002.
Laskaridis, K., Papaioannou, N., Papatrechas, Ch., Kouseris, J. & Christou, G. 2000. *Marble Atlas of the Macedonia and Thrace Divisions.* IGME, LITHOS Laboratory.
Milazzo, G. & Blasi, P. 2004. *An introduction to dimensional stone quarrying in ACP countries,* Internazionale Marmi Macchine Carrara S.p.A., in press.
National Statistics Service of Greece (NSSG), 2003.
Jurado, F., Cano, A. & Carpio, J. 2003. Modelling of Combined Cycle Power Plants Using Biomass. *Renewable Energy* 28: 743–753.
www.wartsila.com/powerplants/english/index/jsp?cid = en_pp_port_Bio, WARTSILA website, 2003.
Thermie, B. 2000. *Innovative Techniques for Large Scale PV Power Stations, Environmental Impacts from the Use of Renewable Energy Technologies.* Project No STR-1000-96-HE.
www.shell.com/solar, SHELL website, 2004.
ENERCON website, 2003, http://www.enercon.com.
European Commission, 2003. *Wind Energy – The facts. Volume 2, Costs – Prices and Values.* Directorate – General for Energy.
Greek Ministry of Development, 2002. *Guidelines for Energy Investments.* Investing Program for Development (IPD).

314

Dimension Stone 2004, Přikryl (ed)
© 2004 Taylor & Francis Group, London, ISBN 90 5809 675 0

Slate wastes role on the production of mullite-rich ceramics

S.C. Vieira & M.T. Vieira

ICEMS, Department of Mechanical Engineering, Pinhal de Marrocos, Polo, Coimbra, Portugal

ABSTRACT: Mullite naturally appears as a product of the phase transformation of slate wastes at about 980°C (primary mullite). The aim of the present research work was to study the improvement of the content of this phase in the sintered product, through the reaction of aluminium hydroxides, from anodizing sludges, with the silica-rich phases in the slate wastes (secondary mullite). Slate wastes and aluminium sludges mixtures were prepared with alumina:silica ratios ranging from 1:1 to 2:1 and sintered up to temperatures of 1300°C. Particularly for the mixture 2:1, it was observed an important increase of the mullite content with the increase of sintering temperature, simultaneously with a significant decrease of silica phases, as well as the formation of important quantities of α-alumina. Final products were mullite-alumina composites with improved mechanical properties, compared to those of traditional clay products, and were characterized namely by XRD, SEM and flexural strength tests.

1 INTRODUCTION

Mullite ($3Al_2O_3.2SiO_2$) is very rare in nature and for many years was mistakenly identified as belonging to the sillimanite group of minerals (Aksay et al. 1991). However, it is commonly found in industrial ceramics and its excellent high-temperature mechanical properties, as well as promising optical and electronic properties (Aksay et al. 1991, Ohira et al. 1996), instigated the interest in its formation, either from natural or from synthetic precursors (Pask 1996, Chen et al. 2000, Li & Thomson 1990). Among the natural precursors, clay minerals played an important role in the production of mullite ceramics, and both clays and mixtures of clays with alumina precursors, as aluminium hydroxides or alumina polymorphs, have been extensively used to produce mullite and mullite composites (Liu & Thomas 1994, Chen et al. 2000).

The formation of mullite is not exclusive of clay systems but it is very common in most aluminium silicates-containing ceramic materials at high temperatures (Aksay et al. 1991), and recent research efforts have shown the formation of mullite in Portuguese slates (Valongo) at temperatures above 980°C (Catarino 1999). From the mineralogical point of view Valongo slate wastes are mainly composed of quartz, muscovite and chlorite and slate powders have been submitted to thermal treatments under several atmospheres and using different heating rates (Catarino 1999). XRD studies showed no phase transformation up to 730°C,

when the powders were heated at 10°C/min in static air atmosphere. Above mentioned temperature was identified the presence of talc and at 800°C was detected the presence of olivine. Traces of hercynite are present at 950°C and mullite, hematite and glassy phase were identified above 1050°C (Catarino 1999).

The quarrying and transformation processes of slate rocks generate just about 30% of useful product, thus resulting in a great quantity of rock wastes that can assume the form of large dimension no use blocks or pieces, cut and polishing muds, but also of very fine powders aspirated during transformation operations. Besides the economic overload resulting from the useless product exploitation, slate wastes have a deleterious effect on the environment as they are responsible for a strong visual impact, when piled up in dumps, and for an important atmospheric and water pollution, in their finer grades (Catarino 1999). Therefore, the conversion of the wastes to a valuable raw material by using them in other industries would be beneficial not only from a socio-environmental point of view, but would simultaneously represent a more efficient use of mineral resources. Through the years several attempts of recovery of these wastes have been performed, but none had a significant industrial impact in the recovery of slate wastes, especially when regarding the enormous amount of existent wastes (Catarino 1999). More recent studies faced the use of slate wastes in ceramic floor and wall tiles industry as a potential high consumption application

315

(Catarino et al. 2003). Bending strengths as high as 75 and 95 MPa have been determined for slate powders uniaxially pressed samples after sintering respectively at 1150° and 1170°C for 30 min, and industrial scale-up of the forming and sintering processes gave rise to products with better properties than those obtained in ceramic tiles produced from traditional raw materials (Catarino et al. 2003). Calcinated powdered slate wastes were also found to be adequate for powder injection moulding, providing products free of defects and with excellent surface finishing (Barreiros 2002). Similarly to the additions of alumina precursors to kaolinite, alumina precursors, namely gamma alumina, have also been added to slate wastes powders in order to react with silica-rich compounds and improve the mullite content in final sintered product (Martins et al. 2002). Slate powders compacts sintered at 1150°C presented a bending strength of about 63 MPa, higher than that observed for slates-gamma alumina mixtures. Nevertheless, gamma alumina addition caused a general increase in the amount of mullite and prevented the swelling observed for slates above 1150°C, therefore inducing higher strengths for the mixture when the sintering proceeded at higher temperatures (1250°C, 6h – 28 MPa for slates and 43 MPa for the mixture sintered compacts) (Martins et al. 2002).

In this study a new approach has been made to the production of mullite-rich ceramic from slate wastes-based mixtures. Mullite precursor mixtures, with alumina-silica ratios ranging from 1:1 to 2:1, were prepared using both natural and industrial wastes as raw materials. At the same time that slate wastes provided the mixture mainly with silica and aluminosilicate compounds, aluminium-rich sludges were used as the source of additional alumina. Aluminium alloys are submitted to anodizing treatments, namely in order to increase corrosion and abrasion resistance, improve decorative appearance or to prepare the surface for subsequent painting operations. Anodizing involves the conversion of the aluminium surface to aluminium oxide and takes place through dipping of the profiles into a sequence of tanks containing different chemical solutions. Aluminium-rich sludges result from these solution-baths treatment in industrial wastewater treatment facilities, previously to the discharge of anodising and lacquering plants effluents into municipal sewerage system. They are removed from wastewaters using flocculation, followed by mechanical dewatering, and are mainly composed by water (75–85 wt.%) and colloidal aluminium hydroxides. The presence of sodium or calcium (generated from neutralising solutions) and aluminium sulphates (used as flocculating agent) (Ribeiro et al. 2002, Oliveira et al. 2003), and some minor elements like iron, copper, zinc, tin, silicon and magnesium, has also been reported (Delmas et al. 1997). In spite of being classified as a non-hazardous waste (Seabra et al. 2000), the growing awareness by the EU relating to the development of acceptable waste disposal practices, improved the research in order to find the most effective long-term solution for this kind of wastes. The introduction of such sludges in ceramic products seems to have very good perspectives, particularly due to the high content of alumina in heat-treated aluminium sludges and to the fineness of its particles (Vieira et al. 2002, Ribeiro et al. 2002), that in the most recent years have acted as a driving-force for the attempts of re-use of aluminium sludges as a raw material in the ceramic industry (Delmas et al. 1997, Pereira et al. 2000, Seabra et al. 2000, Ferreira & Olhero 2002, Ribeiro et al. 2002, Martins et al., in press).

The present paper reports the production of mullite-alumina composites from slate wastes-aluminium sludges mixtures. Uniaxially pressed samples were sintered for 30 min between 1000 and 1300°C and characterized namely by XRD, and density and mechanical strength determinations.

2 EXPERIMENTAL DETAILS

Slate wastes (Empresa Lousas de Valongo, Lda, Valongo, Portugal) and as-received aluminium-rich sludges (Tecnilaca, Mem-Martins, Portugal) were used as raw materials in this study. Mineralogical characterization of Valongo slate wastes has been reported elsewhere (Catarino 1999) and allowed the identification of muscovite, chlorite and quartz as the main phases in the waste powders. Table 1 shows the chemical composition of these rock wastes. As-received aluminium sludges were mainly formed by aluminium hydroxides and water. Sludges were dried at 110°C and their chemical composition was evaluated through atomic absorption spectroscopy and gravimetry (Table 2). Sulphate ions and constitutional water (hydroxyl groups), not removed by low-temperature thermal treatment, are the main responsible for the other ~38 wt.%. In order to determine the amount of as-received sludges required to prepare the desired $Al_2O_3:SiO_2$ ratio, sludges were previously calcinated to evaluate alumina content.

Slate waste powders and as-received aluminium sludges were weighed in order to prepare three batches with the desired 1:1, 1.5:1 and 2:1 $Al_2O_3:SiO_2$ ratios. As-received aluminium sludges were sieved through a 4–5 mm sieve, mixed with water and agitated at 5500 rpm, until a consistent and uniform pulp was formed. The required amount of slate waste was added and the mixture was performed using a turbomixer at 6000 rpm for 60 s. The pulp was then filter-pressed, desagglomerated, dried at 105–110°C, and sieved below 1 mm. The obtained powders were calcinated to 900°C, desagglomerated through a 250 μm

Table 1. Chemical composition and loss on ignition of slate waste powders.

Oxide	Content [wt.%]
SiO$_2$	53.30
Al$_2$O$_3$	24.03
Fe$_2$O$_3$	1.42
FeO	7.37
CaO	0.34
MgO	20.8
Na$_2$O	1.64
K$_2$O	3.06
TiO$_2$	0.89
MnO	0.07
P$_2$O$_5$	0.23
LOI	5.57

Table 2. Dried aluminium sludges chemical composition.

Specie	Content [wt.%]
Al$_2$O$_3$	55
Fe	0.15
S	4.6
Na	1.20
Ca	0.30
Zn	0.07
Cr	0.23

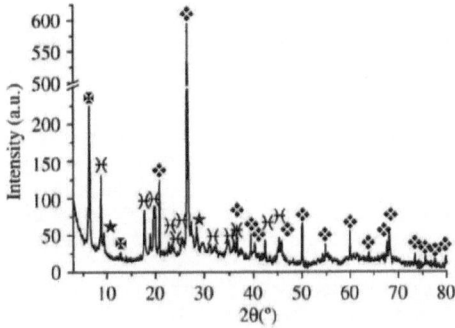

Figure 1. XRD patterns of the as-prepared 2:1 powdered mixture (✤-chlorite, ★-talc, ⅀-muscovite, ✤-quartz).

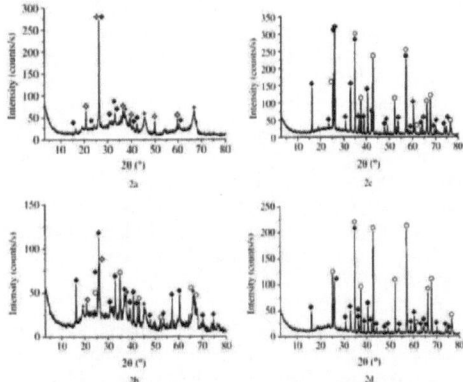

Figure 2. XRD patterns of the slate wastes-aluminium sludges presses bodies sintered at 1000° (a) 1170° (b) 1250° (c) and 1300°C (d) (O -alpha alumina, ✦-gamma alumina, ●-hematite, ✹-hercynite, ◆-mullite, ✤-quartz).

sieve, and then uniaxially pressed at 45 MPa. Pressed samples (72 × 13 × 5 mm) were sintered in air muffle type furnace (heating/cooling rate – 5°C/min, sintering temperatures – 1000 to 1300°C, soaking time – 30 min).

The physical, chemical and mechanical properties of the sintered samples were characterized through shrinkage calculation, density determination (conventional water displacement method) three-point bend tests (ASTM B528-76 Standard). Phase composition was investigated by X-ray diffraction (XRD) using a Rigaku difractometer with copper anticathode. Polished samples were sputtered with gold and their microstructure was studied using a Philips XL30 field emission scanning electron microscopy (FESEM). Coupled EDS was used for chemical analysis.

3 RESULTS AND DISCUSSION

3.1 Phase development

Figure 1 shows the XRD pattern for the 2:1 (Al$_2$O$_3$:SiO$_2$ ratio) as-prepared slate wastes-aluminium sludges mixture. The main crystalline phases identified were muscovite and quartz. Both main phases and minor crystalline phases, like chlorite, talc, olivine and feldspar, result exclusively from the presence of slate wastes in the mixture (Catarino 1999).

After uniaxial pressing, the 2:1 powder compacts were sintered for 30 min at temperatures between 1000° and 1300°C. Quartz was the only crystalline phase from as-prepared mixture that remained in the sample sintered at 1000°C; mullite, hematite, and gamma alumina appeared after sintering at this temperature (Fig. 2a). Both mullite (primary mullite) and hematite resulted from the phase evolution of the slate wastes with the temperature, at the same time that the aluminium hydroxides in the aluminium-rich sludges evolved to gamma alumina.

With the increase of the sintering temperatures, hematite disappeared and mullite diffraction peaks became more intense (Figs 2b,d). Below 1100°C aluminium sludges-slate wastes mixture was largely non-reactive and the dominant reactions were those observed in slate wastes and aluminium sludges reaction series respectively. However, at 1170°C (Fig. 2b)

317

and above it was observed the increase of the intensity of mullite XRD peaks associated to the formation of secondary mullite from the reaction of silica-rich phases, in slate wastes, with gamma alumina from the aluminium sludge.

Alumina alpha is clearly identified in samples sintered at 1170°C and coexists with gamma alumina, which disappears at 1200°C. In opposition to what has been observed for kaolinite-alpha alumina mixtures (Liu & Thomas 1994, Chen et al. 2003), the quantitative interpretation of mullite formation was not easy to perform in the present waste mixture.

Once the formation of secondary mullite from the reaction of slates silica and sludges alumina occurs in the same temperature range that the phase evolution between gamma and alpha alumina polymorphs, the dependence of mullite formation on the alumina source consumption is not readily obvious from XRD patterns analysis. In spite of that, it was clear that above 1200°C it occurred a strong increase of the intensity of mullite diffraction peaks and it was observed a slight broadening of the (130) peak (d = 2.4280 Å) of mullite (Fig. 2c), related with the appearance of the well defined hercynite (311) peak (d = 2.4597 Å) for higher sintering temperatures (Fig. 2d). After sintering at 1300°C, mullite and alpha alumina were the main crystalline phases in the sintered body, and hercynite was also present, together with some glassy phase (Fig. 2d).

Between 1250 and 1300°C it was observed a not fully understood decrease of mullite reflections. Nevertheless, this phenomenon may be related either to the presence of impurities in aluminium hydroxides, that has been reported as a driving-force for the dissolution of primary mullite in clay-alumina systems and for the enhancement of the formation of secondary mullite (Viswabaskaran et al. 2003), or to the presence of iron oxides and alkali impurities in mullite, that showed a tendency to 'demullitization' at high temperature.

From the phase evolution point of view, both mixtures 2:1 and 1:1 behave similar (Fig. 3). The XRD patterns in Figure 3 allowed the comparison between 1:1 and 2:1 mixtures sintered at 1000° (Fig. 3a) and 1300°C (Fig. 3b). These patterns confirm the presence of the same kind of phases in both mixtures, but a clear increase of the intensity of the diffraction peaks belonging to aluminium sludges-related phases (gamma and alpha aluminas) in mixture 2:1 relatively to the observed for mixture 1:1, and a decrease of slate wastes-associated phases due to the dilution effect produced by the addition of increasing amounts of aluminium sludges to the system.

3.2 Final products characterization

Figure 4 shows the densification parameters of the mixture 2:1, in terms of a – shrinkage, b – density,

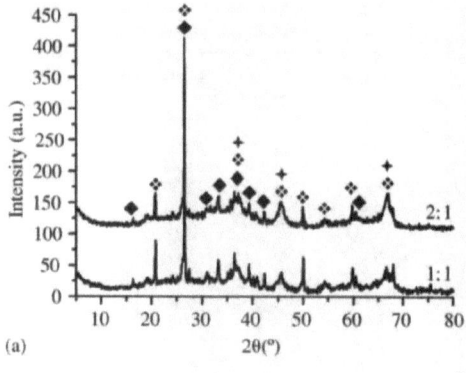

(a)

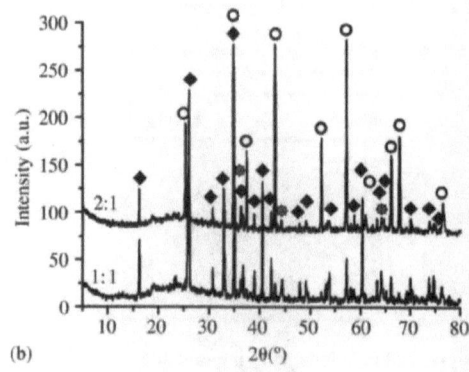

(b)

Figure 3. XRD patterns of the mixtures 1:1 and 2:1 sintered at 1000° (a), and 1300°C (b) (○-alpha alumina, ✚ -gamma alumina, ●-hematite, ✿ -hercynite, ◆-mullite, ✦-quartz).

and c – water absorption, as function of the sintering temperature. It is shown that the densification rate of the sintered bodies was improved by the increasing of sintering temperature.

On raising the sintering temperature, both shrinkage and apparent density values increased, at the same time water absorption values decreased significantly, till ~0% after sintering at 1300°C.

Maximum density was obtained after sintering at 1300°C for 30 min (2.94 g/cm³), and this was also the maximum strength sample – 126 MPa (Weibull modulus – 9) (Fig. 5). In fact, the strength values determined for the mullite-alpha alumina composites obtained by sintering the slate wastes-aluminium sludges mixture between 1250 and 1300°C, are higher than those reported for sintered samples produced either from slate powders (28 MPa) or slate-gamma alumina powder mixtures (37–43 MPa) (Martins et al. 2002) sintered in the same temperature range.

The addition of aluminium sludges to slate wastes lead therefore to a stronger enhancement of the

318

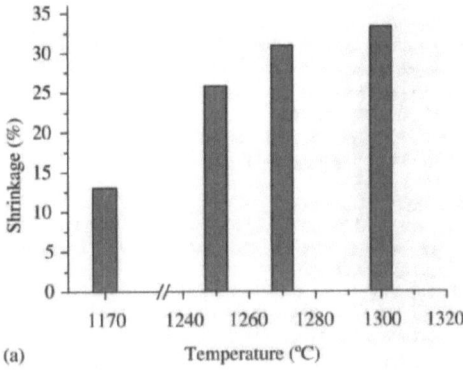

(a)

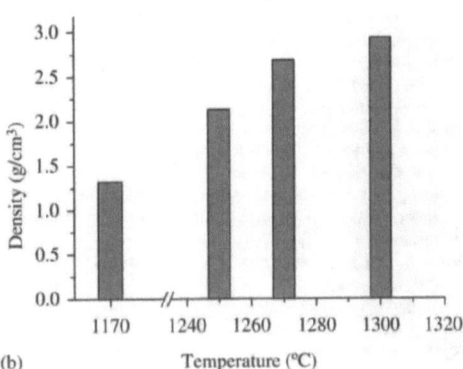

(b)

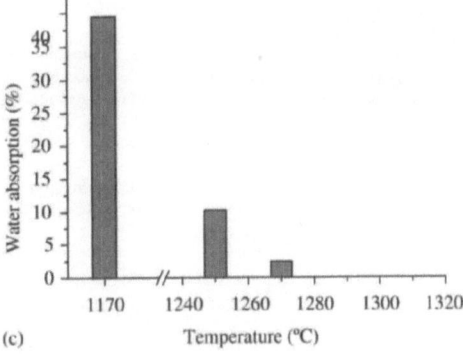

(c)

Figure 4. Properties of the 2:1 sintered bodies as function of the sintering temperature (a – shrinkage; b – apparent density; c – water absorption).

mechanical strength with the sintering temperature than that already observed by the addition of gamma alumina (Martins et al. 2002). Even the strength values reported by Viswabaskaran et al. (2003) for mixtures of calcinated clays with aluminium hydroxides

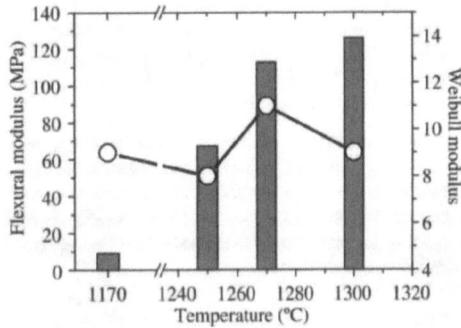

Figure 5. Bending strength (bars) and Weibull modulus of 2:1 sintered bodies.

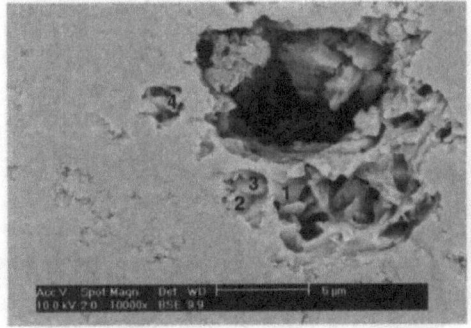

Figure 6. Scanning electron micrography of 2:1 sample heat-treated to 1250°C (1 to 3 – alumina, 4 – mullite).

are as low as, 29 to 38 MPa, for sintered products produced from raw mixtures containing boehmite, or 90 to 99 MPa, for those containing gibbsite, in spite of fully conversion of the samples to mullite.

Figure 6 shows a scanning electron micrography of a 2:1 sample heat-treated non-isothermally up to 1250°C under a vacuum atmosphere. Chemical analysis performed in the defect region presented in the figure allowed the identification of alumina (points 1, 2 and 3) and mullite (point 4), as expected from the XRD results.

4 CONCLUSIONS

The results of the present study are summarized as follows:

– the addition of aluminium sludges to slate wastes lead to an increase of the mullite content in the sintered product, due to the reaction between silica-containing slate wastes and alumina-rich sludges;

319

- the composition of the mixtures lead to a final sintered product composed of mullite and alpha alumina embedded in a glassy phase matrix;
- sintering of 2:1 mixtures at 1300°C produced fully densified products with high flexural strength.

The promising reported results allow identification in floor and wall tiles industries as a possible successful end-user for these wastes mixtures, therefore contributing to the decrease of their environmental impact at the same time final products with improved properties are obtained.

ACKNOWLEDGMENTS

This investigation took place during the POCTI project 35500/CTM/2000 "Inertization kinetics of aluminium sludges by solid state transformation". FCT founding is greatly acknowledged.

REFERENCES

Aksay, I.A., Dabbs, D.M. & Sarikaya, M. 1991. Mullite for structural, electronic, and optical applications. *J. Am. Ceram. Soc.* 74(10): 2343–2358.

Barreiros, F.M. 2002. Optimização da moldação por injecção de residuos industriais inorgânicos. PhD Thesis, University of Coimbra, Coimbra.

Catarino, L. 1999. Xistos ardosíferos: caracterização e recuperação de desperdícios. PhD Thesis, University of Coimbra, Coimbra.

Catarino, L., Sousa, J., Martins, I.M., Vieira, M.T. & Oliveira, M.M. 2003. Ceramic products obtained from rock wastes. *J. Mat. Proc. Tech.* 143–144: 843–845.

Chen, C.Y., Lan, G.S. & Tuan, W.H. 2000. Microstructural evolution of mullite during the sintering of kaolin powder compacts. *Cer. Int.* 26: 715–720.

Chen, Y.F., Wang, M.C. & Hon, M.H. 2003. Transformation kinetics for mullite in kaolin-Al$_2$O$_3$ ceramics. *J. Mater. Res.* 118(6): 1355–1362.

Delmas, F., Gonçalves, L. & Natário, A. 1997. Produção de alumina a partir de lamas de anodização de alumínio. In L.G.Rosa (ed.), *Proceedings of the 8th Meeting of Sociedade Portuguesa de Materiais.*

Ferreira, J.M. & Olhero, S.M. 2002. Al-rich sludge treatements towards recycling. *J. Eur. Cer. Soc.* 22: 2243–2249.

Li, D.X. & Thomson, W.J. 1990. Mullite formation kinetics of a single-phase gel. *J. Am. Ceram. Soc.* 73(4): 964–969.

Liu, K. & Thomas, G. 1994. Time-temperature-transformation curves for kaolinite-α-alumina. *J. Am. Ceram. Soc.* 77(6): 1545–1552.

Martins, I.M., Sousa, J., Catarino, L., Vieira, M.T. & Oliveira, M.M. 2002. The formation of mullite from rock wastes containing alumina and silica. *Key Eng. Mat.* 230–232: 380–383.

Martins, I.M., Vieira, S., Livramento, V., Sousa, J., Delmas, F., Oliveira, M.M. & Vieira, M.T. in press. Manufacture of ceramic products using inerized aluminum sludges. Mat. Sci. For.

Oliveira, A.P., Gomes, V., Hotza, D., Monted. O., Picoli, R., Pereira, F. 2003. Aluminium rich sludge as raw material for the ceramic industry. *Interceram* 52(1): 44–46.

Ohira, H., Ismail, M., Yamamoto, Y., Akiba, T. & Somiya, S. 1996. Mechanical properties of high purity mullite at elevated temperatures. *J. Eur. Ceram. Soc.* 16: 225–229.

Pask, J.A. 1996. Importance of starting materials on reactions and phase equilibria in the Al$_2$O$_3$-SiO$_2$ system. *J. Am. Eur. Soc.* 16: 101–108.

Pereira, D.A., Couto, D.M. & Labrincha, J.A. 2000. Incorporation of alumina rich residues in refractory bricks. *Cer. For. Int.* 77(7): 21–25

Ribeiro, M.J., Tulyaganov, D.U., Ferreira, J.M. & Labrincha, J.A. 2002. Recycling of Al-rich industrial sludge in refractory ceramic pressed bodies. *Cer. Int.* 28: 319–326.

Seabra, A., Pereira, D., Bóia, C. & Labrincha, J. 2000. Pretreatment needs for the recycling of Al-rich anodising sludge as a ceramic raw material. *Proceedings of the 1st Latin American Clay Conference.* II: 176–181.

Vieira, S., Vieira, M.T. & Catarino, L. 2002. Recovery of sludges from aluminum surface treatments. In B. Björkmam, C. Samuelson & J-O. Wikström ed. Proceedings of TMS Fall 2002 Extraction and Processing Division Meeting on Recycling and Waste Treatment in Mineral and Metal Processing: Technical and Economic Aspects 2: 489–498.

Viswabaskaran, V., Gnanam, F.D. & Balasubramanian, B. 2003. Mullitisation behaviour of calcined clay-alumina mixtures. *Cer. Int.* 29: 561–571.

Dimension Stone 2004, Přikryl (ed)
© 2004 Taylor & Francis Group, London, ISBN 90 5809 675 0

Exploitation of marble wastes through pulverization in a prototype ring-mill

G. Zannis, P. Makris & M. Founti

National Technical University of Athens, School of Mechanical Engineering, Laboratory of Heterogeneous Mixtures and Combustion Systems, Athens, Greece

K. Laskaridis

Institute of Geology and Mineral Exploration, Department of Economic Geology, LITHOS Laboratory, Athens, Greece

ABSTRACT: The ability of a horizontal ring-mill for ultra-fine pulverization of $CaCO_3$ and exploitation of marble wastes is examined. The experiments are conducted in batch mode with different mill charges and operating parameters. The influence of grinding aid is examined, resulting to finer products with lower energy consumption requirements. Ultra-fine powder of mean diameter $3.5 \sim 4.5\,\mu m$ can be efficiently produced with the horizontal ring-mill. The size of the final product is independent of the grinding conditions or the charge.

1   INTRODUCTION

Calcium carbonate ($CaCO_3$) can be derived from a variety of sources such as limestone, chalk or marble and to a lesser extent from carbonate, vein calcite, travertine, shells, aragonite sand or dolomite. Limestone and dolomite used for industrial applications hold the seventh place in the list of the world's 50th most important raw minerals with respect to value and the fifth in terms of production (Housa 1999).

Originally, extender and filler pigments were low cost minerals that have been used as additive to paints, paper, plastics, rubber and sealants. Today, in Europe, the most used fillers and extenders are ground calcium carbonate (GCC), kaolin, talc and precipitated calcium carbonate (PCC). Extender and filler applications for these four major pigments have grown from an estimated 5.3 million tons in 1972 to nearly 12 million tons in 1995, representing an average annual growth of 3.6% (Housa 1999). GCC was the strongest performer with an annual growth of around 7%. This preferential increase in consumption of $CaCO_3$ is mainly due to the material's advantages such as reduced cost, ability for partial/total replacement of more expensive fillers (kaolin, talc etc.), particle size, particle size distribution and brightness (Laskaridis 1995). The major end-user applications of GCC include: paper, paints, plastics, rubber, adhesives and sealant carpets, and animal feed industries (Schneider 1997).

Today's European GCC market needs quality products with specific physical properties, hence imposing increasingly stringent particle size requirements. Although limestone is widely available throughout Europe, high purity and high brightness chalk, limestone and marble deposits are not as ubiquitous. With customers demanding higher whiteness products than they used to, securing high-grade deposits is becoming increasingly important. GCC is now forecasted to grow at about the demand rate of its consuming industries, with an average of 2 to 3%.

In Greece, the raw materials used for carbonate filler production are: pure white friable microcrystalline limestone, dolomitic and calcitic marbles. The Greek carbonate filler industry consists mainly of 6 companies, producing fine-grained carbonate fillers. Among them, the three major companies are I. Kalafatis – Ionian Kalk SA., Petrochem K. Zafranas SA. and Dionyssos Pentelikon Commercial S.A. & Industrial Marble Co (Laskaridis 1995).

Ultra-fine calcium carbonate resulting from pulverization of marble wastes, can be used as filler, whitener, extender or reinforcing agent aiming to improve performance characteristics of end products and reduce manufacturing costs in various industrial processes. For example, powdered calcium carbonate is used in the plastics industry to improve properties of plastics such as: increased density and Young's modulus, flexural modulus and surface hardness of the finished article, reduced shrinkage, improved flame retardation etc. (O'Driscoll 1993, Goldman et al. 2000, Omya).

In the paper industry it is used as filler or as coating material improving opacity, printability, bulk and offers reduced cost (Ionides 1995, Harben 1998).

In the pharmaceuticals industry, it can be used as tablet and capsule diluent, therapeutic agent and dissolution aid in dispersible tablets (Bolger 1995).

The most important reason for the widespread use of $CaCO_3$ as colouring pigment (with 30–98% $<2\,\mu m$) is the practically unlimited availability of carbonate deposits. This, in conjunction with the large number of production sites, reduces the distances to consuming industries and the transport costs. At present, $CaCO_3$ is the cheapest colouring pigment and can be obtained at a relatively low cost (Gysau 2000).

Due to its extensive range of applications, ultrafine $CaCO_3$, free of impurities and with pre-specified size distribution is considered to be a high added value product, and its production cost is greatly influenced by the selection of the processing system. In most cases a mill coupled to an air classifier are needed to guarantee the desired particle top size and cost-effective processing of GCC in the ultra-fine regime for commercial applications.

The aim of the present work is to demonstrate the ability of a low-cost horizontal ring mill to pulverize marble wastes ($CaCO_3$) in the ultra-fine region with a pre-determined particle top-size. The experiments are conducted in batch mode with different mill charges and operating parameters. The influence of grinding aid is also examined, the presence of which results in finer product with less energy consumption.

The raw material used in the experiments is $CaCO_3$ powder commercially available as filler for the preparation of mortars, originating from marble wastes, a by-product of marble quarrying. The wastes are shapeless pieces of various dimensions which cannot be used for the production of flagstones and when accumulated can lead to environmental and aesthetic problems especially when the quarry is close to urban or tourist areas.

The paper demonstrates that such marble wastes can be exploited producing ultra-fine $CaCO_3$ powder aiming at: (a) the reduction of the amount of waste blocks at the quarry and (b) the production of a high-value added product.

2 DESCRIPTION OF THE RING-MILL

Figure 1 shows the main components of the ring mill (Makris 1995, Makris et al. 1996, Spitas et al. 1999). A cylinder (1) forms the main body of the mill. A rectangular metal frame is positioned along the principal axis of the cylinder and is attached to the main shaft (4). The main shaft can be rotated with the help of an

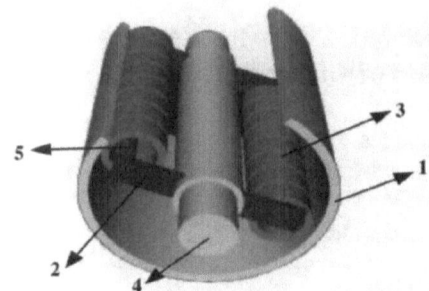

Figure 1. Components of the ring-mill.

electric motor. The rectangular metal frame comprises two arms (2) which are held parallel with the help of two ring-connecting rods (5). A variable number of annular metal rings (3) is inserted along each connecting rod. Each ring is allowed to move freely along and around each rod.

During rotation of the orthogonal metal frame the metal rings, due to centrifugal forces, are compressed against the inner surface of the cylinder of the pulverizer. Simultaneously frictional forces are developed between the metal rings and the inner surface of the machine. Thus, the rings rotate freely around their symmetry axis and at the same time slide along the arms of the frame and roll on the inner surface of the cylinder. The particles are pulverized in the space left between the inner surface of the mill and the annular metal rings. Consequently, particles are under the effect of normal (compression) and shear forces and the mill operates combining comminution with attrition principles.

The mill offers considerable advantages when compared to commonly used comminution processes, especially in the fine and superfine milling of minerals. The originality of the pulverizer lies mainly in that it allows the variation of the ratio of the shear to the compression forces acting on a particle during its pulverization. The ratio is adjusted to the mechanical requirements of the material to be pulverized. Instead of a single roller or balls, the pulverizer employs several annular rings, thus allowing better grinding of the finer fractions of the original material.

3 EXPERIMENTAL PROCEDURE

The experimental set-up is shown in Figure 2. The shaft of the mill is rotated with the aid of an electrical motor, driven through a computer controlled frequency inverter. The frequency inverter supplies an analogue signal of the motor electric power that is digitised and read constantly by the computer. During

322

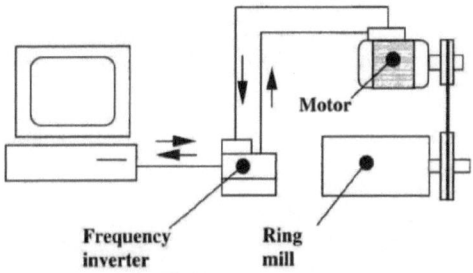

Figure 2. Schematic diagram of the experimental setup.

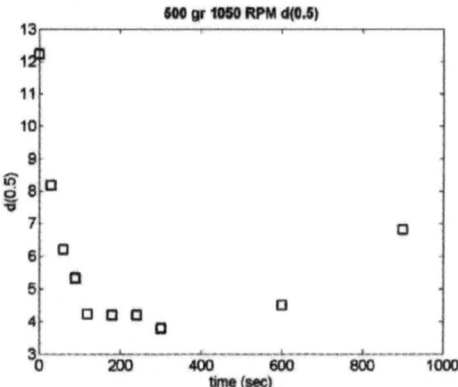

Figure 3. Variation of d(0.5) versus time (mill load 500 g, mill rotated at 1050 rpm).

the experiments, the power-signal was recorded and integrated in time using the LabVIEW platform.

For each experiment, the ring mill was filled with the appropriate mass of material, of initial mean diameter d(0.5) = 45 μm (size range 2–100 μm). For certain experiments, an amount of grinding aid, ca. 2~6 g depending on the feed charge (1% per weight), was added in the mass of material. The particular grinding additive was an aqueous solution of a complex compound of amines and polyalcohols, commonly used in the relevant $CaCO_3$ processing industry. The effect of the grinding additive has been examined as part of the experiments and the results are presented below.

During the experiments, the grinding process was stopped at predetermined time intervals and a representative small sample of ultra-fine ground material was removed for further analysis. Subsequently, the mill was again set in operation. Repeating the start-halt procedure several times demonstrated that it did not affect the mill performance and its operational characteristics.

For the experiments, the mill was rotated at 650 and 1050 RPM with fixed quantities of material mass, corresponding to 250, 500 and 750 g.

The particle size distribution of the various samples was measured with a Master-Sizer 2000 (Malvern) that utilizes the low angle laser scattering (LALLS) method. Size distributions were measured after dispersing the samples in a water supersonic bath. The sample was dispersed in a water bath and agitated with the aid of supersonic waves for 30 sec. The refractive index for the measurements was set equal to 1.690. The total particle-size measurement error was estimated at 5% at 0.95 confidence level.

4 EXPERIMENTAL RESULTS

Figure 3 presents the variation of median particle size d(0.5) versus time for the case of 500 g load and mill rotational speed of 1050 RPM. The trends shown in Figure 3 are representative for all examined cases. It

Table 1. Measured particle size distribution for batch grinding experiments, for various mill loads and rotational speeds-quoted times are those required to achieve minimum particle size.

Mass [g]	Time [s]	d(0.5) [μm]	P < 7 μm [%]	Mean [μm]	Span
650 RPM					
250	–	–	–	–	–
500	840	4.587	62.47	8.212	3.997
750	1200	4.221	63.72	8.231	4.515
1050 RPM					
250	100	4.478	62.02	8.287	4.402
500	300	3.781	67.59	7.315	4.708
750	600	3.415	69.53	7.635	4.991

can be clearly observed that for the first 120 sec the particle size is reduced rapidly, it remains almost stable for the next 180 seconds and then it increases till the end of the experiment after 900 sec. The latter is considered as a result of strong agglomeration occurring during the pulverization and which couldn't be eliminated by the ultrasonic bath during the size-distribution measurements.

The same was observed in all the cases of 250 and 750 g mill load, rotated at 1050 RPM (relevant figures are not shown here, but mean diameter results are tabulated in Table 1). However, this trend was not observed in the cases of 500 and 750 g mill loads when the mill was rotated at 650 RPM (Figure 4), probably because of the milder grinding conditions (and the weaker forces acting on the particles). From the theoretical analysis (Spitas et al. 1995) it is known that the magnitude of the forces acting on the material is proportional

323

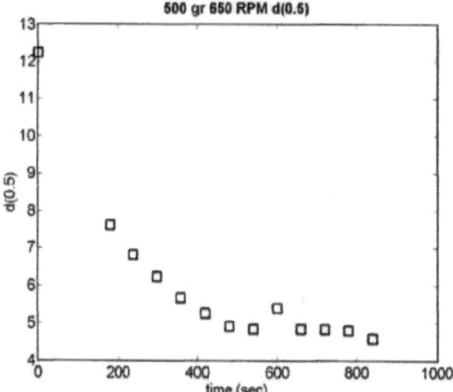

Figure 4. Variation of d(0.5) versus time (load 500 g at 650 RPM).

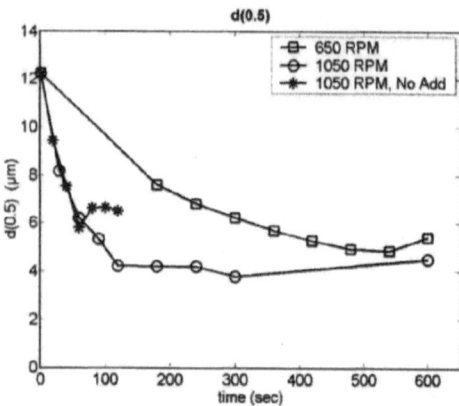

Figure 6. Comparison of variation of d(0.5) with and without additive for various mill rotational speeds.

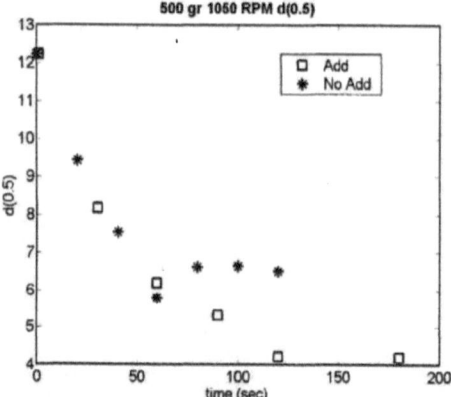

Figure 5. Comparison of variation of d(0.5) with and without additive.

to the square of the angular velocity, implying that the forces acting on the material when the mill is rotated at 1050 RPM are much bigger than those at 650 RPM. As a result intense creation of electrostatic charge is expected, mainly due to the slipping of the rings on the material under increased force conditions.

The grinding additive – injected in the form of a spray – is absorbed by the surface of the particles reducing the electrostatic forces and consequently the formation of agglomerates. From Figures 5 and 6, comparing achieved mean particle diameter versus time with and without grinding-aid, is concluded that the effective grinding process has an obvious end which for the 1050 RPM is about after 120 sec of grinding time while for the 650 RPM is after 420 sec approximately.

The particle size measurement indicated that in the case of rotating the mill at 1050 RPM the d(0.5) is smaller than in the case of 650 RPM, independent of mill load. This is clearly shown in Table 1, tabulating all the particle-size distribution measurements for various mill loads and rotational speeds. For example, for the case of 750 g mill load the measured mean diameters d(0.5) were 3.415 μm, and 4.221 μm for 1050 and 650 RPM, respectively.

With grinding aid (additive), Figure 5, the minimum achieved value of d(0.5) is 3.78 μm while the corresponding value without the additive is 5.79 μm. Also the energy consumption is much higher without additive. It was visually observed that without additive, the material created a crust-layer that covered the active mill surfaces (outer ring surfaces and cylinder internal surface), having the form of "orange peel". This resulted in a dramatic increase of the shear forces between the rings and the internal mill surface and consequently in increased demand of motor power. The results of Figure 5, confirm the need of using an appropriate grinding additive in grinding-pulverization of marble wastes.

The extreme values (minimum or maximum) of d(0.5) (median diameter), of P(d < 7 μm) (cumulative finer than 7 μm), mean diameter (D[4.3]) and span – defined as [d(0.9)−d(0.1)]/d(0.5) – for all the batch experiments are presented in Table 1. The recorded differences in particle size are not remarkable, either for the different quantities of material or the different rotational speeds investigated, indicating that the limiting values (i.e. final particle sizes) lie relatively close. Nevertheless, significant differences in achieved final particle size are observed in relation to the grinding *time*, necessary for the achievement of the above values, and the *consumed energy*.

324

Table 2. Calculated values of mass flow less than 7 μm and corresponding specific energy.

Calculated values	Mill charge [g]		
	250	500	750
t_{60} [s]	48	131	223
$\dot{m}$ [kg/h]	5.3234	3.9459	3.4709
$\dot{E}$ [kWh/kg]	0.1733	0.2104	0.4624

Using the representative value of t_{60} (grinding time to achieve $P(d < 7\,\mu m) = 60\%$), the mass produced per unit time (in kg/h) of material with size smaller that 7 μm was calculated (Austin et al. 1984). The produced mass per unit time is calculated as the difference between the initial (0.32) and final (0.60) values of the experimentally measured percentages of particle size less that 7 μm multiplied by the mass of the material over t_{60} (equation 1):

$$\dot{m} = \frac{0.6 - 0.32}{t_{60}} \cdot 3600 \cdot m \qquad (1)$$

From the mass production per unit time $\dot{m}$, and the measured consumed energy, the specific energy $\dot{E}$ (kWh/kg) was calculated. The specific energy refers to production of material with particle size less than 7 μm. The results are presented in Table 2.

It is obvious that the most favourable case is that of 250 g which not only gives the highest mass production per unit time but also the energy consumption is the lowest. This result is directly related to the operating mode of the mill: the pulverization is conducted on the inner surface of the mill and it is not a volumetric process. The thin layer of material formed on the inner surface of the mill, with a load of 250 g, favours fast pulverization. On the contrary, a thick material layer acts in a "relieving" manner, absorbing the stresses exerted from the rings to the material.

These results are useful hints and can be used as criteria for the selection of operating conditions, during operation in a grinding circuit.

5 CONCLUSIONS

$CaCO_3$ powder commercially available as filler for the preparation of mortars, originating from marble wastes, was further pulverized in the ultra-fine region using a horizontal ring mill in batch mode, under various charges and mill operating conditions. Special attention was given to the production of material finer than 7 μm. It has been experimentally established that in all cases there is a limiting pulverization value, associated with the mill type and operating conditions.

The maximum value of $P(d < 7\,\mu m)$ is approximately 65%. Beyond this point, formation of agglomerates takes place, probably due to electrostatic forces, which results to coarser material. Among the three different examined mill charges at a mill rotational speed of 1050 RPM, the case of mass load equal to 250 g is the optimum resulting to the maximum production with the minimum energy consumption.

REFERENCES

Austin, L.G., Klimpel, R.R. & Luckle, D.T. 1984. *Process Engineering of Size Reduction: Ball milling*. New York: Society of Mining Engineers.

Bolger, R. 1995. Industrial minerals in pharmaceuticals. *Industrial Minerals* 335: 52–63.

Goldman, A.Ya. & Copsey C.J. 2000. Polypropylene toughened with calcium carbonate. *Mat. Res. Innovations.* 3: 302–307.

Gysau, D. 2000. Anti-corrosion properties of paint extender minerals. *Industrial Minerals* 393: 41–49.

Houssa, C.-E. 1999. Filling the gap – A review of European GCC. *Industrial Minerals* 384: 73–81.

Harben, P. 1998. $CaCO_3$ in paper: PCC versus the competition, *Industrial Minerals*. 246: 39–49.

Ionides, G., 2000. Paper pigments – market dynamics and outlook, *Industrial Minerals* 393: 29–39.

Makris, P. 1995. Optimale Konstruktion von Feinstzerkleinerungs-maschinen, *Proc. of ICED 95*, ed. V. Hubka, Praha: 104–109.

Omya A.G. Mineral fillers in the plastics industry – A review. *Technical note*, 184.

O'Driscoll, M. 1993. Minerals in European plastics; polypropylene in the driving seat. *Industrial Minerals* 307: 19–35.

Schneider, B. 1997. Processing of ultra-fine minerals. *Industrial Minerals*: 91–99.

Spitas, V., Makris, P. & Founti, M. 1999. A novel dry pulverizer for low cost production of powders. *Particulate Science and Technology: An International journal* 17(3): 217–228.

Makris, P. & Founti, M. 1996. Constructional aspects of dry pulverizers for energy efficient production of powders, *Proc. of the 5th World Congress of Chemical Engineering, 2nd Particle Technology Forum, AIChE*. San Diego, California. 5.

Laskaridis, K. 1995. Post-quarrying exploitation of wastes from Greek white marble quarries for use as fine filler. *1st Intern. Congress on Natural Stones. 15–18 June*. Lisbon.

Dimension Stone 2004, Přikryl (ed)
© *2004 Taylor & Francis Group, London, ISBN 90 5809 675 0*

Author Index

DIMENSION STONE is one of the few durable building materials that fulfils the criteria of sustainable utilisation of natural resources and that is, at the same time, of high aesthetic value. Stone has drawn attention from times unmemorable and played a significant role in many human activities. The dimension stone evaluation developed to a real scientific discipline during the 20th century. At the beginning of the 21st century, the demand for dimension stone is still growing, although it must compete with other - artificial - building materials. Many countries face massive imports although local historical stone resources that can be employed in a traditional manner are still available.

The Dimension Stone 2004 Conference held in Prague, Czech Republic, aimed to bring together experts from many fields of research including geology, rock mechanics, geotechnics, stone extractive industry, restoration and architecture. This book contains peer-reviewed papers of this multidisciplinary gathering and provides state-of-the-art information on recent developments in the use and application of dimension stones throughout the world.

Themes covered comprise:

- geological studies of traditional local stone types,
- advanced rock fabric and rock mechanics studies applied to dimension stone research,
- application of dimension stone databases for historical research and for stone marketing,
- GIS application to quarry planning,
- aspects of dimension stone deterioration,
- bowing of natural stone cladding and prevention,
- processing and benefits of waste from stone industry.

This book is intended for geologists, mining and civil engineers, restorers and architects.